FIELD IONIZATION MASS SPECTROMETRIE

von

Prof. Dr. Hans-D. Beckey

Institut für Physikalische Chemie
der Universität Bonn

Mit 150 Abbildungen und 38 Tabellen

AKADEMIE-VERLAG · BERLIN
1971

Lizenzausgabe
des Verlages Friedr. Vieweg + Sohn, GmbH, Braunschweig

Erschienen im Akademie-Verlag GmbH, 108 Berlin, Leipziger Straße 3—4
Lizenznummer: 202 · 100/624/71
Gesamtherstellung: VEB Druckhaus „Maxim Gorki“, Altenburg
Bestellnummer: 5797
EDV-Nr.: 761344 5

CONTENTS

PREFACE

In 1967 the area of Field Ionization Mass Spectrometry had developed to a point where it seemed to the author to justify writing a book on this field. No monograph on this subject was available but all information came from original articles spread over several journals. Field Ionization Mass Spectrometry was treated only extremely briefly in some of the existing books on general mass spectrometry. On the other hand, after the introduction of combined electron impact and field ionization sources, the interest in this area of those mass spectrometrists who are engaged in organic analysis grew quickly. Furthermore, the community of field ion microscopists in many countries grew steadily year by year. There seemed to exist some need of these people for information on this parallel branch of field ionization. Moreover, some overlap between the areas of field ionization mass spectrometry and microscopy developed recently with the introduction of combined techniques by E. W. Müller and J. Block and their associates. Finally, physico-chemists working on gas-phase reaction kinetics and on catalytic reactions became interested in the area.

Although the field emission microscope had originally been developed in Germany by E. W. Müller, a careful analysis showed that the number of people working on field electron and ion microscopy in English-speaking countries is at present more than ten times larger than in Germany. In the area of mass spectrometry the situation is not too different, although the corresponding ratio is not as great as indicated above.

For this reason the author decided to write his book in English. Greater efficiency in filling the existing gap in literature was the only reason for this decision. The author was faced with two possibilities: Either to order an English translation of a German text from an English interpreter, or to write the book in English and then ask an expert in this area from an English-speaking country for revision in style and grammar. The author has chosen the second way, hoping to minimize scientifically incorrect formulations which could arise by translation of a text into a foreign language by an interpreter who is not an expert in that field. This procedure would not have been realizable without the invaluable help of the author's British colleague, Dr. A. J. B. Robertson, who revised the text of this book. Dr. Robertson, a distinguished expert in this field for a number of years, has contributed to this book also by many valuable discussions with the author. Many good suggestions for improvement of the manuscript have also been made by D. F. Barofsky who spent one year in the author's laboratory.

The main emphasis of this book is on mass spectrometric problems. Therefore, it was felt that it would be outside the scope of the book to derive the basic equations of the theory of field ionization up to the highest degree of approximation that has been achieved at the present time. Concerning this, the reader is referred to the original publications dealing with tunneling probability, particle supply, field desorption, etc. In most cases, derivations are given which lead to equations on a first approximation level, in order to introduce the mass spectrometrist to the theory of field ionization.

Since the author has been active in the area of field ionization mass spectrometry for over ten years, a great deal of the book is based on his personal research work and that of his associates. Necessarily, problems more closely related to mass spectrometry than to field ion microscopy are treated in greater detail in this book. However, the attempt was made to give a fairly balanced review of all of the more important aspects of field ionization mass spectrometry.

My thanks are expressed to Mrs. Marquardt who did the difficult work of typing the hand-written manuscript and who kindly pushed me to finish the manuscript. Thanks are also given to Mrs. Barofsky who made the numerous drawings, and to all the authors who have given us permission to reproduce figures or tables of their publications.

Finally, I wish to express my special thanks to all my former and present associates who contributed a great deal to this book by their work and by their numerous discussions with me: J. Dahmen, H. Heising, H. Hey, K. G. Hippe, H. Knöppel, H. Krone, K. Levsen, H. G. Metzinger, M. D. Migahed, H. W. Röllgen, G. Tenschert, D. Schütte, P. Schulze and G. Wagner.

INTRODUCTION

FIELD ionization*) mass spectrometry is a rather new branch of mass spectrometry which originated historically from FI microscopy as founded by Erwin Müller in 1953. The first quantitative and comprehensive investigation on FI mass spectrometry was published in 1954 by Inghram and Gomer [1, 2]. The ion source for the mass spectrometer constructed by these authors could be used as an FE microscope as well.

In another paper, Müller and Bahadur [3] reported on visual observation of ions which had been separated according to their mass by a magnetic field, on the luminescent screen of an FI microscope. The close relation between FI or FE microscopy and FI mass spectrometry still remained obvious in these two papers.

The author of this book started with a systematic investigation of the new discipline "Field Ionization Mass Spectrometry" in 1957, in a manner which did not make use of simultaneous or alternative microscopic observations. It was entirely concentrated on field ionization in connection with mass spectrometric problems in order to lead the new method to numerous applications in the area of physico-chemical investigations and quantitative or structural chemical analyses [4–6].

At the beginning, the number of authors working in this new field was very small. Nearly simultaneously with Inghram and Gomer, Drechsler [7] had pointed out the principle of an FI source for mass spectrometric analyses which was realized experimentally a number of years later; Schmidt [8] reported on the design, and on experiments, with an improved version of this source. Block [9] started relatively early with applications of the FI mass spectrometric method to problems of adsorption and catalysis. Robertson [10] contributed on FI mass spectra taken with sharp metal edges. The names and the work of all authors having published on FI mass spectrometry after the authors named so far will be discussed in the course of this book. (References are given

*) FI is used throughout this book for "Field Ionization", FE for "Field Electron" and EI for "Electron Impact".

up to the closing date of the book manuscript in April 1968. Some additional references were introduced during corrections of the proofs in spring 1970.)

The main reasons for the initially rather slow development of the new field were among others:

The number of spectra available was still too small to obtain an insight into the general properties of FI mass spectra; the theory was not yet sufficiently worked out for a good understanding of regularities observed in the FI spectra of homologous series of organic substances; the instrumental techniques were much less developed than today; there were only some rough ideas on what the main applications of the new method would be in the future; and last but not least, there were no commercial FI sources for mass spectrometry available.

The number of users of FI mass spectrometers has been increasing, however, very quickly during the last few years, and a further quick extension of the area is to be expected after the recent introduction of commercial FI sources, especially sources which are combined with electron impact facilities.

At present, it is still relatively easy to obtain a complete survey on all publications in this area because of the exactly defined and relatively short period during which work on FI mass spectrometry has been done.

At the beginning, there will be given a short review on some of the most important directions of research of FI mass spectrometry. One can differentiate between two main fields: Physico-chemical investigations, and applications of the method to the area of quantitative analysis or structural determination of organic substances.

In order to exemplify this, some typical problems may be put into the form of questions.

I. PHYSICO-CHEMICAL PROBLEMS

1. *Probability of field ionization:* How does the FI probability depend on different parameters? (Field strength, ionization potential, polarizability, dipole moment, vapour pressure, mobility on solid surfaces, etc.)
2. *Field dissociation:* What is the mechanism for dissociation of a molecular ion by an electric field of the order $10^7 - 10^8$ V/cm? How can one qualitatively predict the bond where field dissociation occurs most easily?
3. *Charge distribution:* How does the charge distribution of a positive ion change if the ion is brought into a high electric field?

4. *Temperature dependence of FI mass spectra:* How can the temperature dependence of FI mass spectra observed experimentally be interpreted theoretically?
5. *Kinetics of ion decomposition in high fields:* What are the mechanisms and the decomposition times of ions in high electric fields?
6. *Ion clusters, charge transfer complexes, field condensation:* What ionic species are observed in connection with these phenomena, and how can their formation be interpreted?
7. *Field evaporation, field desorption and field corrosion:* What are the mechanisms of these processes and what is the number of unit charges of field evaporated ions?
8. *Adsorption, catalytic processes:* How can the FI method — especially the impulse method — be used for the study of these processes?
9. *Equilibria:* How could a high electric field shift multiphase or chemical equilibria?
10. *Negative ions:* Which mechanisms are possible for the formation of negative ions in a high field, and which ions are formed?
11. *Surface diffusion:* What is the mechanism of particle supply to the FI centres by surface diffusion?
12. *Activation of FI emitters:* What is the reason for the improvement of FI emission of different field anodes with time? (Whisker formation, etc.)

This list is, of course, by no means complete. It should give an impression of how numerous the physico-chemical problems are which have a close relation to this new field of research. The most general question in connection with these problems would be: What is the behaviour of matter — especially of organic molecules, but also inorganic solids — under the influence of extremely high electric fields? One could collect all these problems under the heading "High Field Physico-Chemistry".

The author and his co-workers being physico-chemists devoted the major part of their work to such problems as those just mentioned, and the main weight of this book is directed to this part of field ionization mass spectrometry.

On the other side, it became evident relatively early that the FI mass spectrometer offers interesting new applications to the chemist in the field of quantitative analysis of organic mixtures, and on structural analysis of complicated organic compounds. Again, some typical problems may be indicated.

II. PROBLEMS OF CHEMICAL ANALYSIS

1. *Quantitative Analyses:* Methods for making use of the simplification of FI as compared with EI mass spectra for analysis of complicated mixtures (gasoline samples, for example). Discussion of the advantages and disadvantages of this method. Range of detection of free radicals by FI techniques and its accuracy.
2. *Structure elucidation of organic compounds:* Use of the generally large, relative parent ion intensity for determination of molecular weight and elemental composition of those unknown compounds which do not unambiguously show parent ions after electron impact. Comparison with low voltage electron impact techniques.

   Comparison of the different relative fragment ion intensities in the FI and EI mass spectra for structural information. Use of the "primary metastable fragment ion maps" for determination of structure of unknown molecules in mixtures.
3. *Elucidation of mechanisms and kinetics of decomposition of organic ions:* Use of the different kinetic behaviour of ions in FI and EI sources for differentiation of rearrangement, multi-step and single-step direct bond rupture processes. Use of the strongly preferred formation of primary metastable field ions for the study of reaction paths.

Besides these two main groups of themes, the problems of improving instrumental techniques will be treated in this book. Electrodes of very small radii of curvature are used for production of the extremely high field strengths necessary for field ionization (about $10^7 - 5 \times 10^8$ V/cm). Fine metal tips, thin wires and sharp metal blades have been used for field production. The methods for production of these emitters and their advantages or disadvantages will be treated in the second chapter, together with the main areas of application of these emitters. The ion optics, which are different from those of EI sources because of the special geometry and potential distribution at the FI emitter, will be discussed in some detail.

A brief comment will now, in the remaining space of this introduction, be given on the overall design of FI mass spectrometers. The most specific part of these instruments is the FI source which strongly differs in design and function from EI sources. On the other hand, the mass spectrometers connected with these FI sources are of conventional design. Principles and design of the mass spectrometers themselves will therefore not be described in this book. There exists a number of excellent monographs on mass spectrometry which the reader may use as a general introduction to the principles and the design of mass spectrometers.

All mass spectrometers could be used, in principle, in connection with FI sources which are used in combination with EI sources, i.e. magnetic mass spectrometers or time of flight, high frequency, quadrupole, monopole or other mass spectrometers. Practically, however, there exists a certain restriction at least for the field of normal chemical analysis of organic molecules. The voltage applied to the FI emitter for production of the high field is usually of the order of 10 kV. Mass spectrometers normally using low energy ions (about several hundred eV, for example, with time of flight or quadrupole mass spectrometers) require extremely strong retarding of the ions produced in the FI source at 10 kV. This results in strong diverging of the ion beam due to lens effects. This effect increases the already strong divergence of the ion beam generated on strongly curved FI emitters. This additional divergence due to retardation of the ion beam strongly reduces the transmission of the mass spectrometer. A large transmission is absolutely necessary to achieve accuracy in normal chemical analysis, as will be discussed later. Therefore, one would avoid such strong retardation of the ion beam in the case of normal chemical analysis. On the other hand, one is sometimes interested in observing the processes occurring within very small ranges of area of the emitter (well defined single crystal surfaces of a tip, for example). The ion beam has a very small divergence in that case, and the restrictions with respect to retardation of the ions no longer exist. Time of flight or quadrupole mass spectrometers are even preferable in this case. By far the majority of all FI mass spectrometers used so far has been of the magnetic sector type, and therefore this book will preferentially refer to such instruments.

The general principle of an FI mass spectrometer is shown schematically in figure I. The FI sources will be treated in detail in chapter 2. The pressure of the gas to be analysed is usually several times 10^{-4} Torr in the space surrounding the FI emitter. The voltage between field anode and cathode is usually variable between a few and 20 kV. The ions enter the mass analyser with energies of several kV. Ions of different mass form a uniform beam until they enter the first boundary of the magnetic sector field. This field splits the beam into several ion bundles of different masses. Magnetic sector fields are frequently used having an angle of deflection of 60° and a field strength of several kilo-Gauss.

A simple Faraday-cup detector may be used for ion detection because ion currents of the order of 10^{-10} A may be obtained after mass separation. In many cases, however, the ion current will be several orders of magnitude smaller, and therefore it is customary to use a secondary electron multiplier for ion detection. The mass spectrum is registered on an automatic recorder.

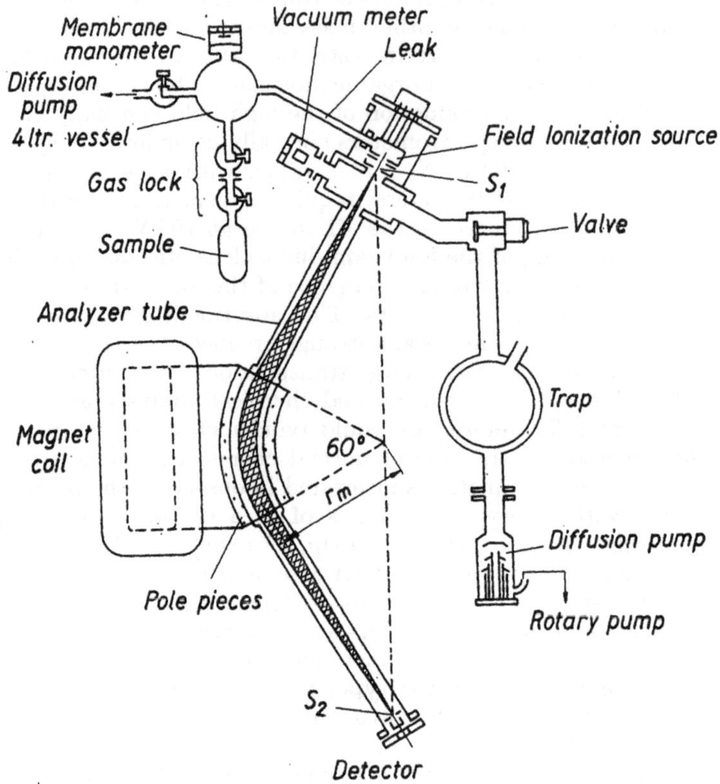

FIG. I. Schematic diagram of an FI mass spectrometer.

1. THEORY OF FIELD IONIZATION (FI) AND FIELD EMISSION (FE)

1.1. FI PROBABILITY

Atoms or molecules are field ionized if high electric fields of the order 10^7 to several times 10^8 V/cm are acting on them. The process of field ionization can be explained theoretically on the basis of the wave properties of matter. The first experimental evidence for the field ionization effect was provided by the observation of Rausch von Traubenberg [11] that certain Stark effect lines in the optical spectrum of highly excited hydrogen atoms disappear if an electric field of several times 10^5 V/cm is applied to the gas. These experiments led to the conclusion that the lifetimes of highly excited electronic states are greatly reduced by high electric fields.

Field strengths of the order 10^8 V/cm are required if field ionization is to occur not from highly excited states but from the ground state of an atom or molecule. Field ionization from the ground state — calculated first by Oppenheimer [12] for the hydrogen atom — has been treated theoretically by Kirchner [13], Inghram and Gomer [2] and by Müller and Bahadur [3].

The theory of field ionization as developed by these authors [2,3,13] may be discussed by means of a potential diagram, fig. 1.1.a). The potential of an electron is drawn as a function of the distance from the ideally smooth surface of a metal. The metal extends over the region $r < 0$ · At $r > 0$ the space is thought to be filled with a very dilute rare gas like helium. The surface of the metal is assumed to be ideally clean, i.e. with no adsorbed layers. The conductivity band of the metal is filled with electrons up to the Fermi level μ. An energy of an amount equal to the work function Φ has to be transferred to an electron at the Fermi level to remove the electron from the metal to infinity.

A helium atom may be at a distance d from the metal surface. The position of the electronic ground level of the H atom is shown in the figure. The ionization energy I of H must be overcome in order to remove the valence electron to infinitely large positive values of r. The electron potential for the free H atom is shown by the solid curve of the figure. An image potential as specified below has to be considered at small distances from the metal. For field ionization the potential of the external field, $F \cdot r$, has to be superimposed. The resulting potential effective

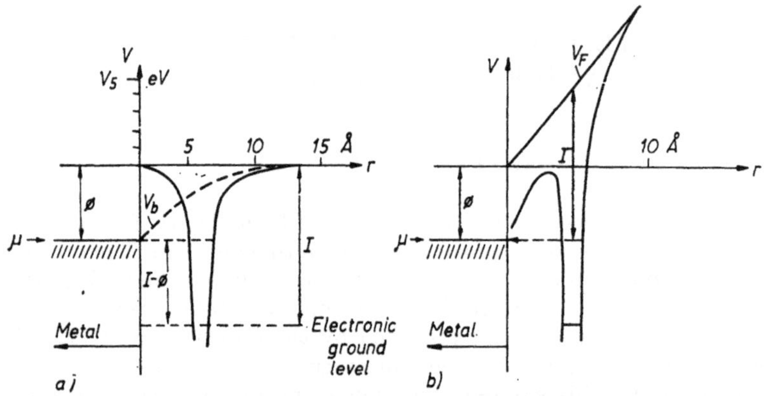

FIG. 1.1. Potential (V) of an atom near a metal surface: a) without field b) with field. I = ionization energy; Φ = work function of the metal; V_b = image potential (dashed curve); V_F = external electric potential $= e.F.r.$ (F = field strength); μ = Fermi level.

for an electron at a distance r from the surface along line perpendicular to the surface and through the ion was assumed by Müller and Bahadur [3] to be approximated for a H atom by

$$V_{(r)} = \frac{-e}{|l - r|} + Fr - \frac{e}{4r} + \frac{e}{l + r} \tag{1.1}$$

This relation would be more complicated for a He atom.

The first term represents the potential of the ion, l being the distance of the atom from the surface. The third term is the image potential of the electron, and the fourth term accounts for the repulsion of the electron by the image of the ion on the metal. This potential is shown by fig. 1.1.b.

At a certain minimum distance of the He atom from the surface, the valence electron of the atom is raised to the Fermi level by the external field. The potential barrier between the atom and the metal surface has then a width of only a few Å and a height of a few eV. Therefore, there exists a certain probability for penetration of the electron through this barrier due to the quantum mechanical tunnel effect.

The mean life time, τ, of atoms or molecules with respect to field ionization can be calculated from the tunneling probability,

D, and the frequency, ν, representing the motion of the electron inside the atom in the direction normal to the potential barrier:

$$\tau = \frac{1}{D \cdot \nu} \tag{1.2}$$

By applying the Wentzel, Kramers, Brillouin (WKB)-Method [14] for one-dimensional barrier penetration one obtains

$$D = exp\left(-\int_{r_1}^{r_2} \sqrt{\frac{8m}{\hbar^2}(e\,V_{(r)} - E) \cdot dr}\right) \tag{1.3}$$

where $V_{(r)}$ is the potential as given by eqn. (1) and E the total energy of the valence electron, $E = -eFl - eV_I + e^2/4l$, (1.3a). V_I = ionization potential. The integral has to be extended over the width of the potential barrier at the energy level E.

Recently, Boudreaux and Cutler [36] have calculated quantum-mechanically the relative ionization probabilities in a three-dimensional model. This was done by considering the process of field ionization as a rearrangement — type collision of an atom with a metal surface and applying the formalism of quantum — mechanical scattering theory. The calculated results are found to be in better agreement with experiment than previous one-dimensional theories.

The frequency of the electron approaching the potential barrier is of the order 10^{15} to 10^{16} sec^{-1}.

Table 1.1. gives the lifetimes of free H atoms at different field strengths as calculated by Inghram and Gomer [2].

TABLE 1.1.

Field strength (MV/cm)	(sec)
50	1.3×10^{-1}
100	1.6×10^{-10}
150	1.6×10^{-13}
200	1.7×10^{-14}
250	2.0×10^{-16}

Gomer [15] pointed out that for approximate calculations, the potential barrier (1) could be replaced by a triangle of height $I - \Phi$ and base I/F.

Inserting this potential into (1.3), one obtains from (1.2) with $\nu = 10^{16}$:

$$\tau \simeq 10^{-16} \times exp\left(\frac{0.68 \times I^{3/2}}{F}\right) \qquad (1.4$$

At present, it seems to be very difficult to calculate the life time τ of a complicated organic molecule with respect to field ionization quantitatively. However, it may be concluded from eqn. (1.4) that τ should decrease exponentially with the field strengths and increase exponentially with some power of the ionization potential.

The calculation of field ion currents at given τ-values has been discussed in detail by Müller and Bahadur [3] and by Gomer [16].

A brief summary of these calculations only will be given here.

The ion current produced in a small volume element is given by

$$i = e \cdot \dot{n} \, (1 - e^{-t/\tau}) \tag{1.5}$$

where e is the electronic charge, $\dot{n}$ the number of particles arriving in the volume element and t the residence time of a particle in that element if no ionization would occur. This approximation holds under the assumption that τ is virtually a step function of the distance l from the tip surface and is constant over t (see below).

Two limiting cases may be discussed.

Very high field strengths

The lifetime τ becomes extremely small at very high fields (see tab. 1.1.). Therefore, eqn. (1.5) reduces to

$$i = e \cdot \dot{n}, \quad t \gg \tau \tag{1.6}$$

In other words, each particle arriving in the ionization zone under discussion, will be ionized. $\dot{n}$, the number of particles arriving will be derived below.

Moderate field strengths

Eqn. (1.5) reduces at moderate field strengths to

$$i = e \cdot \dot{n} \cdot t/\tau = e \cdot c_t \cdot v/\tau \qquad t \ll \tau \tag{1.7}$$

where c_t is the gas concentration (molecules/cm^3) at the tip, and v is the volume where field ionization takes place.

Strictly speaking, ions are formed with continuously varying lifetimes at different distances from the field anode. Müller and Bahadur [3] have shown, however, that most of the ions are formed within a very thin shell of only about 1—2 Å thickness quite near to the surface of the field anode (fig. 1.2.). More recently, Tsong and Müller [17] have shown that for the rare gases and hydrogen, the energy spread of the field ionized atoms was even below 1 eV. Thus the effective depths of the ionization zone at a field strength of a few volts per Å must be below 1 Å.

Therefore, formulas (1.5) and (1.7) may be used, assuming a depth of 1 Å for the ionization volume and a constant value of τ, for order of magnitude calculations of the ion current. (For more exact calculations see [3, 16].)

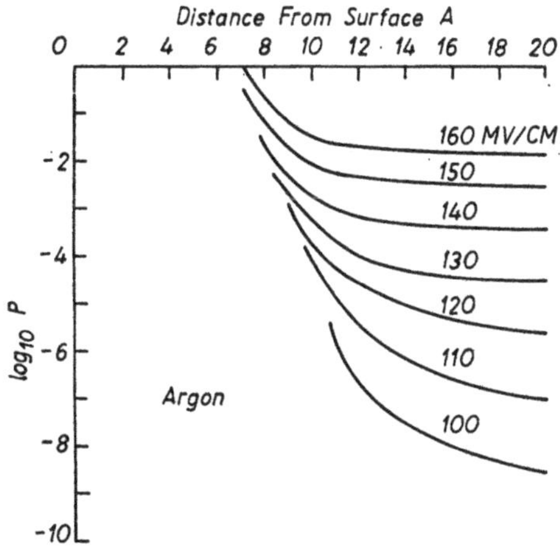

FIG. 1.2. Logarithm of probability P for an argon atom to be ionized while travelling from l to $l - 1$ Å toward the tip surface with a velocity determined by the dipole attraction. (According to Müller and Bahadur [3].)

1.2. PARTICLE SUPPLY

The number of particles $\dot{n}$ approaching the ionization zone has not yet been derived during the discussion of eqn. (1.5)—(1.7). This number is called the "supply function". Its value changes with field strength. At zero field strength, the number $\dot{n}_0$ of particles approaching an area A_0 is

$$\dot{n}_0 = A_0 \cdot P\,(2\pi m k T)^{-1/2} \tag{1.8}$$

where P is the gas pressure in (dyn/cm²), m the mass of the particles, k the Boltzmann constant and T the temperature.

In the case of production of a very high field F by a tip, the particles are attracted by the tip due to polarization. Müller [18] was the first to point out that 10—100 times more particles can be attracted by the tip with high fields as compared with zero field. The field strength F_0 at the surface of a tip can be approximated by

$F_0^{tip} \simeq \dfrac{F_0^{sphere}}{5}$, where F_0^{sphere} is the field strength at the surface of a small free sphere,

$F_o^{sphere} = \frac{U_o}{r_o}$, where U_o is the potential of the sphere, and r_o its radius.

Putting the tip radius r_t equal to r_o, one obtains

$$F_o^{tip} \simeq \frac{U_o}{5 \cdot r_t} \tag{1.9}$$

The potential energy of a particle in an electric field is

$$- V_{(F)} = \bar{\mu} F + 1/2\, \alpha F^2 \tag{1.10}$$

where $\bar{\mu}$ is the average permanent dipole moment in the field direction and α the polarizability.

The effective cross section A of the tip for capture of gas particles is increased by the electric field by a factor

$$A/A_o = \sigma = 1 - \frac{2}{3}\, V_{(F_o)}/kT \tag{1.11}$$

with V given by (1.10). (V has negative values.)

The cross section for particle supply to a tip may be approximated by a hemisphere, $A_o = 2\pi r_t^2$. The effective cross section in the presence of a field would then be $A = 2\pi r_t^2 \cdot \sigma$. The exact tip geometry has to be considered for more accurate calculations of the particle supply.

The supply function $\dot{n}$ has been replaced in eqn. (1.7) by the equilibrium concentration, c_t, near the tip. This concentration is different from the concentration c_g far from the tip if a field is applied. c_t is larger than c_g due to polarization:

$$c_t = c_g \cdot (T_g/T_t)^{1/2} \cdot exp\,(- V_{(F)}/kT_g) \tag{1.12}$$

if the gas and tip temperature (T_g and T_t) are different and if full thermal accommodation of the particles occurs at the tip. The equilibrium concentration c_t given by eqn. (1.12) is maintained only if the number of particles drawn off by ionization is small as compared with the number of neutral particles $\dot{n}$, leaving and approaching the tip. This condition is met by relatively low field strengths.

The increase in density due to polarization of the gas according to eqn. (1.12) can be very large. This increase may be several orders of magnitude at high field strengths and with molecules having a high permanent dipole moment. The gas pressure may exceed the vapor pressure of gaseous polar substances of low volatility so that condensation occurs on the tip. This phenomenon, called "field condensation" will be discussed in section 3.6.2. in greater detail.

Two limiting cases only have been discussed so far, ionization at extremely high fields which is controlled by the supply function only, and ionization at relatively low fields which is controlled by the equilibrium concentration c_t near the tip and the mean lifetime τ with respect to field ionization.

The situation is more complicated at field strengths intermediate between these two limiting cases. A fraction of the particles approaching the tip will be ionized, another fraction will hit the tip. Assume full thermal accommodation of the particles to the tip at the first collision. The velocity of the incoming particles will be higher than that of the rebounding particles, due to polarization. Therefore, the ionization probability in the ionization zone in front of the tip will be larger for rebounding particles than for incoming particles.

Müller [19] first pointed out that a fully accommodated particle having the energy $3/2\ k \cdot T_{tip}$ will be unable to leave the tip in one step if the polarization energy is higher than $3/2\, k \cdot T_{tip}$.

The particle will then perform a series of hops until it will be either ionized or diffuse to the shank of the tip where it may leave the emitter. Thermal accommodation is a function of temperature. Therefore, the ion current depends on temperature in a complicated way. (For more detail see [16, 19].) This will not be discussed here because by far the greater part of this book will be concerned with field ionization mass spectrometry of polyatomic molecules. The hopping mechanism plays a smaller role with these molecules than with rare gases like helium or neon because of physical or chemical adsorption of the gases.

The hopping due to the electrostatic polarization of the particles by the external field should not be confused with the hopping of adsorbed molecules from one adsorbing site to the next, in the absence of an external field. The first relatively far ranging effect may lead to hopping heights up to several Angströms, whereas the second, short ranging effect leads to hopping heights of the order of 1 Å only. Supply of particles by surface diffusion of physically adsorbed particles on the shank of a field ion emitter plays an important role. It may considerably exceed the particle supply from the gas phase. This surface supply will be discussed in chapter 3.6.6.

1.3. MINIMUM DISTANCE FOR FI

It can be seen from fig. 1.1.b that field ionization of an atom or molecule cannot occur at distances below a certain minimum distance d_{min} from the tip surface. The reason for this is that the levels below the Fermi level of the metal are occupied, and the level of the valence electron of the atom must be raised at least to the Fermi level for a transition of the electron to the metal to occur.

It may be seen from the figure that the energy of the electron must be raised by an amount

(1.13a) $I - \Phi - \frac{3.6}{d_{min}}$ for a singly charged particle, where $-\frac{3.6}{d_{min}}$ is an image term $\left(= -\frac{e^2}{4d}\right)$ accounting for the interaction of the ion with the surface after tunneling of the electron. V and Φ are given in volts, d_{min} in Å. The electron level is raised by this amount by the external field potential $F \cdot d_{min}$.

Solving for d_{min} and neglecting the image potential term leads to

$$d_{min} \simeq \frac{I - \Phi}{F} \tag{1.13b}$$

Inserting $I = 10$ eV for an organic molecule, $\Phi = 5$ eV and $F = 0.7$ V/Å (which is a typical value for mass spectrometric analyses) leads to

$$d_{min} \simeq 7 \text{ Å}$$

Much higher field strengths must be applied an atom should not be field ionized in the gas phase 7 Å distant from the tip, but in an adsorbed layer of about 1 Å thickness, for example ($d_{min} \simeq 1$ Å). This process is called "field desorption". Very high field strengths have also to be applied if atoms of the metal substrate itself are to be field ionized at room temperature. This process is called "field evaporation". The phenomena of field desorption and field evaporation will be treated in chapter 2.4.

1.4. FIELD ELECTRON EMISSION

A theoretical treatment of field electron emission and discussion of its numerous applications is not within the scope of this book. However, mass spectroscopists sometimes want to check the properties of tips used in the mass spectrometer. A convenient check consists in measurement of the field electron current as a function of the applied voltage V. For this reason, the basic current — voltage relation for field electron emission, the Fowler-Nordheim equation, will be discussed here very briefly. This equation can be written in the form [38]

$$\log I/V^2 = \log a - b \cdot \Phi^{3/2}/V \tag{1.14}$$

where I is the total ion current in Amperes and

$$a = A \times 1.54 \times 10^{-6}\,\Phi^{-1}\,(krt_{(y)})^{-2} \tag{1.14a}$$

$$b = 2.97 \times 10^{7}\,krv_{(y)} \tag{1.14b}$$

A is the total emitting area, $t_{(y)}$ and $v_{(y)}$ are correction terms depending on the argument $y = \frac{(e^3 F)^{1/2}}{\Phi}$

t and v are tabulated (see, for example, ref. 39)). $t_{(y)}$ is nearly independent of the field strength within limited ranges of applied voltage, the value being about 1.05 for typical working conditions.

k is given by $F = \frac{V}{k \cdot r}$, where k is a constant depending slightly on the geometry of the tip. $k \simeq 5$ may be used in many cases. r is the tip radius, and Φ the work function in volts.

A plot of $\log I/V^2$ versus $1/V$ from experimental data results in a virtually straight line, as shown in fig. 1.3. This figure gives

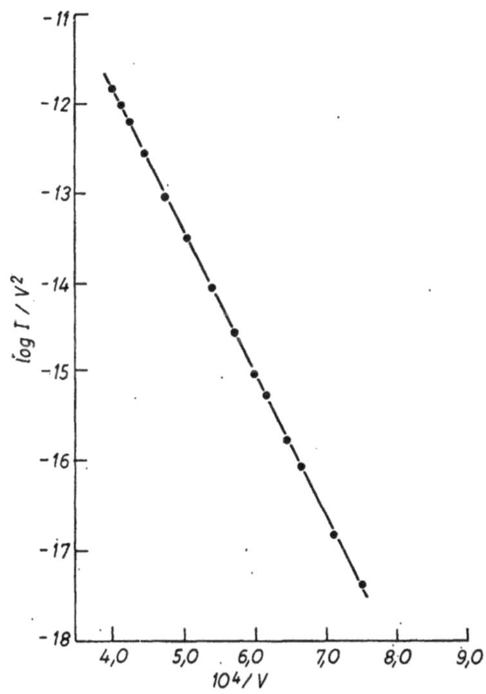

FIG. 1.3. Fowler-Nordheim plot log I/V^2 versus $10^4/V$ [114]. (Field electron currents as a function of applied voltage.)

some indications as to what total electron currents are to be expected at given voltages for a tip with $r \simeq 1000$ Å.

The slope S at any point of the Fowler-Nordheim curve (eqn. 1.14) is given by

$$S = -2.97 \times 10^7 \cdot \Phi^{3/2} \cdot k \cdot r \cdot s_{(y)} \tag{1.15}$$

where $s_{(y)} = v_{(y)} - \frac{y}{2}\left[\frac{dv_{(y)}}{dy}\right].$

The values of $s_{(y)}$ are also tabulated [39]. $s_{(y)}$ has a fairly constant value close to unity. For typical working conditions, $s \simeq 0.95$. It will be shown in section 2.2.2. how the tip radius can be derived from the measured slope S of the Fowler-Nordheim plot.

2. FIELD IONIZATION SOURCES

2.1. GENERAL DESIGN OF FI SOURCES

2.1.1. Non-Focusing FI Sources

There are two principal types of field ionization source which have been developed so far:

a) Non-focusing FI sources, working as sources for both field ionization microscopes and FI mass spectrometers, and
b) focusing FI sources for mass spectrometric analyses.

The first FI source realized experimentally by Inghram and Gomer [2] was of type *a*).

The source is shown schematically in fig. 2.1. It consists of a bulb of "Kovar" glass which is connected with a glass tube of 2.5 cm diam. This tube is sealed to a tube of "Vacon" metal, and this in turn is welded to a flange of "Inconel" metal which serves as a connection to a mass spectrometer of the usual design (single focusing 60° magnetic sector field instrument with secondary electron multiplier detector).

The ion source is made of glass in order to use it as a field electron microscope prior to mass spectrometric measurements. A luminescent screen (*L* in fig. 2.1.) permits visual observation of the field electron emission pattern of the tip. This observation enables one to judge the physical state of the tip, its coverage with adsorbed gases and its crystallographic orientation with respect to the entrance hole of the mass spectrometer. Taking a field electron current curve as a function of voltage (Fowler-Nordheim plot) leads to an approximate determination of the tip radius (see section 2.2.2.). The emission pattern can be observed through a flat glass window W. A tungsten tip of about 1000 Å radius of curvature is used as emitter. The tip is made of 0.1 mm tungsten wire by chemical etching. It is welded to a loop of tungsten wire of 0.25 mm diameter which serves for mechanical support of the tip and for heating it by passing a current through it. The temperature can be measured by two thin tungsten wires (0.075 mm diam.), welded to the loop a few mm from the tip and serving for measurement of the resistance of the loop. The resistance is calibrated as a function of temperature.

A ring of 0.5 mm tungsten wire is mounted in front of the tip. This ring serves as anode with field electron emission and as cathode with field ion emission. This ring prevents secondary

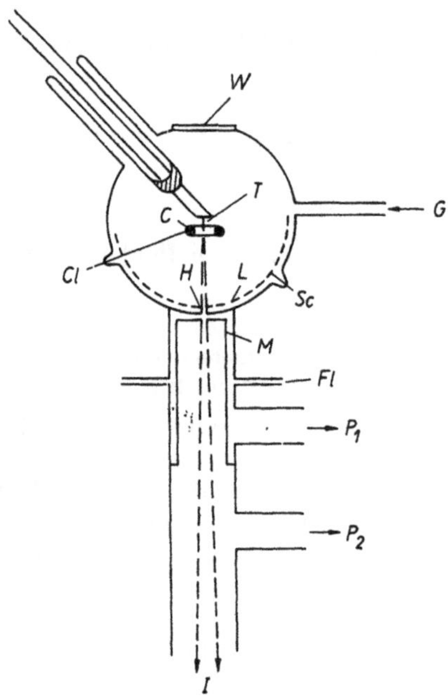

FIG. 2.1. Schematic diagram of field ionization source for mass spectrometer (according to Inghram and Gomer [1]) T = tip; C = cathode ring; CL = cathode lead; L = luminescent screen; SC = screen lead; H = aperture to mass spectrometer; M = entrance slit to mass spetrometer; W = observation window; P_1, P_2 = diffusion pumps.

electrons, produced by ions impinging on the luminescent screen, and field electrons released from the screen, from striking the tip. Moreover, Müller [18] has pointed out that reduction of the anode — cathode distance permits use of higher gas pressures without production of glow discharges.

The luminescent screen consists of zinc silicate deposited on electrically conducting glass. The method of producing conducting glass described by Gomer [20] is now widely used. A small hole is made in the glass carrying the luminescent screen, so that a small fraction of the ions emitted from the tip can pass through this hole. Usually the tip is adjusted carefully so that the (100) face (or another face) of the tungsten emission pattern covers the hole in the screen. Only a small part of the (100) face (about 20 × 20 $Å^2$) depending on tip radius is "seen" by the mass spectrometer.

A metal cylinder (M in fig. 2.1.) with a hole of about 1 mm diam is inserted into the glass tube of the ion source, the hole being close to the screen of the source. (The pressure is rather high in the source during field ionization experiments, up to 5×10^{-3} Torr.) Differential pumping in two stages (P_1 and P_2, fig. 2.1.) reduces the pressure to about 5×10^{-7} Torr in the mass analyser tube. The source is evacuated at the beginning of an experiment to about 10^{-8} Torr. The tip is cleaned by heating, and then a negative potential of several kV is applied to the tip and the field emission pattern is observed. After this, the polarity of the tip is reversed, and the gas to be field ionized is introduced into the ion source through adjustable leak valves.

Müller [22, 223] has described an FI mass spectrometer combining a non-focusing FI microscope source with a time of flight mass spectrometer. This instrument will be described in section 2.9.3. In section 2.9.2. other instruments will be described the properties of which are intermediate between an FI microscope and a simple focusing mass spectrometer [3, 23].

Rotatable tips

Ions of only a very small face of the tip (a few Å in diameter) are collected by the mass spectrometer, as mentioned above. Therefore, one of the most interesting applications of a non-focusing field ionization source would be the observation of ions produced at different single crystal faces. This should be done by use of a rotatable tip.

Rotatable tips for field emission work have been described in the literature [21]. Müller et. al. [22, 223] first used a rotatable tip in connection with a mass spectrometer in their "Atom-Probe FI Microscope".

2.1.2. Focusing FI Sources

Focusing field ionization sources are used for qualitative and quantitative mass spectrometric analyses in order to increase the ion intensity and to reduce the statistical fluctuations of the resolved ion beam. No visual observation of the field emission pattern is provided with these ion sources, in most cases, because it is unnecessary with most of the standard analytical mass spectrometric problems. The author reported first in 1958 on experiments with a field ionization source of this type [4, 24].

Many other FI sources similar to this source in principle, but different in the details of the ion optics, have been developed since then. Before describing of these sources, some general ion optical considerations will be given.

The ion optics used in the FI source depends on the geometrical shape of the field ion emitter. Either tips, thin wires or edges of sharp blades may be used as emitters. The use of tips will be considered first.

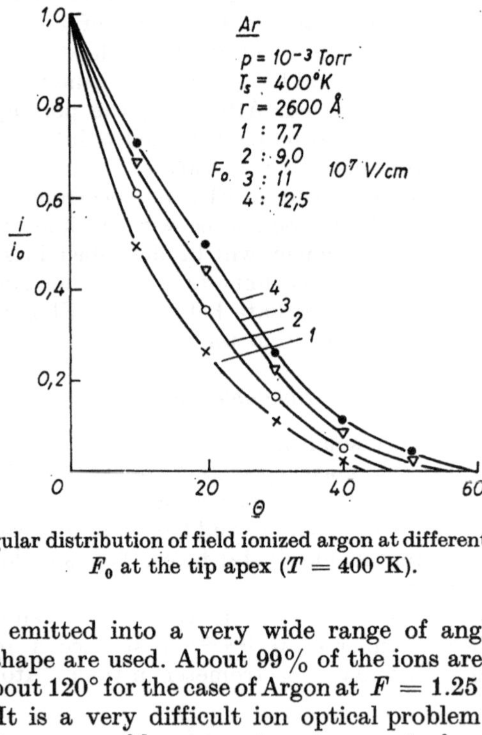

FIG. 2.2. Angular distribution of field ionized argon at different field strengths F_0 at the tip apex ($T = 400$°K).

Ions are emitted into a very wide range of angles if tips of parabolic shape are used. About 99% of the ions are emitted into a cone of about 120° for the case of Argon at $F = 1.25 \cdot 10^8$ V/cm[25] (fig. 2.2.). It is a very difficult ion optical problem to focus ions emitted into a cone of large aperture to a spot of small diameter. The largest image defect arises from the spherical aberration of the electrostatic lens system which increases with the third power of the aperture.

An aperture which can be easily handled with FI sources in connection with a mass spectrometer has a half angle of the order of 1° only, if no special care is taken in the design of the lens system. The angular distribution of the emitted ions as shown in fig. 2.2. may be approximated by a straight line for order of magnitude calculations. The relative intensity is put equal to one at $\vartheta = 0°$ and it drops linearly to zero at $\vartheta_{max}/2 = 60°$.

Then the fraction $\alpha = \frac{\vartheta_{foc}}{\vartheta_{max}} \times 2 = 1/60$, where the angle for focused ions, $\vartheta_{foc} = 1°$, and $\vartheta_{max} = 120°$. The "source transmission" T is defined as the fraction of the ions focused onto an aperture at the exit of the ion source as compared with the total ion current.

Evidently $T_v \approx \alpha^2$, for small angles ϑ_{foc}.

In the numerical example given above, $T = 1/3600$. Assuming that a factor of 3 in ion intensity is lost between the entrance and the exit aperture of the mass spectrometer, an "overall transmission" of about 10^{-4} is obtained. The overall transmission is defined as the number of ions collected behind the exit slit of the mass spectrometer divided by the total number of ions emitted from the field ion emitter. This figure has been reconfirmed with the first focusing ion source [4] where no special efforts were made in optimizing the performance of the lens system.

An experimental determination of the overall transmission can be made by measuring the total ion current emitted by the tip by means of a sensitive dc amplifier which is put into the electrical connection to the tip. The tip is usually at a positive potential of several kV with respect to ground in FI mass spectrometers. Therefore, the amplifier must be separated from the mains by a highly insulating 1:1 transformer, and it must be placed in a well insulated box. Total tip FI currents to be measured are of the order 10^{-10}—10^{-8} A.

The secondary electrons released from impact of the ions on the cathode have to be considered. The sum of emitted ions and collected secondary electrons is measured by the amplifier in the tip admission line. The number of secondary electrons released by impact of a 10 kV ion is of the order of 4, depending on the kind of the ion and the cathode. Therefore, the measured total emission current has to be divided by a factor of about 5 in order to obtain the true ion current. More exact factors regarding the ion energy, the cathode material and the kind of ions have to be taken from the literature [26].

In the case of a lens system focusing only ions emitted into an aperture of 1—2° one could argue that one could completely omit the lens and use the magnetic sector field itself as a lens. The tip could then replace the entrance aperture of the mass spectrometer. However, the magnetic field focuses the ion beam only in a plane perpendicular to the magnetic field lines (R — plane) but not in the direction parallel to the field line (z — direction). So, there should be placed at least a slit lens focusing the beam in the z — direction. Moreover, total voltages of more than 10 kV total often have to be applied between tip and cathode. Many mass spectrometers are operated with positive potentials not exceeding 3—5 kV with respect to ground. Thus a negative voltage of at least 5—7 kV has to be applied to the cathode. After this, the ions have to be decelerated to ground potential. The potential

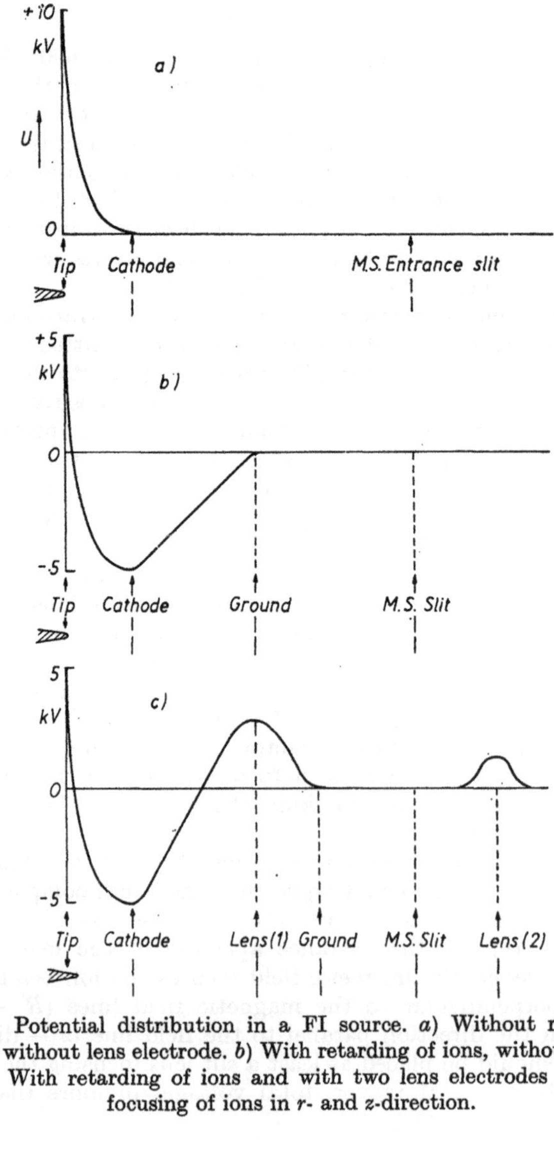

FIG. 2.3. Potential distribution in a FI source. *a*) Without retarding of ions and without lens electrode. *b*) With retarding of ions, without lens electrode. *c*) With retarding of ions and with two lens electrodes (L_1, L_2) for focusing of ions in *r*- and *z*-direction.

distribution without and with deceleration of the ions is shown in fig. 2.3. The cathode and the next electrode on ground potential together are acting as a lens system focusing the ion beam. It cannot be expected that a fixed positive and negative voltage — necessary to produce a certain required field strength — will by chance satisfy the optimum focusing conditions of the beam. In other words, the beam will be generally defocused under these conditions. Robertson and Viney [27] have calculated approximately the loss of transmission with retardation of the ions. 90% of the ions could be lost if the ions are accelerated by a potential difference of 11 kV and then decelerated by a retarding potential of 10 kV. It is therefore advisable to introduce another electrode in order to obtain a complete electrostatic lens system, the focal length of which can be adjusted by the potential of this new electrode, L, independently of the positive voltage of the tip and of the negative voltage of the cathode. The situation is illustrated by fig. 2.3.c. The whole lens system now consists of three discs with spherical apertures. The potential of the lens electrode L_1 is variable between the ground and the positive tip potential, but is usually nearer to the tip potential for optimum focusing. The ion beam is focused onto the entrance slit of the mass spectrometer. The slit has a width from about 0.3—1.0 mm depending on the required resolving power of a single focusing mass spectrometer. It is not meaningful to use much narrower slits with single focusing instruments for the analysis of organic substances because the resolving power is limited to a value of $M/\Delta M$ of about 600 (with 10% valley definition) due to energy inhomogeneity of the ions, as will be discussed later. (The data are given for a mass spectrometer having a 60° magnetic sector field with a radius of 20 cm).

A spherical aperture could be used instead of the entrance slit of the mass spectrometer because of the spherical symmetry of the focused ion beam originating from a tip. However, a slit seems to be more adequate because some of the ions not perfectly focused by the lens system in the direction of the slit may then contribute to the ion current collected by the mass spectrometer.

The path of the ions in the y — direction (contained in the "R — plane") as well as that in the z — direction perpendicular to this is shown in fig. 2.4.a—c. A pair of deflecting electrodes is placed between lens L_1 and the entrance slit (fig. 2.4.a). After crossover within the entrance slit, the ion beam diverges again into a cone. It will be focused again by the magnetic sector field only in one direction (fig. 2.4.a, in the R-plane as mentioned above), but not in the z — direction (see fig. 2.4.b). Therefore, it is advisable to place an additional lens behind the entrance aperture which focuses the beam in the z — direction (see fig. 2.4.c). This must be a slit lens (L_2), in contrast to the spherical lens (L_1). Lens L_2 consists of three electrodes, the outer two being at ground potential the central one on a variable positive potential of a few kV maximum. The two parts of the central electrode may be supplied with separate, slightly different voltages so that the ion beam may be adjusted in the z — direction.

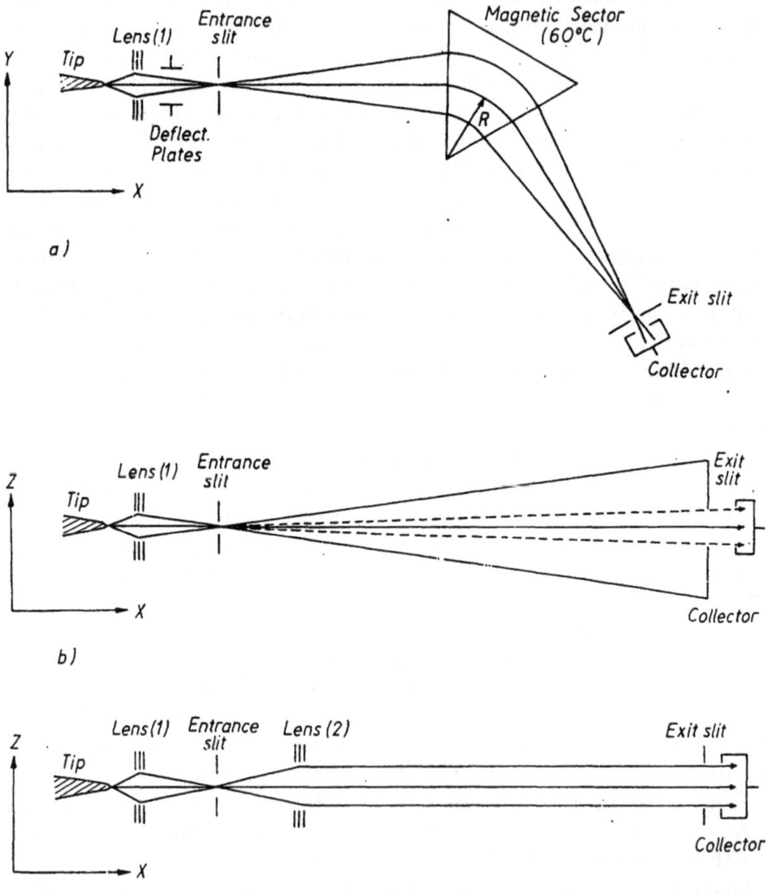

FIG. 2.4. Path of the ions in a FI mass spectrometer. *a*) Path of ions in the $x-y$-plane ("R-focusing" by lens L_1). *b*) Path of ions in the $x-z$-plane (without "z-focusing"). *c*) Path of ions in the $x-z$-plane (with "z-focusing" by lens L_2).

2.1.3. Lens systems

The question arises which of the different possible lens systems would be most suitable for use in a field ionization source. The answer depends in part on the geometry of the emitter system. Tip emitters will be discussed first.

It was suggested by Drechsler [7] that a tube lens should be used for focusing of the strongly diverging ion beam emitted from a

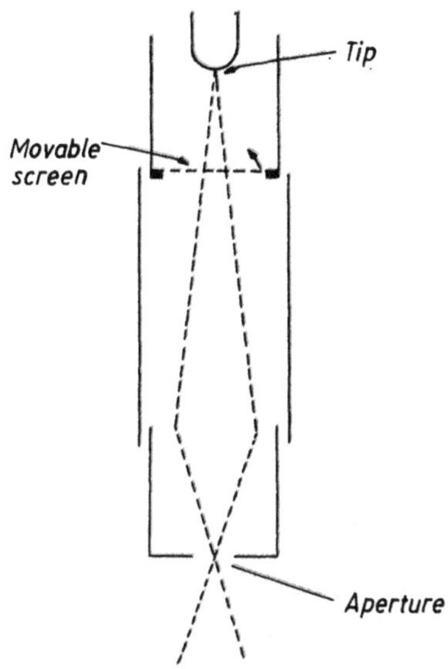

Fig. 2.5. Schematic presentation of a FI source as used by Schmidt [8, 28]. Electrostatic lens consisting of three concentric tubes.

tip. A tube lens was used in FI sources as constructed by Schmidt [8] [28] and by the author [5]. The advantage of using a lens consisting of three tubes of relatively large diameters (a few cm) lies in the fact that a movable luminescent screen can be placed in the first tube, thus permitting the field emission pattern of the tip to be observed alternatively to the mass spectrometric measurements. The essential parts of such a source as used by Schmidt are shown schematically by fig. 2.5.

The focusing properties of tube lenses and apertured disc lenses in connection with FI sources have been compared theoretically [29]. The results are as follows. It is well known that a beam emitted into a large aperture cannot be focused indefinitely sharply. This is due to the image defects of the lens system. The object to be depicted is a point lying on the axis of the tube lens in the case to be discussed here. Therefore, the only image defect among possible other defects will be, in this case, the spherical aberration of the third order. A beam leaving the plane of the object under the angle ϑ with respect to the axis arrives in the image plane at a distance r from the axis:

$$r = c_3 \cdot V \cdot \vartheta^3 \tag{2.1}$$

Here, V is the magnification and c_3 the constant of spherical aberration. The product $c_3 \cdot V$ should be made as small as possible in order to make the radius r of the image small. c_3 is a function

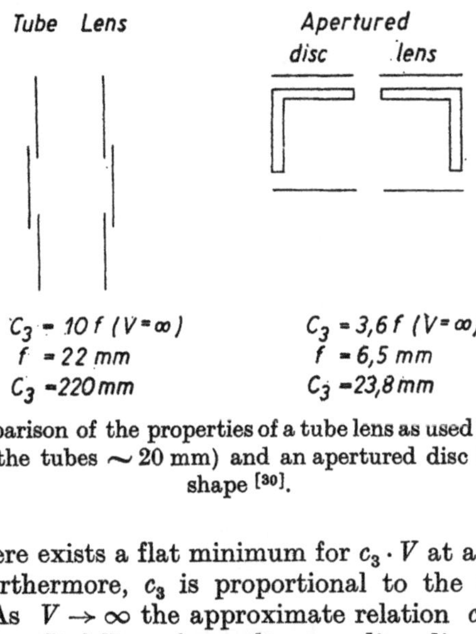

FIG. 2.6. Comparison of the properties of a tube lens as used in a FI source [29] (diameters of the tubes ~ 20 mm) and an apertured disc lens of optimum shape [30].

of V, and there exists a flat minimum for $c_3 \cdot V$ at a magnification $V = 3.5$. Furthermore, c_3 is proportional to the focal length f of the lens. As $V \to \infty$ the approximate relation $c_3 = 10\, f$ holds for tube lenses. It follows from the preceding discussions that c_3 can be further reduced by minimizing the focal length, or in other words, by reducing the dimensions of the lens system. The other possibility of reducing c_3 consists in reducing the factor A of the relation $c_3 = A \cdot f$ by suitable electrode geometries, other than that of the tube lens. Seeliger [30] has compared apertured disc lenses of different shape and found that the factor A could be made three times smaller as compared with a tube lens for a special lens geometry as shown in fig. 2.6. The focal length of this lens system could also be reduced by a factor of three as compared with the tube lens. The resulting data are collected in fig. 2.6. It can be shown [29] that the transmission T — defined as the ion current passing through an aperture of diameter d after focusing by a lens, divided by the total emission current of a tip — is given by

$$T = \left(\frac{d}{2 c_3 \cdot V}\right)^{2/3} \tag{2.2}$$

The value c_3^t of a tube lens is about ten times larger than that of the optimum disc lens c_3^d. Therefore,

$$T^d/T^t \simeq 10^{2/3} \simeq 4.6$$

Summarizing, an optimized disc lens is preferable to a tube lens because a very high transmission is required for most of the analytical problems, as will be discussed below.

Correction of image defects by octopole systems

Another possibility of increasing the transmission is to correct the spherical aberration of a tube or disc lens by an octopole system. Seeliger [31] has realized such a system in electron microscopes. However, the beam apertures are much larger in FI sources than in sources for electron microscopes. Therefore, the spherical aberration of 5th order plays a larger role in FI sources. The corresponding constant c_5 is much larger for octopole systems than for tube or disc lenses. Therefore, the transmission is increased only about 10% by correction of a lens with an octopole system and will not be considered further.

Use of quadrupole systems

Giese [32] has shown that a system of two quadrupole lenses can be used for producing an image having the form of a slit, the original beam having a quadratic cross section. This obviously has advantages in connection with a mass spectrometer containing a

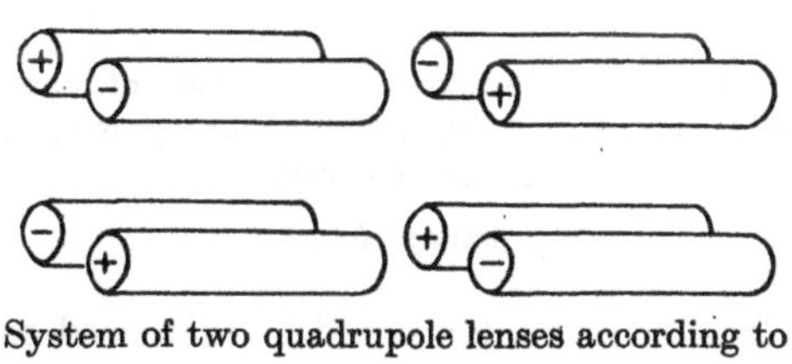

Fig. 2.7. System of two quadrupole lenses according to Giese [32].

series of slits. A beam leaving the tip of an FI source with rotational symmetry could also be made slit-shaped by a quadrupole system. The system consists of eight cylindrical pole pieces of alternating potential, as shown in fig. 2.7. The system focuses positive ions in the + direction (convex lens) and defocuses them in the − direction (concave lens).

There are two different possibilities for using quadrupole lenses for focusing an ion beam leaving an FI tip:

1. Correction of the spherical aberration of a tube or disc lens by the quadrupole system.
2. Focusing by two quadrupole lenses only.

1. Combination of a spherical lens with two quadrupole lenses

Fig. 2.8. shows the path of the beam in a combined spherical and quadrupole lens system in a schematic manner. Two quadrupole lenses behind the spherical lens are focusing the pre-focused beam such that it just covers the area of the mass spectrometer

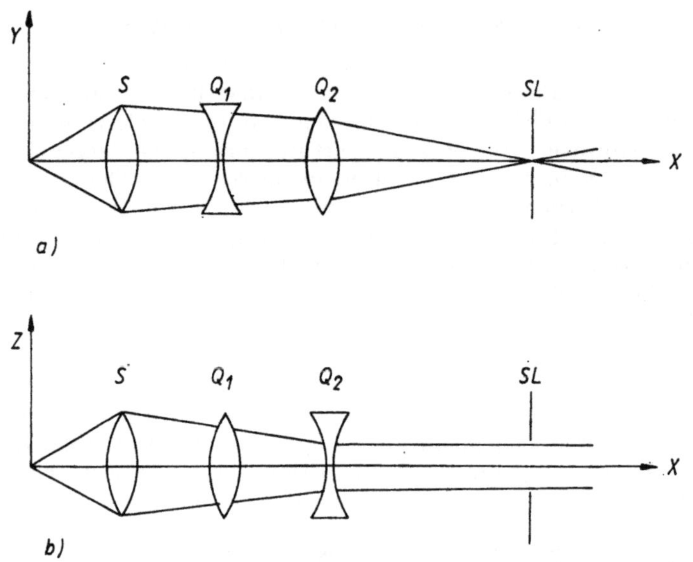

FIG. 2.8. Path of ion beam through a combination of a spherical and two quadrupole lenses, represented by a light beam passing through a set of focusing and defocusing lenses.

a) $x-y$-plane; *b)* $x-z$-plane.

entrance slit. The transmission could be increased experimentally — by optimizing all lens parameters — by a factor of 2.3 as compared with use of a spherical lens only. The quadrupole system itself has a spherical aberration of the same magnitude as the spherical lens, but in the combination of the two lens systems both image defects are compensating, as can be shown theoretically [29]. Focusing with a quadrupole lens was more perfect in the case of Giese's system [32]. The difference lies in the fact that in that case, the ion beam was homogeneous in energy and exactly parallel, whereas in the case of a FI tip the ion beam is strongly diverging, and consequently the gain in transmission is restricted to a factor of 2.3.

2. Use of two quadrupole lenses only

Two quadrupole lenses can be used without previous focusing by a spherical lens. The tip can be placed directly in front of the quadrupole system. It can be shown experimentally that the quality of the focusing is not better than with a tube lens alone. The constant of spherical aberration is two powers of ten larger for the second quadrupole lens (convex lens) than for the first lens (concave lens). Therefore, the image defects could not be decreased below those of a tube lens, which in turn are larger than those of an optimized disc lens as discussed earlier.

Summarizing, the combination of a spherical lens with two quadrupole lenses seems to improve the transmission of the FI source by about a factor of two. It seems to be worth while to make use of this improvement for special investigations like that of Jason, Burns and Inghram [33] (structure in FI energy distributions). However, for routine investigations with a FI mass spectrometer the need for adjustment of four additional potentials to optimum value seems to be a drawback which is not compensated for by a gain in transmission by a factor of two. A well designed apertured disc lens system seems to be quite adequate for normal FI sources.

Lenses for use with wires or sharp edges as FI emitters

Thin wires or sharp edges of metal blades may be used instead of tips for field ionization. The ion beam is then of cylindrical instead of spherical symmetry. Clearly, a cylindrical lens consisting of a series of slits is then more adequate for focusing of the ion beam. In the case of a sharp edge, the emitter usually has about the same length as the slit system, and the edge is placed parallel with and near to the first slit [27]. The beam leaves the edge as a parallel beam in the direction of the longer axis of the slit (z — direction) and is strongly diverging in the direction perpendicular to this (y — direction). The slit lens focuses a small fraction of the ions spread over the y — direction.

The situation is somewhat different with thin wires. It is advisable to make the emitting part of the wire as short as possible in order to reduce the chance of rupture of the thin wire. Usually, a length of about 3 mm is etched down to the characteristic diameter of a few times 10^{-4} cm if Wollaston wires are used. About 2 mm of this length are effectively contributing in field ionization.

It is then preferable to place the wire perpendicular to the slit as shown in fig. 2.9. which shows the emission pattern of the ions which can be observed on the first slit electrode after several hours ion bombardment. The emission pattern fits well with the slit. The ion beam is nearly parallel in the y — direction and can be easily focused onto the entrance slit of the mass spectrometer which has a width of 0.1—0.3 mm. The beam is strongly diverging in the z — direction, but the focusing is less critical in that

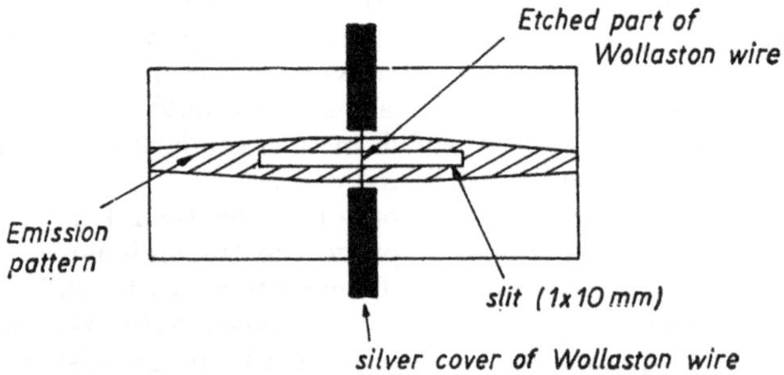

FIG. 2.9. Coloured spot "burnt" by ions imping on the cathode.

direction because the beam may have a width of about 5 mm even at the narrowest part of the mass spectrometer tube between the magnetic pole pieces. In this case, a lens with slits parallel to the z — direction is placed near the FI wire and another lens with slits parallel with the y — direction is placed behind the first lens.

2.1.4. Fluctuations of the FI Currents

A serious problem in FI mass spectrometry consists of the relatively strong, statistical, short-time fluctuations of the ion intensity. This has to be differentiated from the long time changes in the sensitivity coefficients for the substances to be analysed, as will be discussed in section 3.6.5. The author and Schütte carried out investigations on the nature of these fluctuations at a relatively early stage of development of FI mass spectrometry [29]. It turned out that these fluctuations are of a completely statistical nature. Two cases have to be differentiated. First, at extremely high field strengths such as are produced with tungsten tips (5×10^8 V/cm) field evaporation of atoms of the tip material takes place. This phenomenon was investigated in detail first by Müller [34]. The effect will be discussed in section 3.6.4. Field evaporation of tip surface atoms influences the FI current of the organic molecules to be investigated in the following manner. Assume that the transmission of the mass spectrometer would be 10^{-4}, i.e. only a fraction of 10^{-4} of the ion current produced at the tip is collected behind the exit slit of the mass spectrometer. The diameter of the effective emitting area of the tip may be 1000 Å. Then only a very small spot within the tip area, having a diameter of about 10 Å is contributing to the collected ion current. In other words, this emission spot has a diameter of a few atoms only. The micro field strength distribution at this spot may be greatly changed if only one or a few atoms of the spot are removed. Very small changes in field strength can cause very large changes in ion current because the ion current increases with a high power of the field strength at field strengths where the ionization probability is smaller than one. At higher field strengths, where the gas supply limits the ion current, the FI current still increases with about the third power of the field strength.

Removal of atoms of the tip material can also occur by field corrosion in the range of $1 - 5 \times 10^8$ V/cm. Field corrosion consists of chemical reactions of atoms like nitrogen, hydrogen or oxygen with the substrate atoms of the tip and subsequent field evaporation of the surface complex [35, 28]. This process will be discussed in section 2.2.5. Field evaporation of metal surface complexes may occur at lower field strengths than field evaporation of the pure tip material. Therefore, new tips inserted into an FI source which have initially a radius of curvature of 1000 Å or less are quickly transformed to blunter shapes in the presence of most of the substances to be analysed mass spectrometrically, with high voltage applied. Schmidt [28] has shown that traces of water which are usually present in the residual gas of evacuated mass spectrometers are sufficient to produce quick blunting of tungsten tips. Blunting of very sharp platinum tips can be carried out by applying about +10 kV to the tip and by introducing methanol at a pressure of a few microns during a period of several hours. The equilibrium radius of curvature of tungsten or platinum tips after extended mass spectrometric analyses is of the order of 10000 Å.

This blunting of the tips has great analytical advantages. It was shown above that the effective emission spot at the surface of a 1000 Å tip has a diameter of only about 10 Å at a transmission of the mass spectrometer of 10^{-4}. Removal of only a few atoms leads to considerable changes in micro field strength. If the radius of the tip is increased to 10000 Å, the area of the emission spot increases by a factor of a hundred, the aperture of the focused ion beam remaining constant. Removal of the same small number of atoms as considered above from the emission spot causes a smaller local perturbation. The emitted FI current should therefore be more stable than in the case of a very sharp tip. This is in accordance with the experimental experience. The relative fluctuations of the FI current emitted from very sharp tips may go up to the order of $\pm 50\%$ in the presence of most gaseous substances. However, these fluctuations may be reduced to about $\pm 5\%$ with well rounded tips having radii of the order of 10000 Å.

Increase of the stability of the emitted FI current may also be obtained by increasing the angle of aperture of the tip which can be focused by the lens system. More atoms of the tip are then active in producing ions which can be collected by the mass spectrometer, and hence removal of a few surface atoms causes a less drastic change of the ion current. It is for this reason that much attention was directed to constructing ion optical systems having a high transmission. The reduction of the fluctuations of the FI current is a more important aim than the gain in intensity of the collected ion beam, since sensitive multiplier detectors may be used for obtaining a sufficiently high signal from FI sources with only moderate transmission.

The collector current of the mass spectrometer is about 10^{-12} A with a total ion current of 10^{-8} A emitted by the tip and a transmission of 10^{-4}. This corresponds to a collection of about 0.6×10^7 ions per second. This number is too large to explain the observed fluctuations of the ion current of about $\pm 20\%$ by statistical scattering of the number of collected ions around the average value.

The statistical fluctuations of the collected ion current are caused by changes of the micro structure of the tip as mentioned above. The number of micro surface elements changing their structure and emission properties independently of each other per second may be equal to N. The instrument recording the ion current has a time constant t, and it averages over Nt events. Then a scattering of the events is to be expected — as with radioactive decay — which corresponds to counting of Nt events

$$dI/I = (Nt)^{-1/2} \tag{2.3}$$

where dI/I is the fluctuation of the total emission. If only a fraction v of the total emission is collected, only $n = vN$ events are recorded. The fluctuation of the corresponding collected ion current i is enhanced to

$$d\,i/i = (nt)^{-1/2} \tag{2.4}$$

With the relation $n = i \cdot N/I$ it follows

$$d\,i/i = (N \cdot t \cdot i/I)^{-1/2} \tag{2.5}$$

The relative fluctuation of the collected ion current i should be proportional to $(i/I)^{-1/2}$, or in other words, inversely proportional to the square root of the transmission T

$$d\,i/i \approx 1/\sqrt{T} \tag{2.6}$$

This relation was tested experimentally by changing the transmission by using apertures which could be changed continuously in size [29]. The result is shown in fig. 2.10. It is in qualitative agreement with the equation given above.

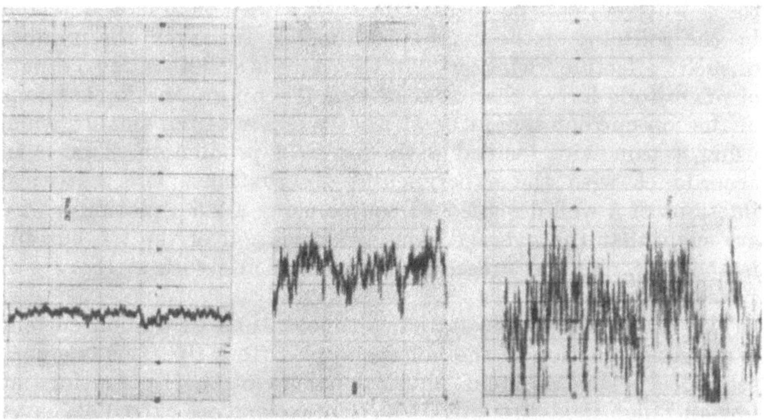

Fig. 2.10. Fluctuations of the FI current at different transmissions i/I (i = collected ion current, I = total ion current).

a) $i/I = 10^{-2}$; b) $i/I = 7 \times 10^{-4}$; c) $i/I = 6 \times 10^{-5}$.

The second type of phenomena causing fluctuations of the ion current occurs at relatively low field strengths. It will be shown later that whiskers or other micro needles grow on the smooth emitter surfaces of tips or thin wires at field strengths of about 1×10^7 V/cm at the smooth emitter surface. For simplifying the nomenclature, micro needles consisting of organic semiconducting polymers will be termed also "whiskers".

These whiskers grow rather quickly on wires of 2.5 μ diameter if a high voltage of about + 12 kV is applied to the wires and acetone at about 10^{-3} Torr is introduced into the FI source. Whisker growth on tips occurs at about + 2 kV with a tip of about 1000 Å radius of curvature. Whisker growth on tips and wires has been proved by taking electron micrographs (see section 2.2.3.). The whiskers are torn off at higher field strengths. A dynamic equilibrium between whisker formation and removal is established at the voltages indicated above. The appearance and disappearance of whiskers is connected with fluctuations of the ion current. Again it is to be expected that the relative fluctuations of the current are inversely proportional to the square root of the number of whiskers. The number of whiskers observable on the tip apex is of the order of ten, whereas the number of whiskers on a 2.5 μ wire of about 2 mm length is of the order of 10^5-10^6. (The average whisker length is of the order of 1000 Å and the average diameter about 100 Å if the whiskers are grown by acetone.) The emission current of only a small fraction of these whiskers can be focused in such a way that it is detected by the collector of the mass spectrometer. However, the number of such "effective" whiskers of a thin wire is still about two orders of magnitude larger than that of tips. Therefore, the fluctuations of the ion current should be at least a factor of ten smaller when using a thin wire instead a tip for field production. This is in accordance with the experimental observations. The statistical fluctions of a well designed FI source using a 2.5 μ platinum wire are such that the average standard deviation of the ion current is about 1% for ten subsequent measurements with a recorder of 0.3 sec rise time.

The magnitude of the statistical fluctuations of the ion current is a strong function of the voltage applied to a tip. This is a consequence of the different physical nature of the fluctuations at low and high field strengths. The fluctuations are extremely large (up to the order of ± 50% of the average value of the current) at low voltages (and hence low field strengths). This is due to the very small number of whiskers on the tip apex changing with time. At high voltages, the whiskers are torn off, and FI takes place at a smoother tip surface. Fluctuations due to field evaporation or corrosion of tip surface atoms can be reduced to about ± 5% of the average value of the ion current at high voltages.

The fluctuations of the beam emitted from thin wires are not so strongly dependent on the voltage as in the case of tips. This is a consequence of the distribution of the whisker size on wires, as will be discussed in section 2.3.3.

2.2. TIPS AS FI EMITTERS

Methods for production of tips suitable for field ionization and their emission properties will be discussed in this section.

2.2.1. Methods for Production of FI Tips

Methods for production of fine metal tips suitable for FI microscopy have been described by Müller [37]. These methods are also suitable for production of tips as used in FI mass spectrometers, the only difference being that larger tip radii are usually required in FI mass spectrometry.

There are two main methods for tip production. The first consists in dipping a wire of the metal concerned, with a diameter of about 0.1 mm, periodically into a molten salt such as $NaNO_2$ or into a solution as indicated in tab. 2.0. The periodic dipping can be made simply by hand or by a device driven by a motor if etching times of several minutes are required. The second method consists in electrolytic etching of the wires in molten salts or in

TABLE 2.0. TIP ETCHING PROCEDURES (E. W. MÜLLER [37])

Metal	Etchant	Remarks
W	Molten $NaNO_2$	Dip wire into melt
	Molten $NaNO_2$	0.5—1 v ac
	Aqu. sol. NaOH	1—5 v ac
Ta	Molten $NaNO_2$ or KOH	Up to 8 v ac
Re	Conc. HNO_3	10 v dc
Ir	Molten NaCl	2—3 v ac
	Aqu. Sol. KCN	2—3 v ac
Nb	Molten $NaNO_2$	6 v ac
Mo	Same as W	
Pt	Molten NaCl	5.5—6 v dc
	Aqu. Sol. KCN	1 v ac
Zr	10% HF	Dip into solution
Be	Conc. H_3PO_4	30—50 v dc
Rh	Aqu. sol. KCN	1 v ac
Si	Sol: 45 parts HF (40%), 60 parts HNO_3 (conc.), 20 parts acetic acid, 3 parts bromine	Dip into fresh solution
Au	50% HCl, 50% HNO_3	10 v ac
Fe	1% HCl	0.5—1 v ac
Co	10% HCl	4—6 v dc
Ti	40% HF	4—12 v dc
Pd	30% HCl, 70% HNO_3	3 v ac
Ni	40% HCl	1—2 v ac
Cu	Conc. H_3PO_4	1—5 v ac
Zn	Conc. KOH	10—15 v dc
Sn	40% HF	1—6 v ac

aqueous solutions of salts with a few volts *ac* or *dc* applied. Müller points out that there is a large variation in the treatment possible. One has to find out the optimum conditions of temperature, applied voltages and currents, composition of the etching bath and etching time, because of the different behaviour of a specific sample.

The tip wires have to be inserted as nearly vertically as possible into the etching bath. The etching process is examined by microscopic inspection from time to time. Optical microscopes at about 500 times magnification are generally used. The actual end of a tip with a radius of about 1000 Å can be observed in detail only with an electron microscope. However, one can judge the quality of a tip after having gained some experience from observations in the optical microscope. The tip should show a long, slender taper.

2.2.2. Measurement of Tip Radius and Shape Factor

Different methods have been described in the literature for determination of tip radii. One has been described by Gomer [16] making use of the Fowler-Nordheim equation (eqn. 1.14, chapter 1.)

A graph of log I/V^2 versus $1/V$ is taken and the slope $S = -b \times \Phi^{3/2}$ is determined.

$$b = 2.97 \times 10^7 \cdot k \cdot r \cdot s_{(y)}$$

If $s_{(y)}$ is taken to be unity and $k \simeq 5$, a value of r, correct to within 20 percent, can be found at once from S if Φ is known and is uniform.

A more accurate method consists in inserting the lower limit of $k \cdot r$ found from S into the equation $F = \dfrac{v}{k \cdot r}$ by taking $s = 1$ Thus, a mean value of the field strength F for the limited range of voltages of the experiment is found. s is a function of the parameter y, which can be approximated by $s = (1 - y)^{1/2}$, where $y = 3.8 \times 10^{-4} F^{1/2}/\Phi$. y may be found from the mean F value as just determined. Then s can be calculated or taken from a table [39]. s is inserted in b, and an upper limit of kr is thus found, from which a new F and hence s can be found. Iteration leads quickly to a good value of kr. r can then be determined if the "shape factor" k is evaluated. This factor can be assumed to be approximately

$$k = 5$$

for a certain range of tip geometries. However, its exact value depends on the detailed shape of the tip which can be determined by several methods.

The most direct method of determining the tip shape is, of course, observation of the tip in an electron microscope. Fig. 2.11. shows typical shapes of FE tips as observed in an electron microscope. The shapes change between that of a hyperboloid and that of a sphere on a slander shank. The potential distribution at such tips can be approximated in different ways. Haefer [40] approximated the tip by a hyperboloid, for which the formula

$$k = \frac{1}{2} \ln (4R/r) \tag{2.7}$$

is obtained, where R is the tip to screen distance.

Becker [41] used a paraboloid, for which the formula

$$k = \frac{1}{2} \ln (R/r) \tag{2.8}$$

is obtained. More generally, Drechsler and Henkel [42] approximated different tip shapes by superimposing the potential of a sphere onto the potential of a hyperboloid, the center of the sphere coinciding with the center of the sphere of curvature at the apex of the hyperboloid.

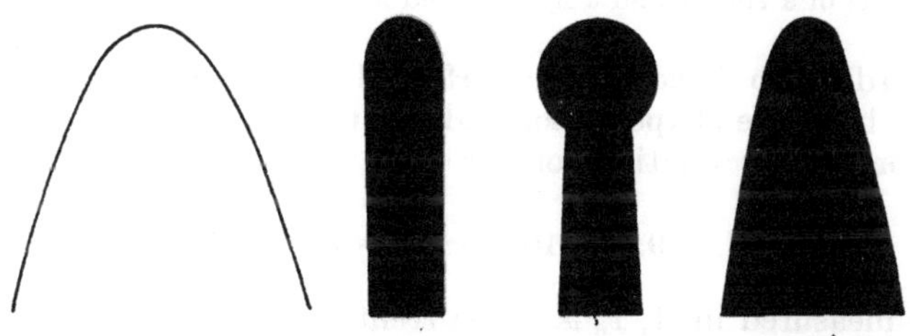

FIG. 2.11. Typical shapes of FE tips as observed in an electron microscope (Drechsler and Henkel [42]).

Fig. 2.12. shows some equipotential surfaces as obtained by superposition of the potential of a sphere and a hyperboloid, at constant tip radius. These different equipotential surfaces change with one parameter only, the "shape factor" α, which is the contribution of the spherical field as a fraction of unity. $1 - \alpha$ is the contribution of the hyperboloid field. The maximum field strength at the tip apex is

$$F_0 = (1 - \alpha) F_{hyp} + \alpha F_{sphere} = V \left(\frac{1 - \alpha}{\frac{r}{2} \cdot \ln \frac{4R}{r}} \right) \tag{2.9}$$

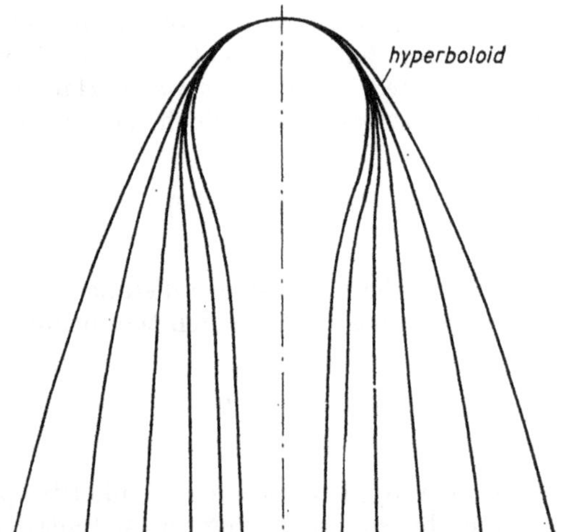

FIG. 2.12. Equipotential surfaces as obtained by superposition of the potentials of a sphere and a hyperboloid at constant tip radius [42].

According to Drechsler and Henkel [42] one method of determining both the shape factor and the tip radius consists in using an approximate relation for the total FE current emitted from a tip:

$$I = 9.5 \times 10^{-10} \pi^2 \cdot r^2 \cdot I_0 F_0 \cdot e^{6.4\alpha} \tag{2.10}$$

I is measured in A, I_0 is the current density at the tip apex in A/cm² as calculated by combination of the Fowler-Nordheim equation (1.14) and eqn. (2.10.), F_0 is the field strength at the apex in V/cm.

Both r and α are unknown. Two pairs of measured values I and V inserted into eqn. (2.9) and (2.10) lead to two equations which may be solved for r and α. The accuracy of this method is about 5—10%. Drechsler and Henkel described also more exact methods for determining r and α which can be carried out if the tip emission pattern is observed in an FE microscope.

Dyke et al. [43] approximated different tip shapes by a sphere mounted on an orthogonal cone. The method is somewhat more complicated than that of Drechsler and Henkel because three shape factors have to be adjusted instead of one. However, the parameters can be adjusted to make the equipotential surfaces conform to almost any tip shape.

A formula for more approximate determination of tip radii, irrespectively of their shape, was given by Müller [44]

$$r = \frac{0.85}{\varphi^2} \cdot V^{*5/4} \tag{2.11}$$

where r is given an Å, φ the average work function in volts and V^* the voltage applied to the anode in order to obtain a total tip emission current of 10^{-5} Amp.

A slightly improved formula given by Drechsler and Henkel [42] is based on the observation that very fine tips tend to have hyperbolic shapes, whereas tips having larger radii tend more to the shape of a sphere on a slender shank. The relation is

$$r = \frac{0.15}{\varphi^2} \cdot V^{*3/2} \tag{2.12}$$

with the same meaning and units of the parameters as with (2.11).

A more direct method of determining tip radii, and especially of the variation of the local radii of curvature on the tip surface, has been discussed by Müller [37]. The method consists in counting the number of net plane rings of known step height in the highly resolved image of a FI microscope.

2.2.3. *Angular Distribution of FI Currents Emitted from Tips*

A knowledge of the angular distribution of field ionized molecules originating from a tip in the source of a FI mass spectrometer is of theoretical and practical interest. The reason is as follows. It is observed that the collector current behind the exit slit of a FI mass spectrometer does not increase with the voltage applied to the tip in the same way as the total FI current produced at the tip. Part of this effect is due to ion optical effects, but the main reason consists in a change of the angular distribution of the FI current with field strength. Thus, the transmission of a FI mass spectrometer is dependent on the tip voltage.

Investigations concerning these problems have been carried out by the author, Dahmen and Knöppel [25]. Platinum tips were used for these investigations. First, the tip radii were determined using eqn. (2.12.). Then, the angular distribution of the field emitted electrons was drawn according to the results of Drechsler and Henkel [42] for different shape factors α. These curves were compared with experimental curves obtained in the following manner. A field emission tip assembly is mounted in an evacuated bulb. The tip is surrounded by 10 ion detectors consisting of concentric rings, forming a hemisphere with the tip in the centre. The difference in polar angle of the collectors is 10°. A small disc is used as central collector opposite the tip. The experimental curves of the relative electron current per unit detector area related to the central collector current, i/i_0 are plotted as a function of the polar angle (see fig. 2.13.).

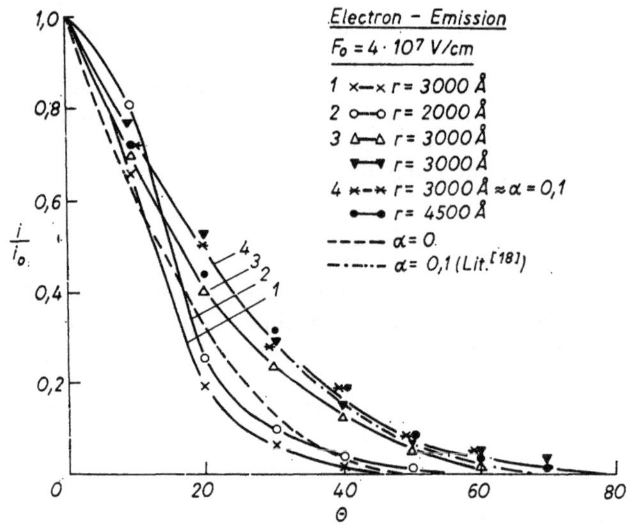

FIG. 2.13. Angular distribution of field emitted electrons for different tip radii and shape factors.

The theoretical curves are also drawn for some different shape factors. The shape factor of a specific tip is determined by selecting the theoretical curve which fits best with the experimental curve. The experimental curves for four of the six tips tested in the experiments underlying fig. 2.13. fit reasonably well with a shape factor $\alpha = 0.1$. The experimental curves 1 and 2 do not agree well with any theoretical curve, which indicates anomalous tip shapes. Such tips were not used in the subsequent field ionization experiments. The platinum tips used here had nearly hyperbolic shapes. This is in agreement with investigations by Schmidt [28]. Shape factors $\alpha > 0.15$ could not be observed. The experimental curves were smooth, i.e. with no maxima due to local changes in work function of the tip. The rather uniform work function is a consequence of a layer of adsorbed residual gases on the tip. This tip condition was an intended one in order to make the status of the tip similar to that in an FI mass spectrometer where the tip is surrounded both by residual gases and the gas to be analysed.

For a theoretical calculation of the angular distribution of field ionized gases like argon, a relation between ion current and field strength must be established. This was done — as discussed in chapter 1. — by Müller [3] for the limiting cases of low and high fields, and also by Gomer [16] for intermediate fields. The formulae

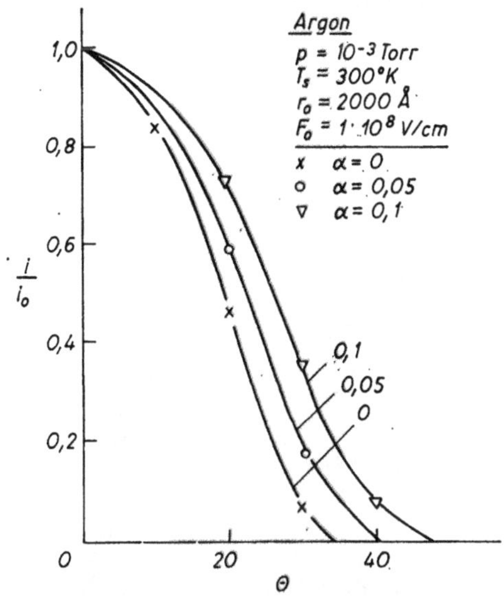

FIG. 2.14. Angular distribution of field ionized argon atoms at three different tip shape factors. Field strength at tip apex, F_0 = const.

given by Gomer were extended here for molecules having permanent dipole moments μ. A closed expression for the ion current as a function of the field strength and the different molecular parameters was used to calculate the total FI current of argon and the angular distribution of the argon ions. Southon and Brandon [48] have derived more refined equations for the FI current (see 2.2.4.), however, these had not become known to the author's group when they calculated the angular distribution of the argon ions. The qualitative agreement of the calculated total FI current of argon with that measured by Müller and Bahadur justified the use of Gomer's current-voltage relation for calculation of the angular distribution.

The calculated relative ion current of argon is plotted in fig. 2.14. as a function of the polar angle for three different shape factors, at constant field strength for the tip apex. The experimental curves related to about the same F_0 — values (curves no. 2 and 3 of fig. 2.13.) drop somewhat faster wit Θ at small angles. However, the general agreement of the experimental and theoretical curves is within the limits which have to be expected if one takes into account the numerous approximations contained in the theoretical equation for the FI current.

The angular distribution of field ionized substances strongly differs from that of argon or other rare gases if larger molecules are

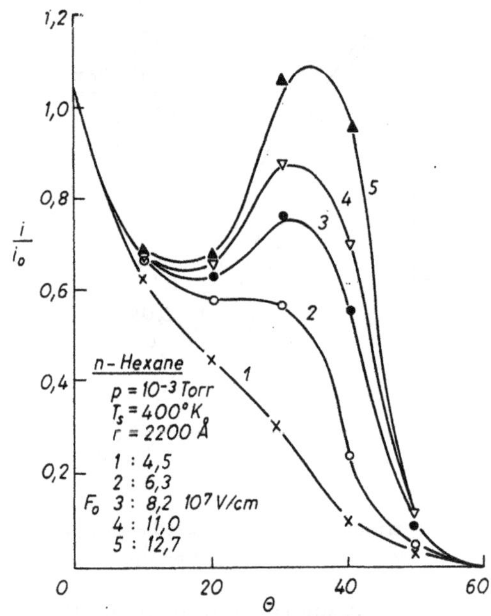

FIG. 2.15.a. Angular distribution of field ionized molecules at different field strengths F_0. — hexane.

concerned, especially if they contain polar groups. Fig. 2.15. shows the angular FI distribution for hexane and acetone. There is a pronounced side maximum in the angular distribution, lying at about 40° at higher field strengths. The maximum shifts from smaller to larger angles with increasing field strength.

The theoretical interpretation of the side maxima is as follows. Molecules like hexane are adsorbed at the tip, thereby forming at most a monolayer at the pressure of about 10^{-3} Torr in the FI source and a field strength of max. value 1.3×10^8 V/cm. Acetone may be adsorbed in the form of a multilayer due to the increase in gas density near the tip as a result of the electrostatic polarization [16].

Adsorption of molecules in a mono- or multilayer changes the supply of particles to the region of field ionization. The relative fraction of particles supplied from the gas phase and from the shank of the tip by surface diffusion changes with field strength.

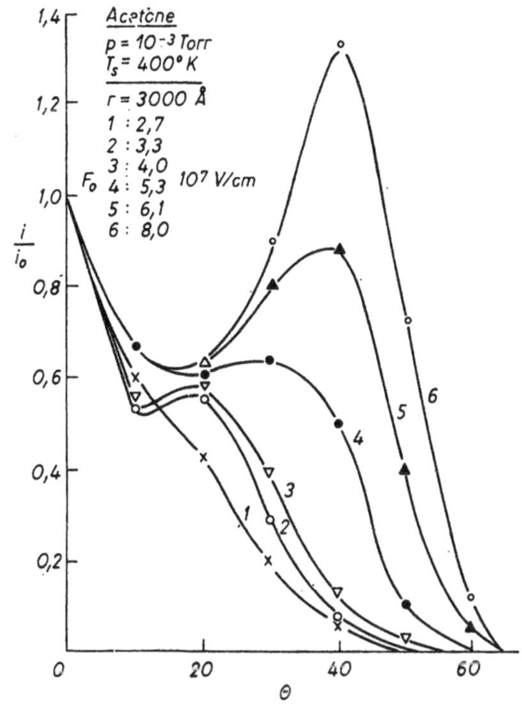

FIG. 2.15.b. Angular distribution of field ionized molecules at different field strengths F_0. — acetone.

a) *Low fields*

A homogeneous coverage of the tip with an adsorbed layer is very probable at low field strengths, slightly above the "threshold field strength" where field ionization becomes just measurable. Ionization takes place only in the vicinity of the tip apex because the field strength is sufficiently high only at small polar angles. Particle supply to the tip apex occurs partially from the gas phase (enhanced by electrostatic attraction of the particles), and partially by surface diffusion from the shank of the tip.

b) High fields

The area where field ionization occurs extends to larger polar angles with increasing field strength. The relative supply of particles from the shank of the tip increases. The shank represents a large reservoir of adsorbed particles as compared with the tip apex. All particles arriving at the apex from the gas phase are field ionized at high fields. Therefore, there is a low coverage with adsorbed particles at small angles $(0 < \Theta < 10°)$ under these conditions. There exists a gradient in concentration of adsorbed particles from the shank to the apex. The low coverage of the tip

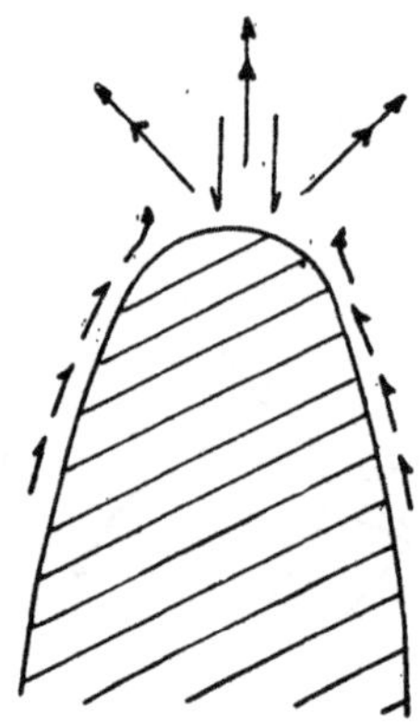

FIG. 2.16. Schematic presentation of the particle supply to the FI tips. → supply from the surface and from the gas phase; ↠ ion current.

at small angles is also partially due to the relatively small heats of adsorption in that area (faces of close packing with platinum).

The tip is covered with adsorbate at large Θ — values because of the lower field strength and the larger average adsorption energies. The surface particle supply from the shank to the apex due to the concentration gradient is still enhanced by the electrostatic attraction of the particles in the direction of higher field strength. The surface particle supply may become larger than that from the gas phase. This is the reason why intensities of the side maxima of fig. 2.15. are larger than that of the central maxima. At high fields, the field ionization probability becomes approximately equal to unity not only at the tip apex, but also at angles corresponding to the side maxima. All particles arriving at angles of about 30—40° by surface supply are ionized there. The surface supply is exhausted at smaller angles, thus the ion current drops to smaller values as determined by the smaller gas phase supply. The final rise of ion current from $\Theta \simeq 20°$ to $\Theta = 0°$ must then be due to the increase of local field strength with decreasing polar angle for a tip of hyperbolic shape. The conditions for particle supply are shown schematically in fig. 2.16. The position of the side maximum of relative ion intensity is essentially the same with all substances investigated ($\Theta \simeq 30-40°$). This range of angles comprises the faces of greatest surface roughness and largest heats of adsorption. The position of the side maximum shifts to larger angles with increasing field strength because the range of the tip where the FI probability becomes equal to one also shifts in that direction.

This interpretation of the observed phenomena is confirmed by investigations with FI microscopes. Schulze [46] investigated the adsorption of nitrogen on tungsten tips at liquid nitrogen temperature in a He—FI microscope. At low fields, the whole tip is covered with adsorbate rather uniformly, whereas at high fields the adsorbate was desorbed from the tip apex, and the structure of the tungsten tip appeared there. The adsorbate remained at the tip in the region of $\Theta \gtrsim 30-40°$. Drechsler and Pankow [47] observed a bright ring at about 30—40° angular distance from the tip pole if argon was adsorbed physically on a tungsten tip at low temperatures.

Summarizing, the transmission of an FI mass spectrometer using tip anodes depends on the angular distribution of the field ionized molecules which in turn is a function of the field strength. The transmission decreases with increasing field strength if the molecules form adsorbed layers.

The side maxima of intensity also depend on the emitter shape (For details see [25, 261]).

2.2.4. FI Current-Voltage Curves for Tips as Emitters

The FI current-voltage curves were calculated and measured first by Müller and Bahadur [3]. Their measured curves agree qualitatively with the curves calculated theoretically. Fig. 2.17. shows the FI current-voltage curve for helium with a tungsten tip as measured by Southon and Brandon [48]. At low voltages, the ion current increases very strongly with the field strength, the slope of the log i versus voltage curve being about 30. The increase of the ion current within this range between points A and B of fig. 2.17. is due to the extremely strong increase of the ionization probability P with field strength. The slope of the curve between points B and C is about 3. The ionization probability is constant and equal to about one in that range. Here, the increase of ion current with field strength is due to the increase of the particle supply function due to increasing polarization of the particles approaching the tip. The supply should vary according to theory as the square of the field [3, 16]. The observed variation with F^3 was discussed by Southon [49].

The extrapolated linear parts of the curves shown in fig. 2.17. intersect in point B. It can be assumed that the ionization probability is equal to about one at the point of inflection B. The

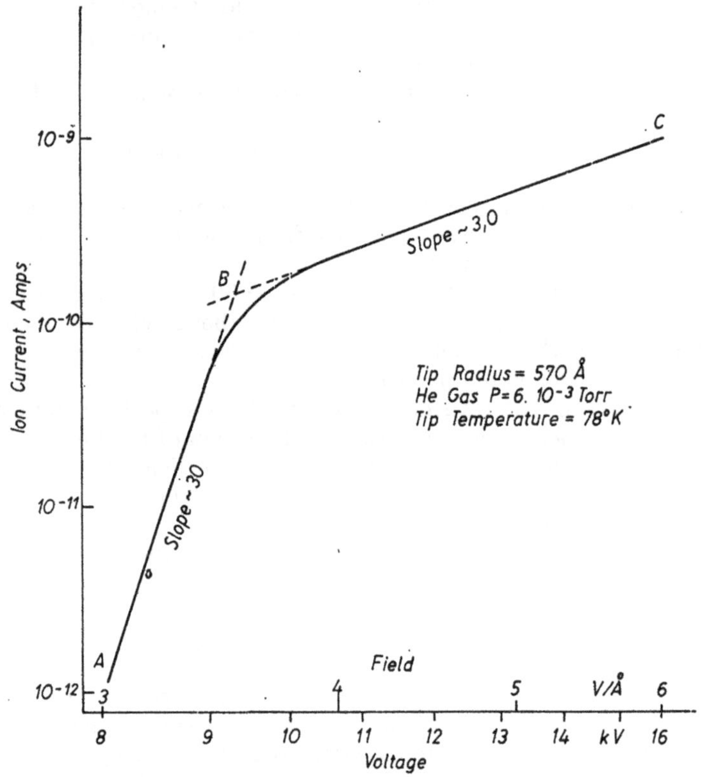

FIG. 2.17. FI current-voltage curve for helium at a tungsten tip [48].

curve ends at point C although it could be extrapolated theoretically to higher voltages. The limit in voltage which can be applied in practice is given by the effect of field evaporation of atoms of the tip material (see chapter 3.6.4.) The evaporation field strength depends on the tip material, as shown by fig. 2.18. The evaporation field is highest for tungsten, and much smaller for other materials like nickel, for example. Correspondingly, a suitable tip material has to be chosen if FI mass spectrometric investigations of molecules are to be made at very high field strengths without field evaporation of the substrate.

Both, the magnitude of the ion current and the evaporation field strength change with temperature, as shown by fig. 2.19. for the case of helium. The evaporation field and the ion current fall with increasing temperature, and the point of inflection of the curves increases. The FI current of organic molecules also

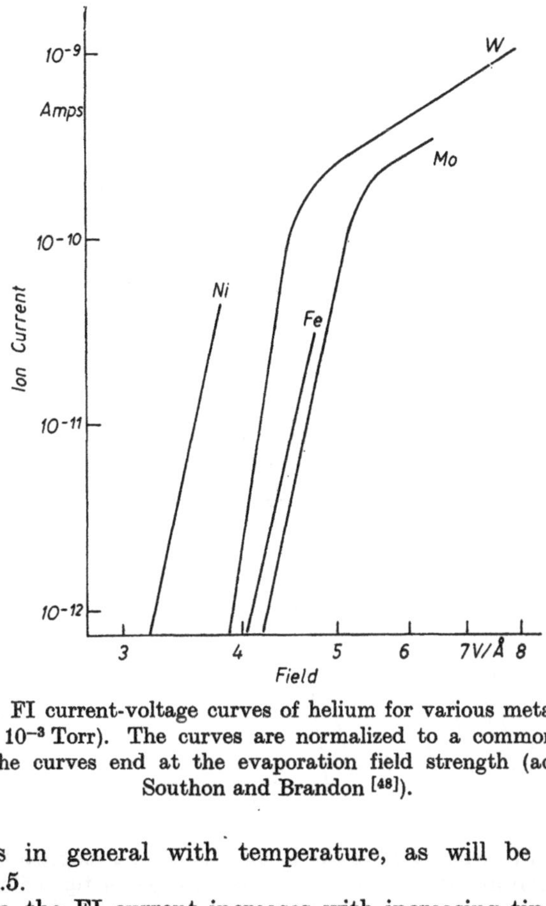

FIG. 2.18. FI current-voltage curves of helium for various metals at 77 °K ($p = 6 \times 10^{-3}$ Torr). The curves are normalized to a common radius of 480 Å. The curves end at the evaporation field strength (according to Southon and Brandon [48]).

decreases in general with temperature, as will be shown in section 3.5.

Clearly, the FI current increases with increasing tip radius at constant field strength because of the larger particle supply. Southon and Brandon [48] found the variation of the ion current to be proportional to a power of the tip radius between 2.5 and 3.0.

FI current-voltage curves for a number of substances, including organic ones, have been measured [25], as shown by fig. 2.20. The curves were taken with the apparatus used for measurement of the angular distribution of field emitted ions. Only the central collector measuring ions coming from the tip within a polar angle of 10° was used for the current measurements. Thus, the change of angular distribution of the ions with field strength has not to be considered in evaluation of the data, in contrast to measurements of the total ion current emitted into a cone of 120°.

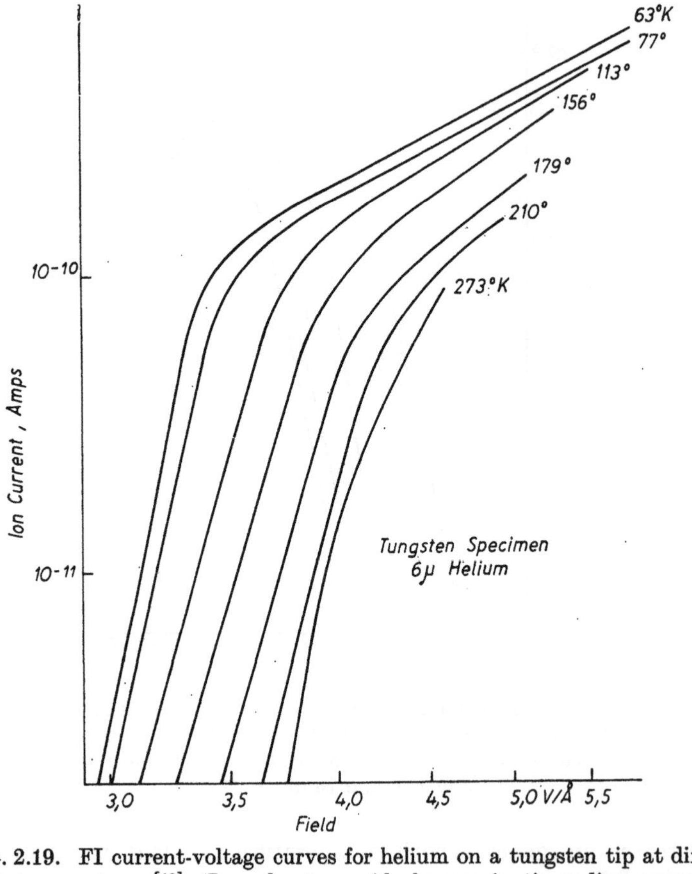

FIG. 2.19. FI current-voltage curves for helium on a tungsten tip at different temperatures [48]. (In order to avoid changes in tip radius, currents were not measured right up to the evaporation field.)

The general shape of the curves is very similar to that shown in fig. 2.17. Again, the large slope of the curves at lower fields corresponds to the strong increase of ionization probability with field strength, whereas the smaller slope at higher field corresponds to the increase of particle supply at constant ionization probability.

Interestingly, it can be seen from fig. 2.20. that the ion intensity of argon hydride, ArH^+, is much larger than that of argon, at the same field strength and argon pressure. The ArH^+ ions are formed if water is adsorbed at the tip surface. Multilayers of water can be formed on the tip surface at room temperature, as will be

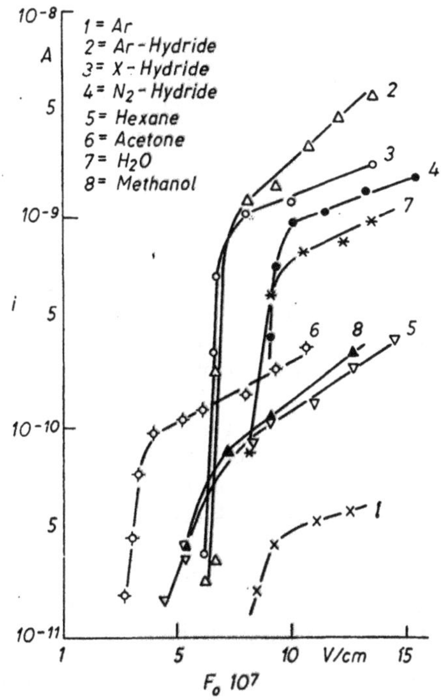

FIG. 2.20. FI current-voltage curves for different substances at a platinum tip [25]. The ions emitted into an aperture of 10° were collected.

discussed later. Protons are easily formed in these multilayers. The ArH^+ intensity is large because of the high value of the proton affinity of Ar.

The curve for ArH^+ was taken at a tip temperature of 300 °K, and that for Ar^+ at 400 °K with careful removal of water traces. The increase of temperature from 300 to 400 °K leads to a general decrease of ion intensity, but this is only by a factor of about two [5]. This cannot account for the observed decrease in ion intensity of argon as compared with argon hydride by a factor of about one hundred. Thus, the effect of water discussed above seems to be the predominant one.

2.2.5. Threshold Field Strength

It is necessary to define a threshold field strength in order to compare the current-voltage curves of different substances. There is no theoretical discontinuity in the field ionization probability comparable to quantized processes like ionization by electron impact. However, field ionization is a very steep function of the field strength, as can be seen from fig. 2.17. Below the point of inflection B, the field ion current plotted linearly is virtually a step function of the field strength. Therefore, in practice there is not much difference in field strength if the threshold field is defined as the field where the FI current becomes just conveniently measurable (10^{-12} A, for example) or as the field strength at the point of inflection B in fig. 2.17. The latter definition is physically more meaningful, and will be used here. The threshold field F^* is then the minimum field strength where the FI probability P_i becomes virtually just equal to one.

P_i has been calculated approximately for n-heptane by Metzinger and Beckey [51] making use of the relation $P_i \simeq (1 - e^{-\nu \mathrm{D} t}) \simeq 1$ at F^*. (A value of $\nu \mathrm{D} t = 4$ may be used.*)) At $T = 300\,°\mathrm{K}$ and $M = 100$, t was calculated to be in the order 10^{-12} sec, where t is the residence time of the molecule within the ionization zone of about 1 Å thickness. ν was taken to be about $10^{16}\ \mathrm{sec}^{-1}$.

Then $\nu \cdot t \simeq 10^4$, and correspondingly $D_{(F^*)} \simeq 4 \times 10^{-4}$. Both the values of ν and t are extremely approximate ones only. However, D is such a steep function of the field strength that even an error of two orders of magnitude in $\frac{4}{\nu \cdot t}$ and hence $D_{(F^*)}$ would cause a change in F^* by only about 20%. The equation given by Müller for the tunneling probability (eqn. 1.3, with eqn. 1.1 for the potential) was used to calculate D as a function of the field strength. The results were plotted as a curve and at $D_{(F^*)} = 10^{-4}$ a threshold field strength $F^* = 6 \times 10^7$ V/cm was found. (The parameters used for the curve were: $I = 10.4$ eV, $\Phi = 6$ eV for heptane on platinum). The use of eqn. 1.3 for calculation of the tunneling probability has the same approximate character as the values assumed for t and ν. However, the procedure is justified by the same argument as given above.

With regard to the numerous approximations made in the calculations of F^* for heptane, an experimental check of the threshold field strength was made. This consisted in measuring the

*) ν and D were defined in section 1.1.

current-voltage curve for n-hexane (the molecular data of which are very similar to those of n-heptane) with the small central collector of the apparatus used for measurement of the angular distribution of field ions. The result is presented as curve 5 of fig. 2.20. Extrapolation of the linear parts of the curve leads to a threshold field strength $F^* \simeq 7 \times 10^7$ V/cm which is sufficiently close to the theoretical value if one considers the approximations underlying the calculation. For acetone, $F^* \simeq 3 \times 10^7$ V/cm was found.

A third method of determining the threshold field strength consists in measuring in an electron microscope the average height h and diameter d of whiskers, grown on the shank of a tip under certain conditions (see 2.3.3.). Knowing the ratio h/d and the field strength at the surface of the smooth emitter before the growth of whiskers, one can calculate the local field strength at the top of the whiskers. It will be shown later that in the presence of whiskers most of the FI current will be generated at the top of the whiskers at the threshold field strength F^*. In this way, F^* can be derived from the narrow voltage range where whisker formation occurs and from the shape of the whiskers and the substrate. For acetone, one arrives at $F \simeq 4 \times 10^7$ V/cm, which is sufficiently close to the value given above.

Schmidt [28] has measured the threshold field for water on tips of different material. He found $F^* \simeq 6 \times 10^7$ V/cm for H_2O on platinum, whereas a value of 9×10^7 V/cm may be derived from fig. 2.19. The difference of the two values is somewhat larger than that given for the other substances. (Schmidt's value seems to be the more reliable one because it is based on a larger number of measurements.) In general, one could expect to obtain values for the threshold field strength to within an accuracy of about 20%, using different methods.

Interestingly, Schmidt noted that the threshold field strength is different for water on tungsten, iridium and platinum tips (2×10^7, 5.1×10^7 and 6.1×10^7 V/cm, respectively). Care was taken that the tungsten tip had not an essentially larger roughness factor than that of iridium or platinum tips, which was achieved by smoothing of the tip surface by tempering. A possible explanation of the dependence of the threshold field for water on the tip material would be the following one [28]. Water is a substance capable of forming multilayers on the tip surface near the threshold field strength. The thickness of this layer is of the order of the minimum distance from the tip for field ionization (eqn. 1.13b). The ionization probability is much enhanced at the surface of this layer, as will be discussed in chapter 3.6.2. The thickness (d) of the multilayer is assumed to be constant in all of the three cases. Then

$$F'_{min} \simeq \frac{I' - \Phi}{d}$$

where F' and I' are the field strength and the ionization potential, modified by the dielectric medium. Since Φ decreases in the sequence Wo, Ir, Pt, the threshold fields F'_{min} should increase in the same order.
(For very recent results by Schmidt etal. see ret. 262).

2.2.6. *Heating of Emission Tips*

Facilities for heating of tips as used for FE or FI studies must be provided for the purpose of cleaning of the tip surface by desorption of gases, for smoothing of a rough tip surface and for investigations on the temperature dependence of FE or FI currents.

For this purpose, the tip of a few mm length is spotwelded to a wire forming a loop. The loop can be heated by passing a current of several Amps through it. Typical values of the diameters of wires used for the tip and the loop are 0.1 and 0.3 mm, respectively. The tips may be heated up to a very high temperature ($\simeq 2800\,°$K with tungsten, for example). The tip temperature may be controlled by an optical pyrometer if a glass window is provided in the ion source for observation of the tip. Otherwise, it is possible to spotweld thin potential leads to the loop near the apex and to determine T from the measured resistance and a previous R — vs. — T calibration [52]. In most cases the nonuniformity of temperature between the potential leads can be neglected. At temperatures above 1000 °K radiation losses become important and therefore, corrections must be made for the temperature drop along the tip itself. Müller [38] calculated temperature differences between the tip apex and the point where the tip is spotwelded to the supporting wire. The result is $\Delta T = 2°$ at $T = 1200\,°$K, and 86° at 2800 °K for a 0.5 mm tungsten tip with a 5° half cone angle.

Smoothing of roughly sharpened points is achieved by surface migration [53, 54] during a heat treatment in high vacuum. The free surface energy is reduced by rearrangement due to surface migration of the atoms. Perfectly smooth tips can be achieved after a few minutes heating at about $^2/_3$ of the absolute melting temperature.

Cleaning and smoothing of tips can also be achieved by field evaporation of the tip material, as will be discussed in section 3.6.4.

2.3. THIN WIRES AS FI EMITTERS

2.3.1. Production of Thin Wires for FI

Thin wires of different origin may be used for investigation of FI phenomena.

2.3.1.1. Wollaston wires

Wollaston wires are drawn wires made of a thin central core of metal x and a covering metal y which has a diameter many times larger than that of the central wire. Central wires of a few μm thickness are of interest in FI work. The diameter of the whole wire is usually 20—100 times larger than that of the core. Wires where metal x is platinum and metal y is silver are often

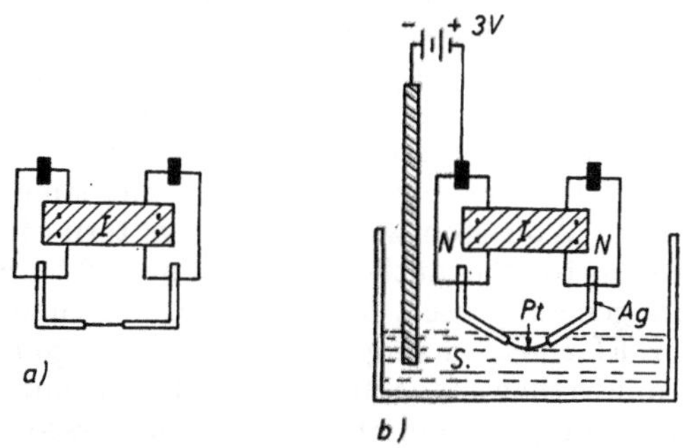

FIG. 2.21. Wollaston wire with wire holder and etching bath. I = insulator; N = nickel sheets, Ag = silver cover of Wollaston wire, 2.5 μm ∅, Pt = platinum core, 0.1 μm ∅, S = 1 molar KCN solution.
a) stretched wire; *b*) bent wire (see section 2.3.2.2.).

used in FI mass spectrometry. The silver cover is etched away over the length which is desired for FI work (usually a few mm). Wollaston wires can be pre-shaped and fixed into wire holders prior to etching, and thus the thin and fragile central wires can be handled relatively easily. Fig. 2.21. shows schematically a wire holder with an etched Wollaston wire.

The typical length of the etched part is about 3 mm. The effective emission length is then about 2 mm which is sufficient if the wire is mounted in a position vertical to the cathode slit of about 1 mm width, as shown in fig. 2.9. A larger length of the wire increases the possibility of rupture of the wire.

For etching of a limited length of the wire the following procedure is suitable. First, the Wollaston wire is bent to the shape shown in fig. 2.21. and spotwelded to the wire holder, and a rod of about 2 mm diameter is heated in a flame and subsequently brought in contact with a wax candle. The Wollaston wire is then painted with the wax on the rod, except for the central part of the wire of 3 mm length. The wire holder is clamped in a position and a beaker filled with 1 molar KCN solution is placed on a small laboratory table adjustable in height by a screw. The beaker is placed below the Wollaston wire and the table is raised until the solution just covers the horizontal part of the wire by about 1 mm. A *dc* voltage of about 3 Volts is applied, with the Wollaston wire as anode. A one mm platinum wire serves as cathode in the solution. The current may be measured during electrolysis (although this is not necessary) by a milli-Amperemeter. The etching is finished after a few minutes, but it is advisable to continue with the electrolysis for about 20 minutes in order to remove traces of silver on the platinum wire. Then the voltage is switched off and the table with the beaker is lowered extremely slowly and carefully. Otherwise the very thin wire may break because of the surface tension which leads to a thin lamella of liquid between the wire and the surface of the solution which is slightly below the wire. The wire may be broken if the lamella suddenly tears away. 1 μm Wollaston wires are always torn off due to this effect, according to the author's experience. 2.5 μm wires are stable and are generally used. 5 μm wires are mechanically stronger, but their FI emission properties are not as good as that of 2.5 μm wires unless they are „activated" with benzonitrile (2.3.1.3).

After this etching process the wire is immersed carefully in a beaker containing a solvent like toluene at a somewhat elevated temperature in order to dissolve the wax. The withdrawal of the wire from the solvent must again be made with care. Finally, the wire is cleaned further by immersion in a beaker containing acetone. Wollaston wires with gold cores were also used for FI work. The diameter was 3 μm and the mechanical strength was lower than that of platinum wires (see table 2.1.).

2.3.1.2. Taylor wires

Taylor wires may also be used for FI work. Taylor wires are made by embedding a metal in a capillary of glass or quartz [55]. The melting point of the metal must be somewhat below that of the material which it is embedded in. The sample is heated up to the melting point of the metal and is then drawn through a hole in a metal plate. The glass or quartz cover is subsequently etched away by hydrofluoric acid. Very fine wires of a few μm diameter may be obtained in this way. The mechanical strength of these wires is not quite as good as that of Wollaston wires (see table 2.1.).

TABLE 2.1. TENSILE STRENGTH OF DIFFERENT METALS IN kgf/mm².
+ THE BULK TUNGSTEN SAMPLE WAS A HAMMERED ROD. * Optimum values.

	Bulk-material	Drawn wire	Whisker [56]		Wollaston wire [57]*		Taylor wire [60]
Diameter:		0.1 mm	1 µm	2.5 µm	2.5 µm	5 µm	
Material							
Fe	17—23 [58]	56—84 [61]	1340	650	—	—	250
Cu	13—35 [58]	42—49 [61]	300	180	—	—	40
Pt	12—15 [59]	35 [51]	—	—	70	125	50
Au	13 [59]	27 [54]	—	—	21	—	17 [57]
Wo	35—150 [59]+	300 [59]	—	—	—		
Wo		410 (5 µm)	—	—			
C		—	—	700 [224]			

2.3.1.3. Tungsten wires

Tungsten wires of about 3 μm diameter are available from industry or may be produced by etching of thicker wires [62] and may be mounted directly onto a holder. The thin tungsten wires were connected with the wire holders by conducting silver glue. Dense whisker growth occurs on tungsten, like that on platinum or gold, as will be shown later. Therefore, FI mass spectra can be obtained with tungsten wires. However, the FI intensities of some substances are extremely small due to the specific interaction between them and the tungsten. This is the case with alcohols, for example. Moreover, tungsten may be easily oxidized by oxygen-containing substances. Further, it seems that twisting of the thin tungsten wire occurs frequently when it is fixed to the two ends of the holder. The internal stress of the very brittle tungsten adds to the external stress by the electric field. By this reason and because of the corrosiveness of tungsten, the average "lifetime" of a 3 μm tungsten wire in a high electric field does not seem to be much larger than that of a platinum wire, unless it is treated in a specific manner.

The situation is different if 10 μm tungsten wires are activated by a specific method developed by Migahed and the author [224]. The activation is not carried out in the ion source of a mass spectrometer because of poisoning of the source during this process. The activation process will be described in greater detail in section 2.3.4.

If benzonitrile is used as the activating gas, the tungsten wire will be surrounded by a solid layer of semiconducting organic material after the activation. This prevents the tungsten from corrosion. Furthermore, very long micro needles of the same semiconducting material grow on the wire surface. Therefore, the field strength on top of these needles is enhanced to a value which corresponds to the field strength on the surface of the much shorter needles grown on a 2.5 μm wire with acetone as activating gas.

The emission properties of the benzonitrile needles are much improved by the following temperature-time program:

Benzonitrile pressure: $3 - 5 \times 10^{-3}$ Torr. 10 μm tungsten wire. Anode — cathode distance about 2 mm, voltage about 15 kV at the beginning. 3 min heating by passing a current of about 35 mA through the wire. 3 hours of activation and gradually reduction of the voltage from 15 kV to about 8—10 kV during that time. 3 min heating of the wire (35 mA) after every one hour. Heating of the wire again after insertion into the mass spectrometer FI source. The resolving power, the intensity of the ion current and the stability of the current are much improved by this treatment. One should check the needle growth by an optical microscope at a magnification of about 1000 fold. Depending on the geometry of the wire holder, its distance from the cathode and on the applied voltage, the needles grow either preferentially on the front side or the back side of the FI wire. The latter case must be avoided, of course, by changing the voltage or the anode — cathode distance to a suitable value. Activation of the wire for more than 4 hours leads to an unfavourable increase of the wire diameter.

The mass resolving power is improved by the heat treatment because the electrical conductivity of the micro needles increases, as pointed out by Röllgen [139]. The voltage drop across the needles is thus reduced, and hence the energy inhomogenity of the field ions.

The cross section of a 10 μm tungsten wire is sixteen times larger than that of the widely used 2.5 μm platinum-Wollaston wire, and its tensile strength is about ten times larger than that of platinum. Therefore, the mechanical force necessary for wire rupture should be about 160 times larger for a 10 μm tungsten wire than for a 2.5 μm platinum wire. Tungsten treated in this specific manner seems to be ideally suitable as an emitter wire in an FI source.

2.3.1.4. Quartz and glass ribbons covered with platinum or gold

Metzinger [51] has covered 5 μm quartz ribbons with platinum by cathode sputtering of platinum wires. It is essential that the platinum cover forms a completely closed layer. Otherwise the platinum will be torn off after applying high voltage to the emitter in the FI source. The quartz ribbons proved to be stable up to about 15 kV. Materials other than platinum, such as chromium for example, may be placed on the quartz ribbon. Gold films turned out not to be very stable on quartz ribbons. The tensile strength of 3 μm quartz filaments was found to be about 330 kgf/mm^2. Glass fibres of 1—3 μm diameter and a high tensile strength (up to 190 kgf/mm^2) have also been coated with chromium and then with gold (chromium was deposited first for better adhesion of the gold film).

The metal-coated quartz and glass filaments could be activated to higher FI emission by the "acetone activation method" (see section 2.3.4.). However, the drawback of quartz and glass filaments as compared with metal and carbon filaments lies in the fact that the thin metal coating is rather unstable in the high electric field, and that it is virtually impossible to raise the filament temperature by resistive heating.

2.3.1.5. Carbon filaments used as FI emitters

Thin carbon filaments of a few μm diameter are now being produced as components for very strong compound materials. "Thornel" filaments, made by the Union Carbide Corporation, of 4.5 μm diameter and a tensile strength of about 300 kgf/mm^2

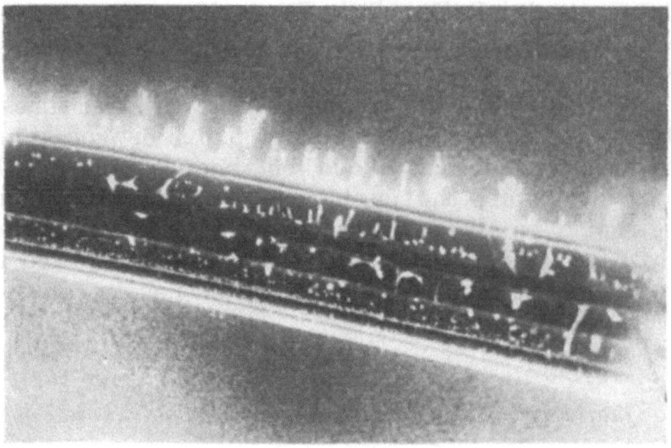

FIG. 2.22. STEREOSCAN foto of a 9 μm carbon filament activated with acetone. Magnification about 20000 times. Preferential micro needle growth along the ridges.

have been tested as FI emitters [224]. They could be activated by the "wire activation method" to approximately the same extent as Wollaston wires, i.e. a very large number of micro needles was produced on the surface of the carbon filament. The cross section of the carbon filaments was not circular but showed several long ridges. A Stereoscan micrograph at a magnification of 20000 times showed preferential needle growth along these ridges (see fig. 2.22.). The FI current of acetone of a 9 μm carbon filament at 15 kV was found to be comparable to the FI current of a 2.5 μm platinum wire at 10 kV at the same acetone pressure. From this observation relatively large field enhancement factors of the micro needles on the filament surface can be concluded.

2.3.1.6. *Whiskers used as FI wires*

It is well known that whiskers can be grown as perfect single crystals having a much higher tensile strength than the bulk metal. The tensile strength of 2.5 μm iron whiskers is reported in the literature to be about 30 times larger than that of the bulk material [56]. This would make whiskers very convenient for use as FI emitters. On the other hand, the tensile strength of thin drawn wires as compared with the bulk material also increases with decreasing wire diameter. So, the net gain in tensile strength of a whisker as compared with a Wollaston wire of 2.5 μm diameter should be expected to be only in the order of a factor of three. Copper whiskers of about 10 mm length and a few μm diameter have been grown for testing their properties as FI emitters [57]. The whiskers can be made by reduction of copper iodide by hydrogen flowing across a quartz boat containing the copper iodide at a temperature of 610 °C [56]. The whiskers found in the boat after about 90 min varied in length and diameter.

A suitable whisker is selected and taken out of the boat by a small pair of tweezers. It is then fixed with conducting silver glue to the wire holder. Acetone FI mass spectra were obtained with copper whiskers as field anodes.

The platinum whiskers which have been described in the literature so far have a length of only about 3 mm maximum. More systematic research has to be done to grow platinum whiskers of greater length.

The highest tensile strength of graphite reported so far in the literature is that of graphite whiskers. Bacon [225] has produced such whiskers by means of a high pressure, high current dc carbon arc. The whiskers are embedded in a cylindrical solid carbon matrix, called "boule", which grows between the arc electrodes by deposition of carbon atoms from the gas phase. Bacon has reported a maximum tensile strength of 3000 kgf/mm^2 for a few whiskers, and an average value of 500—1000 kgf/mm^2. The whisker diameters are about 0.5—5.0 μm, and their lengths range up to 15 mm, virtually independently of the diameter. These values make the carbon whiskers very suitable for FI emission. (An average tensile strength of about 670 kgf/mm^2 was measured in the author's laboratory for whiskers from a boule provided by R. Bacon.) The graphite whiskers were fixed to the holders by conducting silver glue. In spite of the very perfect structure of the graphite whiskers, a dense growth of micro needles on the surface of the basic whisker was observed. Very intense and stable FI mass spectra were obtained with such graphite whiskers as field anodes. (During previous measurements by Schissel [45], carbon whiskers were not used as wire emitters but as point emitters.)

2.3.2. *Strength of Wires and Forces Acting on Them*

2.3.2.1. *Tensile strength of wires*

A micro tensile strength measuring apparatus has been constructed in order to compare the tensile strengths of whiskers, drawn wires, Wollaston wires and Taylor wires [57]. Some of the results are given in tab. 2.1. together with the results of other authors.

The tensile strength of the bulk materials is also included in the table. The tensile strength increases with decreasing wire diameter. It is optimum for 1 μm iron whiskers. However, 1 μm wires are difficult to handle, and a diameter of 2.5 μm seems to be more suitable from this point of view. The tensile strength of 2.5 μm iron whiskers of 650 kgf/mm^2 — as reported by Brenner [56] is still very high. The probability of rupture increases with increasing whisker length. The tensile strength of 6 μm tungsten wires, 410 kgf/mm^2, is also very high, and it can be assumed that it will be at least as high with 2 μm tungsten wires.

The tensile strength of platinum Wollaston wires made by different companies can differ greatly. Specimens may be produced having rather strongly changing diameters. The tensile strength itself may change within a given specimen. The tensile strength of 2.5 μm Pt-Wollaston wires of one manufacturer varied between about 5 and 78 kgf/mm^2. The scatter of tensile strength was lowest with the wire produced by the Siegmund Cohen Corp., Mt. Vernon, N. Y. The average tensile strength of 2.5 μm Pt-Wollaston wires from that company was about 33 ± 5 kgf/mm^2. In contrast to the general finding that the tensile strength decreases with increasing wire diameter, it was found that 5 μm platinum-Wollaston wires had a higher tensile strength than 2.5 μm wires. This might be a property of that specific sample.

2.3.2.2. *Forces acting on wires in high fields*

The surface stress per unit area due to a field F at a conductor surface is given by

$$\sigma = \frac{10^{-8}}{8\pi}\left(\frac{F}{300}\right)^2 \ [\text{kgf/mm}^2] \tag{2.13}$$

where F is given in volts/cm. Taking $F = 7 \times 10^7$ V/cm leads to a value $\sigma \simeq 22$ kgf/mm², which is comparable with the tensile strength of drawn copper or platinum wires. The full stress acts in the direction of the wire axis if the wire is fixed at one end, and the other end is directed to a cathode (wire axis perpendicular to cathode).

The situation is different if a wire opposite to, and with its axis parallel with, a cathode is fixed at both ends. If a thin wire as field anode, having a radius r_1, is surrounded by a concentric cylinder, having a radius r_2 as a cathode, the forces acting on the wire are compensated. However, this is an unstable equili-

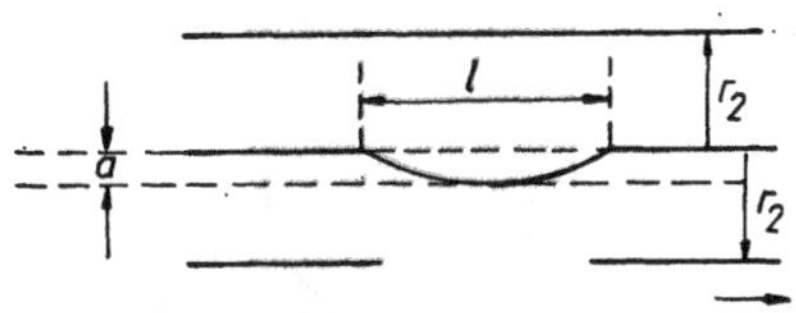

FIG. 2.23.a. Different mountings of FI wires.
FI wire mounted in the centre.

brium. If there is a slight difference in the centres of the two electrodes, a force acts on the wire. Wind [63] has shown that at small distances, a, between the electrode centres (see fig. 2.23.a.), and with the condition $a \ll r_2$, and $r_1 \ll r_2$ the force acting per unit length of the wire is

$$K = p \cdot a \ [\text{kgf/mm}^2] \tag{2.14}$$

where

$$p = \frac{2\pi \varepsilon_0 V^2}{r_2^2 (\ln r_2/r_1)^2} \tag{2.15}$$

where $\varepsilon_0 = 8.86 \times 10^{-12} \left[\frac{\text{A sec}}{\text{Vm}}\right]$. It may be assumed that the wire holders (and thus the ends of the wire) are concentric with the cathode cylinder, but that the central part of the wire is bent so that the wire has the shape of a parabola, $y = \frac{4a}{l^2} \cdot x^2$ (2.16), where l is the length of the wire, x the direction parallel, and y that perpendicular to the wire, and a is the maximum y — value in the centre of the wire. The mechanical stress in the wire is given by [78]

$$S = \frac{K}{d^2y/dx^2} \tag{2.17}$$

where K is the force per unit length and $y_{(x)}$ the shape of the wire (S has the dimension of a force). Substituting $K = p \cdot a$ one finds

$$\frac{S}{\pi r_1^2} = \frac{p \cdot l^2}{8\pi r_1^2} = \frac{\varepsilon_0 F_0^2\, 10^{-2}}{4} \left(\frac{l}{r_2}\right)^2 \tag{2.18}$$

where F_0 is the field strength at the smooth parts of the surface of a wire in V/cm.

Taking $l/r_2 = 3/2$ and $F_0 = 10^7$ V/cm, one obtains $\frac{S}{\pi r_1^2} = 3.3$ kgf/mm².

However, eqn. (2.18) is applicable only at $l > 2r_2$; therefore, a numerical value of the stress between that resulting from (2.13) and (2.18) may be expected. This is independent of the maximum

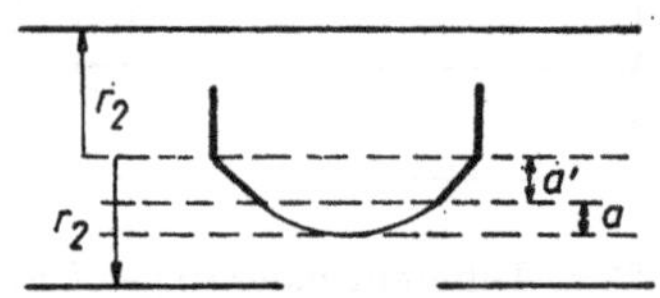

FIG. 2.23.b. Different mountings of FI wires. Out of the centre of the cylindrical cathode.

displacement, a, of the wire, as long as the approximation $K = (p \cdot a)$ is valid. Therefore, theoretically there should be no difference in the stress whether the wire would be taut or parabolically bent. In practice, however, it is found that taut wires mounted as shown in fig. 2.21.a. are more easily broken than slightly bent wires, where the ends of the unetched Wollaston wires have the same direction as shown in fig. 2.21.b. If a wire corresponding to fig. 2.21.a. is broken, rupture usually occurs at the points where the unetched wire ends join the etched core. In this respect, mounting of the wire according to fig. 2.21.b. seems to be more favoured. In general, the FI wire will be mounted out of the centre of the cylindrical cathode, as shown in fig. 2.23.b. The distance between the mounting points of the wire and the centre of the cylindrical cathode is a', and it is assumed that the deflection of the wire, a, is small as compared with a'.

$$\text{Then } K \simeq P \cdot a' \text{ and } S \simeq \frac{P(a' + a)}{8a/l^2} = \frac{Pl^2}{8} \cdot \frac{(a' + a)}{a} \tag{2.19}$$

$$r_2 \gg a', a .$$

It is seen from this equation that the stress becomes indefinitely high if the wire is taut ($a = 0$) and if it is eccentric with respect to the cathode ($a' > 0$). Therefore, the FI wire should have a curved shape, i.e. $a > 0$. On the other hand,

a should not be essentially larger than l ($l \simeq 3$ mm, $a \simeq 1.5$ mm) because the ion beam, being a parallel beam in the y-direction with a taut wire, becomes more diverging with increasing curvature. Further, it is seen from eqn. 2.19 and 2.18 that the stress on the wire is proportional to the square of the wire length. Therefore, l should be kept as short as possible, as mentioned above.

As a practical example, the values $a' = 1.0$ mm, $a = 1.5$ mm, $d = 1.5$ mm (anode-cathode distance) and $r_2 = 4.0$ mm are given. In this case, the condition $r_2 \gg a'$, a is not well satisfied, but the general result that l should be as short as possible and that the FI wire should be slightly curved remains valid.

The supporting wires should be bent by about 30° as shown in fig. 2.21.b. because they then point approximately in the direction of the FI wire under stress. If the supporting wires are horizontal, the FI wire under stress forms an angle of about 30° with the supports. The ends of the supporting wires have irregular sharp edges from the etching process which may then cut the FI wire. In fact, if the supporting wires are horizontal and if rupture of the FI wire occurs, it is usually observed that rupture takes place at the mounting points and not in the centre of the wire.

The maximum stress which a wire can withstand before rupture is $S_{max} = \sigma \cdot \pi r_1^2$, where σ is the tensile strength, the values of which may be taken from table 2.1.

The calculations of Wind lead to the result that the maximum field strength, F_{max}, for producing the maximum allowable stress is

$$F_{max} = 6 \cdot 10^6 \cdot \sqrt{\sigma} \cdot r_2/l \ [\text{V/cm}] \tag{2.20}$$

where σ is given in kgf/mm². Although this formula is valid only if the boundary effects do not intervene, which means $l \gg 2r_2$, and is not well established in real FI sources, where $l \simeq 3$ mm and $2r_2 \simeq 4$ mm, some general conclusions may be drawn from this formula. The maximum field strength does not depend on the radius of the wire, r_1. So the probability of rupture should be virtually the same for a 2 μm wire at 10 kV and a 6 μm wire at 30 kV, for example. In practice, however, a voltage of 10 kV is much more convenient than 30 kV both from ion optical considerations as discussed in chapter 2.1.2. and because of the greater danger of sparking in the ion source at the higher voltage, which would destroy the wire. Further it is seen from the formula that the length of the wire should be made as small as possible because $F_{max} \propto 1/l$. l should, however, not be made smaller than about 3 mm because of the field reducing effect of the unetched ends of the Wollaston wire. Inserting the typical values $r_2 = 2$ mm, $l = 3$ mm, $\sigma = 33$ kgf/mm², one obtains from 2.20: $F_{max} = 2.3$

$\times 10^7$ V/cm, for a smooth wire. The local fields are much higher because of the roughness factor.

The mechanical force which is exerted on the wire in the presence of a high field if a cathode is mounted in front of the wire can be largely compensated by an auxiliary electrode which is at cathode potential and is mounted behind the wire. Because of difficulties with such a construction during the changing of a wire, a nearly closed cathode construction as shown in fig. 2.24. may be used, which partially compensates the force directed from the wire to the front part of the cathode. The wire holder is introduced into the cathode cylinder by a slit in the upper disk. The slit in the lower disk of the cathode is perpendicular to the wire.

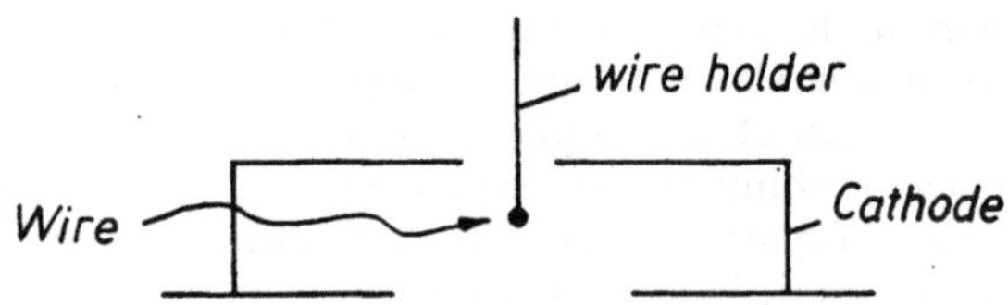

FIG. 2.24. Nearly closed structure of the electrode opposite to FI wire.

2.3.3. *Current-Voltage Curves for FI Wires*

A current-voltage curve for field ionization of n-heptane on a platinum — Wollaston wire which had already been operated for several days is shown in fig. 2.25. The shape of the curve is quite different from that for a tip emitter as shown in figs. 2.17 and 2.18. The point of inflection seen in those curves is much less pronounced in the curves for FI from wires. The shape of this current-voltage curve is a consequence of the very rough surface structure of wires which have been in use for long times, as will be discussed now [51].

The wire is assumed to be surrounded by a cylindrical, coaxial cathode of radius R_0. The radius of the wire is r_0. The potential as a function of the distance r from the wire axis is given by

$$V_{(r)} = V_0 \ln (R_0/r)/\ln (R_0/r_0) \qquad (2.21)$$

The field strength at the surface of the wire is

$$F_0 = V_0/(r_0 \cdot \ln R_0/r_0) \qquad (2.22)$$

Taking the values $r_0 = 1.25 \times 10^{-4}$ cm, $R_0 = 2.5$ cm, $V_0 = 10$ kV one obtains

$$F_0 = 8.1 \times 10^6 \text{ V/cm}.$$

The threshold field strength for n-heptane is about 6×10^7 V/cm as discussed in 2.2.5. In spite of the much lower field strength, $F_0 = 8.1 \times 10^6$ V/cm, the FI current is very large (about 10^{-7} A at $p = 1 \times 10^{-3}$ Torr and $V_0 = 10$ kV). From this it can be concluded that the actual field strength at the wire surface is larger by a factor of about seven as compared with that calculated for a smooth wire. The higher local field strength is due to surface protrusions of the wire, such as edges of crystallites or whiskers which are produced during FI emission.

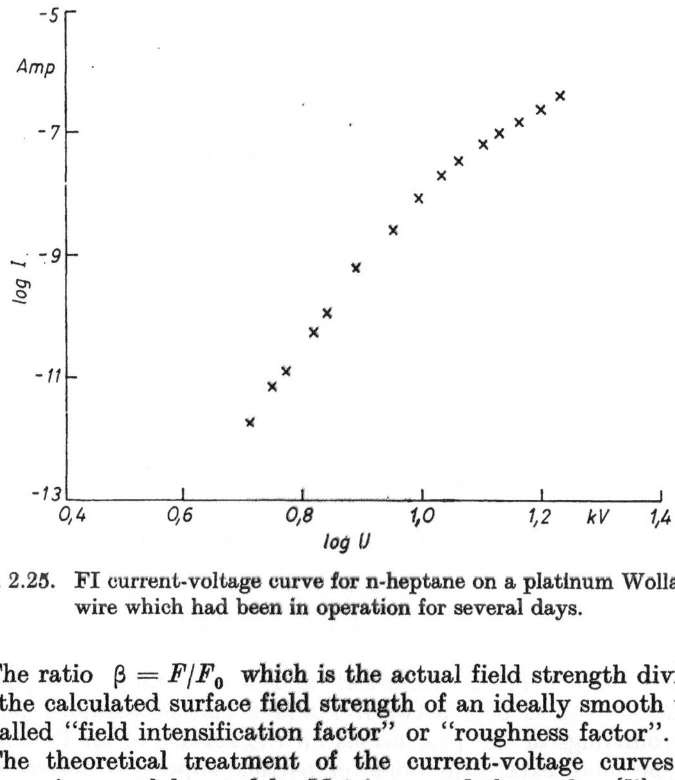

FIG. 2.25. FI current-voltage curve for n-heptane on a platinum Wollaston wire which had been in operation for several days.

The ratio $\beta = F/F_0$ which is the actual field strength divided by the calculated surface field strength of an ideally smooth wire is called "field intensification factor" or "roughness factor".

The theoretical treatment of the current-voltage curves for FI on wires as elaborated by Metzinger and the author [51] starts with the consideration that there is a distribution of surface protrusions on the wire surface having different radii of curvature. Different β-values are attributed to the different protrusions. The measured total FI current of a wire is a superposition of many partial currents of the individual protrusions. The ion current-voltage curve for each protrusion is assumed to be similar to that of a single tip, as shown in fig. 2.17. The threshold field strength F^* where the FI probability becomes equal to one is realized at different voltages $V_{(\beta)}$ for the different groups of protrusions having different β-values. This leads to the equation

$$F^* = \frac{\beta \cdot V_{(\beta)}}{r_0 \cdot \ln (R_0/r_0)} = c \cdot \beta \cdot V_{(\beta)} \tag{2.23}$$

A protrusion having an emission area A may be considered for which $F \geqq F^*$. The FI probability is equal to one, and the ion current is given by

$$i = A \cdot e \cdot S \tag{2.24}$$

where e is the elementary charge and S the particle supply function from the gas phase to the emitter per cm^2. (Supply by diffusion of particles over the emitter surface will be not considered in this section.) The number of particles arriving at the emitter is given by the gas kinetic collision number $n\bar{c}/4$ (n = number of particles per cm^3, $\bar{c}$ = mean molecular velocity) which is to be multiplied by a factor depending on the field strength and the shape of the emitter. This factor was given for molecules with no permanent dipole moment by eqn. 1.10 and 1.11 as

The factor is

$$\left(1 + \frac{\alpha F^2}{3RT}\right) \text{ for hemispherical emitter.}$$

$$\left(1 + \frac{\alpha F^2}{3RT}\right)^{1/2} \text{ for a cylindrical emitter.}$$

In general, the FI current of a single protrusion at $F \geqq F^*$ is

$$i = A \cdot e \cdot \frac{n\bar{c}}{4}\left(1 + \frac{\alpha F^2}{3RT}\right)^{s/2} \tag{2.25}$$

where $s = 1$ for a cylindrical, and $s = 2$ for a hemispherical emitter. Values of s between 1 and 2 can be assumed for the surroundings of the protrusions on the surface of an FI wire.

The more exact equations for particle supply as derived from Southon [49] may be used instead of those given here. However, the latter will be sufficient for the degree of approximation required here.

The FI current is governed by the steep increase of FI probability at $F < F^*$. As can be seen from the curve taken by Southon and Brandon [48] (fig. 2.17.), the FI current can be represented by a factor $(F/F^*)^t$, where $t = 30$ in the case of helium on tungsten. However, t depends on the gas and on the emitter material.

A total current-voltage relation for a FI wire is obtained by integrating over the contributions of the protrusions having different β-values. This relation contains $dA_{(\beta)}$ and β as parameters. The equation can be transformed to $dA_{(\beta)}/d\beta$ as *a* function of i and di/dV, which are observable quantities. In this way the distribution of the micro areas connected with protrusions of different roughness factor can be obtained. This distribution function can be determined for three subsequently improving degrees of approximation.

First approximation

The current-voltage curve for a single protrusion is described in the first approximation by a step function (see fig. 2.26.a.).

$$di_{(\beta)} = 0 \quad \text{at} \quad F < F^*$$

$$di_{(\beta)} = dA_{(\beta)} \cdot e \cdot \frac{n\bar{c}}{4}\left(1 + \frac{\alpha F^{*2}}{3RT}\right)^{s/2} \quad \text{at} \quad F > F^* \quad (2.26)$$

For a given value β_0 of the roughness factor there exists a voltage $V^*_{(\beta_0)}$ where the FI probability becomes equal to one. V^* is determined by the equation

$$V^*_{(\beta_0)} = F^* \frac{r_0 \ln(R/r_0)}{\beta_0} = \frac{K_1}{\beta_0} \quad (2.27)$$

The total current at $V^*_{(\beta_0)}$ is obtained in the first approximation by integrating over the emission areas of all protrusions having β-values $\geq \beta_0$.

$$i_{(\beta_0)} = e \cdot \frac{n\bar{c}}{4}\left(1 + \frac{\alpha F^{*2}}{3RT}\right)^{s/2} \int dA_{(\beta)}$$

$$= K_2 \int_{\beta=\infty}^{\beta=\beta_0} (dA/d\beta)\, d\beta \quad (2.28)$$

The total area of the protrusions emitting at a given minimum value β_0 is

$$A = \frac{i_{(\beta_0)}}{K_2} \quad (2.29)$$

and the distribution of the differential areas is

$$\frac{dA}{d\beta} = \frac{1}{K_2} \cdot \frac{di}{d\beta} = -\frac{K_1}{K_2 \cdot \beta_0^2} \cdot \frac{di}{dV} \quad (2.30)$$

Both i and di/dV can be taken from the measured current-voltage curves.

Second approximation

In the second approximation the increase of the particle supply due to polarization at field strengths greater than the threshold field F^* is taken into account (see fig. 2.26.b.).

Again a certain value $\beta = \beta_0$ may be considered where at a given voltage $V_{(\beta_0)}$ the FI probability becomes just equal to one. Sharper protrusions, having values $\beta_i > \beta_0$ will contribute to the ion current, but no blunter ones will emit ions $(\beta_i < \beta_0)$.

$$\beta_i < \beta_0 \qquad di_{i(\beta_0)} = 0$$

$$\beta_i \geqq \beta_0 \quad di_{i(\beta_0)} = dA_{(\beta_i)} \cdot e \cdot \frac{n\bar{c}}{4}\left(1 + \frac{\alpha F^2}{3RT}\right)^{s/2} =$$

$$dA_{(\beta_i)} \cdot e \cdot \frac{n\bar{c}}{4}\left(1 + \frac{\alpha F^{*2}}{3RT}\left(\frac{\beta_i}{\beta_0}\right)^2\right)^{s/2} \tag{2.31}$$

where $dA_{(\beta_i)}$ is the area of the protrusions having β_i.

The polarization energy is much larger than the thermal energy at room temperature at the usual field strengths,

$$i \cdot e \cdot \frac{\alpha F^{*2}}{3RT}\left(\frac{\beta_i}{\beta_0}\right)^2 \gg 1. \text{ Then}$$

$$di_{i(\beta_0)} = dA_{(\beta_i)} \cdot K_3 \left(\frac{\beta_i}{\beta_0}\right)^s \tag{2.32}$$

where $K_3 = e \cdot \frac{n\bar{c}}{4} \cdot \left(\frac{\alpha F^{*2}}{3RT}\right)^{s/2}$

The total current is

$$i_{(\beta_0)} = \int_{\beta=\infty}^{\beta=\beta_0} K_3 \, (\beta/\beta_0)^s \cdot (dA/d\beta) \cdot d\beta \tag{2.33}$$

One obtains by differentiation

$$di/d\beta = -\,(sK_3/\beta_0{}^{s+1} \int_{\beta=\infty}^{\beta=\beta_0} \beta^s \, (dA/d\beta) \, d\beta + K_3 \, (dA/d\beta)_{\beta_0}$$

$$di_{(\beta_0)}/d\beta_0 = -\, s \cdot i_{(\beta_0)}/\beta_0 + K_3 \, (dA/d\beta)_{\beta_0} \tag{2.34}$$

The distribution of micro areas with β factors is then

$$(dA_{(\beta)}/d\beta)_{\beta_0} = \frac{1}{K_3} \left[s \cdot i_{(\beta_0)}/\beta_0 + di_{(\beta_0)}/d\beta_0\right] \tag{2.35}$$

The measured current-voltage curve must be described by an analytical equation in order to make an integration of eqn. (2.35) possible; this results in calculation of the total area $A_{(\beta_0)}$ which emits ions with a FI probability equal to one.

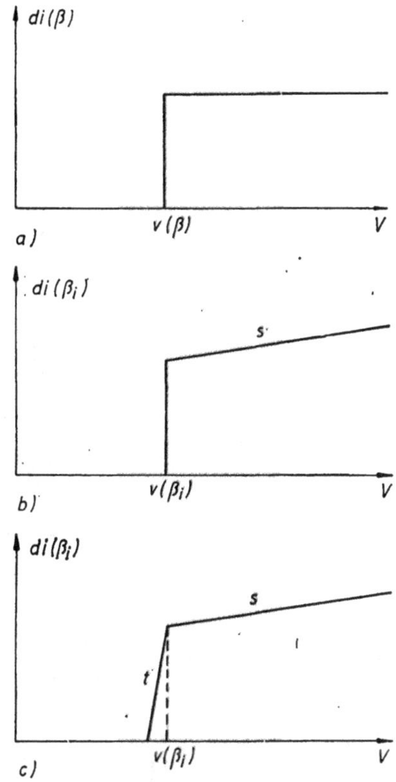

FIG. 2.26. Theoretical FI current-voltage curves for a needle emitter grown on a wire, schematically.

a) first approximation; *b*) second approximation; *c*) third approximation
β = field enhancement factor.

The simplest procedure consists in subdivision of the measured curve into portions where the current changes with a fixed power of the voltage. It is sufficient to describe the curve shown in fig. 2.25. by two portions each having a fixed exponent:

$$V < V(B) \quad i \sim V^{m_1}, \quad \text{with} \quad m_1 = 16$$

$$V > V(B) \quad i \sim V^{m_2}, \quad \text{with} \quad m_2 = 6 \quad \text{(see fig. 2.27.)}$$

B is the point of inflection of the curve. $\beta_0 \sim 1/V_{(\beta_0)}$ because of eqn. (2.27). Therefore

$$i_{(\beta_0)} = i(B) \cdot (\beta_B/\beta_0)^m,$$

$$di_{(\beta_0)}/d\beta_0 = -m \cdot i(B)\, \beta_B^m/\beta_0^{m+1} \tag{2.36}$$

Inserting (2.36) into (2.35), one obtains

$$dA_{(\beta)}/d\beta = (s-m) \cdot i_{(B)} \cdot \beta_B^m/(K_3 \cdot \beta^{m+1}) \tag{2.37}$$

The total emitting area at a given value β_0 is obtained by integrating over all micro areas dA. Three equations are obtained, depending on whether $\beta_0 > \beta_B$, $\beta_0 = \beta_B$ or $\beta_0 < \beta_B$.

$$\beta_0 > \beta_B, \quad A_{(\beta_0)} = \int\limits_{\beta=\infty}^{\beta=\beta_0} \frac{(s-m_1) \cdot i_{(B)} \cdot \beta_B^{m_1}}{K_3 \cdot \beta^{m_1+1}} \cdot d\beta$$

$$= i_{(B)} \cdot \frac{(m_1-s)}{K_3 \cdot m_1}\left(\frac{\beta_B}{\beta_0}\right)^{m_1} = i_{(\beta_0)} \cdot \frac{(m_1-s)}{K_3 \cdot m_1} \tag{2.38a}$$

$$\beta_0 = \beta_B, \quad A_{(B)} = \frac{i_{(B)} \cdot (m_1-s)}{K_3 \cdot m_1} \tag{2.38b}$$

$$\beta_0 \leqq \beta_B, \quad A_{(\beta_0)} = A_{(B)} + \int\limits_{\beta=\beta_B}^{\beta=\beta_0} \frac{(s-m_2) \cdot i_{(B)} \cdot \beta_B^{m_2}}{K_3 \cdot \beta^{m_2+1}} \cdot d\beta$$

$$= A_{(B)} + i_{(B)} \cdot \frac{(m_2-s)}{K_3 \cdot m_2}\left[\left(\frac{\beta_B}{\beta_0}\right)^{m_2} - 1\right]$$

$$= \frac{1}{K_3}\left[\frac{m_1-s}{m_1} \cdot i_{(B)} + \frac{m_2-s}{m_2}\left[i_{(\beta_0)} - i_{(B)}\right]\right] \tag{2.38c}$$

Third approximation

In the third approximation for the FI current-voltage curve for wires which will not be derived in detail here (see [51]), the strong increase in tunneling probability with field strength below the threshold field F^* is taken into account (see fig. 2.26.c.). The ion current is assumed to increase with F^t, where t is equal to about 30 [48] (see fig. 2.17.).

The result of the calculations is:

$$(dA/d\beta)_{3rd\,Approx.} = (dA/d\beta)_{2nd\,Approx.} \cdot \left(\frac{s+t-m}{t}\right) \quad (2.39)$$

with $m = m_1$ at $\infty > \beta > \beta_B + 1$

$m = m_2$ at $\beta_B + 1 > \beta > 1$.

The total emission area for $\beta_0 = \beta_B$ and $\beta_0 < \beta_B$ is given by

$$A_{(\beta_0)3rd\,Approx.} = A_{(\beta_0)2nd\,Approx.} \cdot \left(\frac{s+t-m}{t}\right) \quad (2.40)$$

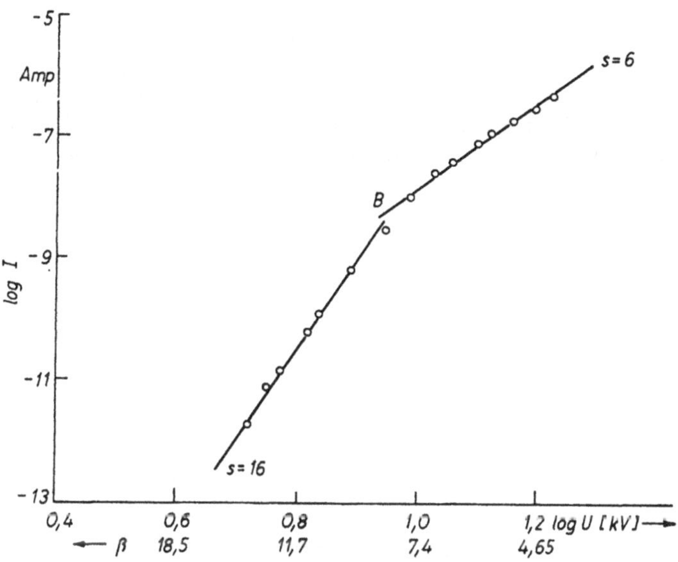

FIG. 2.27. Subdivision of the current-voltage curve of fig. 2.25. into two ranges of different field strength dependence.

The equation will be applied to the case of n-heptane on a platinum wire. The following data are used:

$s = 2$, $t = 30$, $m_1 = 16$, $m_2 = 6$ (see fig. 2.27.)

$F^* = 6 \times 10^7$ V/cm.

The other constants entering K_3 are well known. The total area A_w of the smooth wire of 1 cm length and

$r_0 = 1.25 \times 10^{-4}$ cm is 7.85×10^{-4} cm^2.

Inserting these values into eqn. (2.40), together with (2.38b) or (2.38c) respectively, one obtains:

$$A_{(B)} = A_{(9\,\mathrm{kV})} = 2.2 \times 10^{-7}\,\mathrm{cm}^2 = \frac{1}{3500} A_w \quad (\beta_0 = 8.2)$$

$$A_{(12.5\,\mathrm{kV})} = 9.0 \times 10^{-7}\,\mathrm{cm}^2 = \frac{1}{850} A_w \quad (\beta_0 = 5.82)$$

$$A_{(15.8\,\mathrm{kV})} = 4.2 \times 10^{-6}\,\mathrm{cm}^2 = \frac{1}{180} A_w \quad (\beta_0 = 4.65)$$

Evaluation of the results thus obtained leads to the following consideration of the emission properties of rough wires. At voltages above the point of inflection (fig. 2.27.) change of the roughness factor β by about 15% leads to a doubling of the area emitting ions with an ionization probability equal to one. The change of the emitter area as a function of β is even more pronounced below B. The situation is illustrated schematically by fig. 2.28.

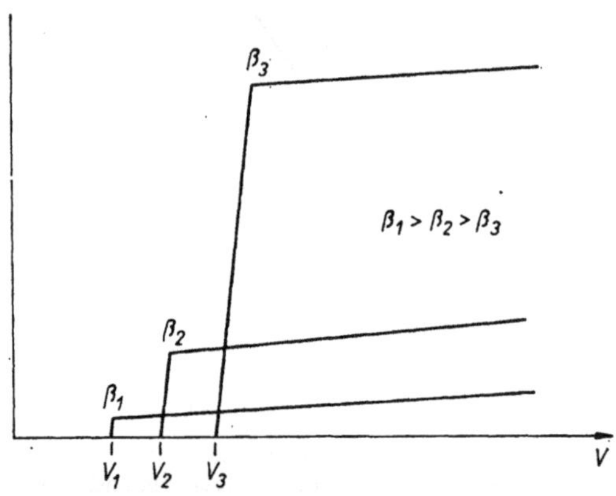

FIG. 2.28. Superposition of the FI currents of several protrusions having different field enhancement factors β. (Schematically.)

There are certain groups of protrusions (whiskers, for example) having different field enhancement factors β_i. The finest protrusions, having the highest field enhancement factor, β_1, will field ionize the organic molecules at the lowest applied voltage, V_1. At a somewhat higher voltage, V_2, the next group of somewhat blunter protrusions having somewhat lower β values, β_2, will be emitting. The total area A_2 of the protrusions (*2*) is much larger than that of the protrusions (*1*). ($A_2 \gg A_1$.) The same holds for the next group of protrusions having increasing radii of curvature, ($A_3 \gg A_2$), etc. At any given voltage V_i, a specific group of protrusions (whiskers) having a specific value β_i will field ionize predominantly, and the field strength at the top of these protrusions will just be the threshold field strength, F^*. The blunter protrusions cannot field ionize at that voltage V_i because of the low tunneling probability for the organic molecules at that local field strength; the finer whiskers will field ionize, but their contribution is negligible as compared to the protrusions having just the value β_i. Consequently, the "effective" field strength (which is the field strength at the protrusions contributing most to the FI current) should be independent of the applied voltage.

The rather surprising conclusion that the "effective" field strength should be independent of the applied voltage was confirmed by mass spectrometric measurements [51]. The ratio of the fragment to parent ion intensity of FI mass spectra is a very sensitive measure of the field strength. The relative intensity of the $C_2H_5^+$ — fragments ($M = 29$) of n-heptane ($M = 100$) from a tip increases by a factor of 100 when the voltage applied increases from 4.4 to 5.5 kV. On the other hand, with a thin wire with a rough surface, the increase of the $C_2H_5^+$ — fragments is only by a factor of 1.36 if the voltage is increased from 10 to 12 kV (see tab. 2.2.).

TAB. 2.2. DEPENDENCE OF THE — $C_2H_5^+$ — FRAGMENT INTENSITY FOR FIELD IONIZATION OF N-HEPTANE ON TIPS AND THIN WIRES

a) tip		*b*) wire	
kV	*M* 29 (% *M* 100)	kV	*M* 29 (% *M* 100)
4.4	1	8	0.84
4.6	10	10	2.56
5.0	50	12	3.5
5.5	100	14	5.8

These results prove that the "effective" field strength of a rough wire is largely independent of the applied voltage. However, there remains a small increase of the fragment ion intensity with the voltage applied to a wire, although very much less pronounced than with tips. This slight increase in fragment ion intensity can be explained by the fact that a small fraction of the protrusions emits ions at a field strength $F > F^*$, whereas the main contribution comes from those where $F \simeq F^*$. The fraction of the protrusions emitting at $F > F^*$ increases with increasing applied voltage, which can be concluded from the decrease of the slope of the $\log i - \log V$ curve.

Finally it should be mentioned that the field intensification factor of a wire could also be determined by a method worked out by Robertson, Viney, and Warrington [65] which is based on investigations of Stern, Gossling and Fowler [66]. This consists in measurement of the temperature and voltage dependence of the field electron emission and comparison with the change of the FI current with these parameters as derived from the Fowler-Nordheim equation. The sharpest protrusions contribute most to the FE current in the whole range of applied voltages because of the characteristic properties of the Fowler-Nordheim equation. Therefore, Robertson et al. find a β-factor of 70. This value will have to be applied if one considers field electron emission. With field ion emission, however, the blunter protrusions contribute more to the total ion current because of the different current-voltage characteristics of FI and FE emission and because of the distribution of the protrusions, as discussed above. Therefore, the lower β-values as listed in tab. 2.1. have to be considered in FI work, with a typical value of $\beta \simeq 7.4$ at $V = 10\,\mathrm{kV}$ and a wire diameter of 2.5 μm.

2.3.4. Activation of thin wires for FI

The following phenomenon is observed if one operates an FI mass spectrometer with a freshly etched platinum — Wollaston wire, introducing acetone at about 10^{-3} Torr as the substance to be analysed. At the beginning, the FI current is rather low (several 10^{-10} A at 10 kV), but increases continuously with time. After a couple of hours, the FI current has increased by up to a factor of a thousand (several 10^{-7} A), and a slight further increase may occur even over days. Some change of the surface properties of the wire must have occured during that time. It is clear from the preceding section that new and sharp protrusions must have grown up during this "activation period". The method for activation of wires for high FI emission by specific organic substances is termed the "wire activation method".

Electron micrographs of the wires taken after this activation showed that a very great number of micro needles had grown at the surface of the wire in parallel with the increase of ion current (fig. 2.29.)*).

*) "Whiskers" has sometimes been used synonymously for "micro needles" in previous publications and in this book. However, the organic semiconducting needles described here are not whiskers in the strict sense of the definition of whiskers. Therefore, the term micro needle seems to be more adequate.

Many experiments have been carried out in order to obtain a better understanding of the activation process for the wires and to find optimum conditions for this process [67, 248]. A number of parameters have been varied systematically, such as the sub-

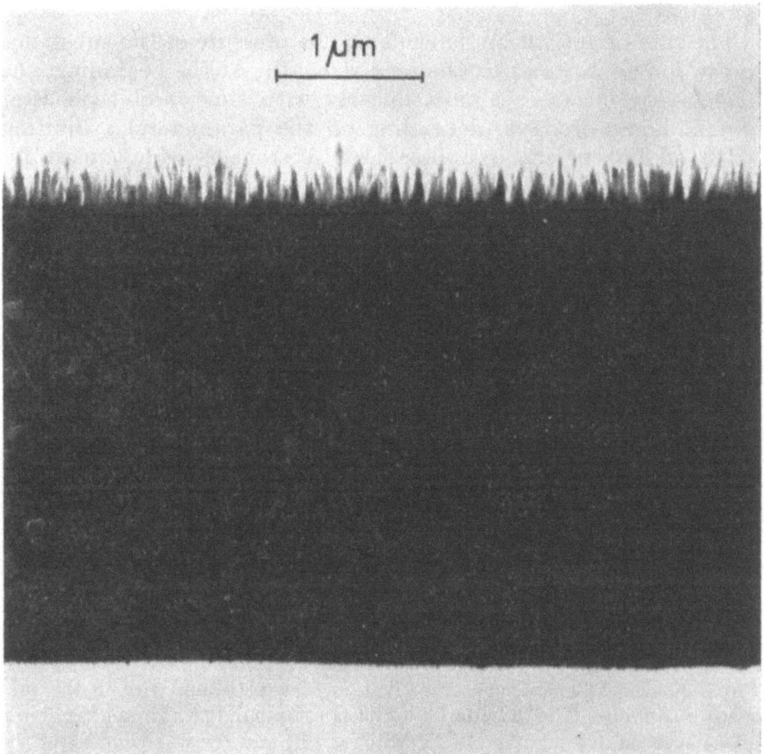

Fig. 2.29. Electron micrograph of a 3 μm platinum wire activated in the presence of acetone at 10 kV, showing protrusions. (Taken by A. Maas, Bonn.)

stance forming the ions, pressure, temperature, time of activation, voltage and others.

The experiments were carried out without mass separation of the ions in a simple, field, anode — cathode ion source. Thin wires at positive potentials of several kV were used as FI emitters. A fine grid at ground potential served as cathode. An ion collector was placed close behind the grid. Collector currents can be measured between about 10^{-10} and several 10^{-5} A, if organic substances more active in the formation of protrusions than acetone are used. The length of the etched part of the Wollaston wire was about 2—3 mm.

2.3.4.1. *Time dependence of activation*

The rate of activation depends on the pressure of the substance in the ion source and on the field strength. At the beginning, the ion current increases almost linearly with time, and then after several hours or days (depending on the parameters) a limiting value of ion current is approached. Acetone is widely used for these experiments because of its high activation efficiency, but

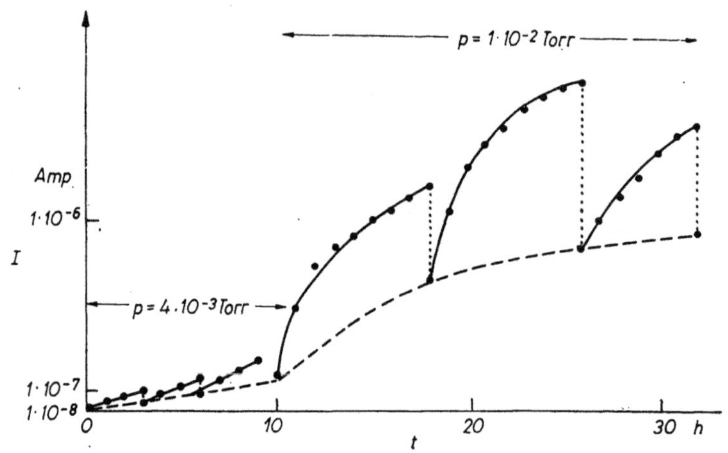

FIG. 2.30. Time dependence of activation of a platinum wire in the presence of acetone. The minima of currents correspond to values after pump off of acetone overnight (voltage switched off) and re-establishment of the original pressure and voltage.

it turned out that other substances like benzonitrile or nitrobenzene have an even larger activation efficiency [224]. The result of a typical experiment showing the time dependence of activation is given in fig. 2.30. A potential of 10 kV is applied to the wire, and acetone at a pressure of 4×10^{-3} or 1×10^{-2} Torr, respectively, is admitted to the ion source. The voltage is applied for several hours a day. The gas is pumped off and the voltage is switched off overnight. Next day, the activation process is continued under the same working conditions. This cycle is repeated for several days. It is seen from fig. 2.30. that the curves are linear at the beginning, but that they tend to saturation with increasing time of activation and with pressure. The ion current after pumping off the acetone overnight is always lower than the last value of the preceding day. This can be explained by assuming two different processes to occur in the following way.

Micro needle growth. The lowest points of the curve for $P = 4 \times 10^{-3}$ Torr can be connected by a straight line. The increase of ion current in this interval corresponds to increasing growth of stable micro needles, as reconfirmed by electron micrographs. The wire area for needle growth is limited, therefore saturation of the curve must occur after long time. This time is reduced as the pressure is increased, because the rate of activation increases with pressure.

Adsorption phenomena. The right part of the upper curve shown in fig. 2.30. resembles Langmuir-type adsorption isotherms. The change of the curves seems to be reversible. This reversible change which is superimposed on the irreversible increase of FI current as shown by the dashed curve (due to increasing growth of stable needles) can be attributed to a change of the FI current with adsorption of organic molecules. Increase of work function due to chemisorption of specific substances increases the FI current, as will be shown in 3.6.5. In addition to this effect, desorption of less tightly bound needles may occur during evaporation overnight and subsequent application of high voltage before introduction of a new charge of acetone.

The speed of activation is different for other organic substances. The highest speed of activation was found with benzonitrile, so far [224]. Almost the maximum activation of a wire can be achieved within about 30 min.

2.3.4.2. *Pressure dependence of activation*

It has been shown by several authors that roughening of the surface of a smooth metal tip occurs at sufficiently high temperatures and field strengths (see for example [254]).

This is due to a certain surface mobility of the metal atoms and their tendency to move in the direction of locally enhanced field strength. The micro needle growth on smooth wires with a ground field of about 1×10^7 V/cm and at room temperature is certainly not due to this effect. This can be shown by applying 10 kV to a 2.5 μm platinum wire for several days without introduction of acetone. The electron micrographs clearly show that there is virtually no micro needle formation under these condition. The acetone itself (or other suitable organic substances) must play an important role in the needle growth mechanism.

The increase of the ion current is linear with time at the beginning of the activation process. The rate of activation increases linearly with pressure. This can be shown experimentally, and the results are shown in fig. 2.31. The pressure is increased in six steps from 5×10^{-5} to 5×10^{-3} Torr, the activation process continuing during 60, 30 or 15 minutes during each of these six constant pressure periods. At the end of each period, a pressure of 1×10^{-3} Torr (upper curve) or 5×10^{-4} Torr (lower curve) is introducted into the ion source. The corresponding points of measurement are connected in fig. 2.31.

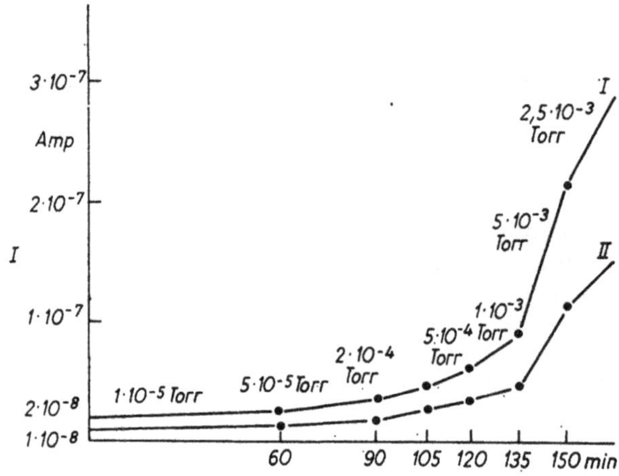

FIG. 2.31. Dependence of the rate of activation on the pressure. Pressures as indicated in the figure are applied during the long activation periods. The pressure during the short test periods was 1×10^{-3} Torr with curve *I* and 5×10^{-4} Torr with curve *II*.

It can be seen from the figure that the rate of activation increases linearly with pressure, which is equivalent to the assumption that the number of new micro needles grown per second increases with pressure. The mass spectrometrist may activate a new wire either in the ion source of the mass spectrometer itself, or in a separate "conditioning apparatus" which is a simple ion source with a pump as briefly described at the beginning of 2.3.4. The latter method has the advantage that no measuring time of the mass spectrometer is lost with activation procedures, and that wires having mechanical defects are ruled out before insertion into the mass spectrometer ion source. High rates of activation at safe working conditions are found at a pressure of 5×10^{-3} Torr, but glow discharges become disturbing at higher pressures.

2.3.4.3. Dependence of activation on field strength

There exists an optimum field strength for activation of thin wires in the presence of substances like acetone. This range is rather narrow and corresponds to about $8 \times 10^6 - 1.3 \times 10^7$ V/cm at the smooth wire before micro needle formation. This corresponds to an applied voltage of 8.5 to 13.5 kV with a 2.5 μm wire and an anode — cathode distance of 0.2 cm. (The field strengths and hence the applied voltages — about 6—10 kV — are lower for optimum wire activation with substances like benzonitrile or nitrobenzene.) The rate of activation within this range increases with increasing field strength. At 1.4×10^7 V/cm no further increase of activation is observed, and at field strengths above 1.6×10^7 V/cm the activation is reduced. The reason for this reduction lies in the pulling off of the needles by the high electric field.

A deactivation process occurs at lower field strengths. At $F_0 \simeq 6 \times 10^6$ V/cm a decrease of ion current with time is observed. If this decrease has proceeded to a certain point, raising of the field strength to the normal value of about 1×10^7 V/cm does not restore the high initial ion emission, but the drop in ion intensity proceeds irreversibly. The deactivation process becomes more pronounced at higher pressures. Electron micrographs of such deactivated wires show surface protrusions similar to needles grown nearly parallel to the wire surface instead of the needle growth perpendicular to the surface. In other words, the wire diameter increases under these conditions by a solid deposit of semiconducting organic material on the wire surface, thus reducing the macroscopic field strength.

The surface seems to be "poisoned" irreversibly by these deposits. Therefore, a freshly etched Wollaston wire should be conditioned in such a way that voltages below 9 kV (for a 2.5 μm wire) should be applied only during short periods in the presence of acetone.

On the other hand, first application of high voltage to a new wire should not occur instantaneously because a layer of adsorbed impurities may desorb spontaneously at high voltages giving raise to a sudden burst of ions which could lead to a spark and thus to a rupture of the wire. Thus a compromise for the initial raise of the voltage with time has to be made. A quick increase from zero to 5 kV can be recommended (within a few seconds) because sparking is not very probable in this range. Then the voltage should be raised to 9 kV within about 10 min. The final increase from 9 to about 12 kV should be made gradually within the activation process during a couple of hours.

The inactive status of the wire after the deactivation process can be removed if a negative voltage of several kV is applied to the wire, or if very high pressures (about 10^{-1} Torr) are established in the ion sources with the wire at about +10 kV or more. The deactivated surface seems to be prepared for re-activation by on bombardment during these two processes. Once a wire has

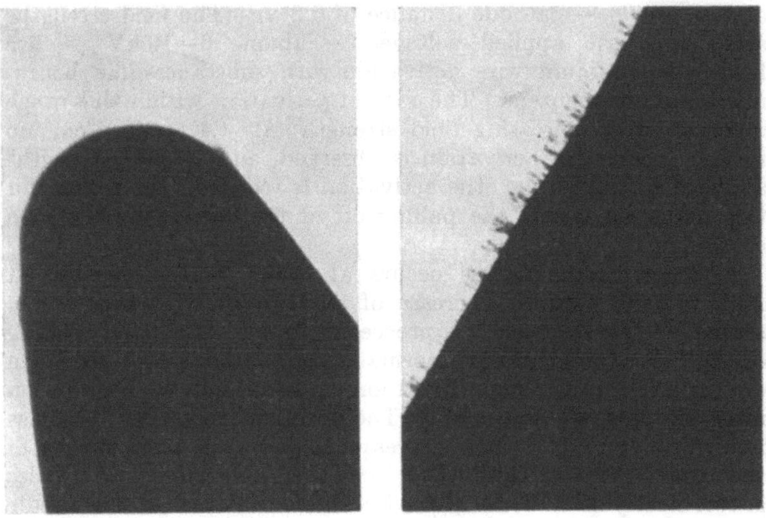

a *b*

FIG. 2.32. Electron micrograph of a platinum tip activated in the presence of acetone. Magnification: 25000 times.

a) Tip apex; $F_0 = 4 \times 10^7$ V/cm; *b*) shank of tip, $F_0 = 8 \times 10^6$ V/cm.

been thoroughly activated by dense needle growth in the suitable voltage range, reduction of the voltage below 9 kV (for a 2.5 μm wire) does not lead to deactivation of the wire. So, care has to be taken to avoid deactivation of the wire only during the first activation period of a new wire.

All three conditions for activation mentioned so far (very high field — no activation; medium fields — activation; low fields — deactivation) can be realized simultaneously with a fine metal tip used as an FI emitter instead of wires [67]. Fig. 2.32. shows a platinum tip which was operated in an FI source at a pressure of 5×10^{-3} Torr at a voltage of +10 kV for a couple of hours. The front part of the tip is very smooth, and no needles are observed. The field strength calculated for this part of the tip is greater than 2×10^7 V/cm. The field strength is somewhat lower at the shank just behind the tip, and micro needles are observed here. Still farther away from the tip the field strength is quite low, and deactivation takes place. In this range, only few needles are found, and they are grown on those places where the local field strength is enhanced by surface roughness. The needles are longer at low macroscopic field strengths. This is explained by the fact that the field strength decreases quickly at small distances from the local protrusion, so that the needles can grow longer until the local field at the top of the needles attains the evaporation field strength which limits further growth. The field strength on top of a needle grown on a wire is (as will be derived later)

$$F_w \simeq F_0 h / r_w \tag{2.41}$$

where F_0 is the field strength of the smooth wire, h the height and r_w the radius of the needle. Field evaporation of metal atoms from the top of the needles occurs at a certain field strength F_w. Hence it follows from (2.41) that the needle height is inversely proportional to the ground field F_0

$$h \approx 1/F_0 \tag{2.42}$$

By the same argument it could be concluded that at constant ground field F_0, the ratio $h/r_w = \beta$ (field enhancement factor) should be constant for a fixed evaporation field strength. In other words, there should be either longer and blunter or shorter and sharper protrusions. Inspection of many electron micrographs shows, however, that there is a distribution of β-values, in accordance with the results of 2.3.3. This seems to be due to a dynamic nature of the equilibrium of the needle growth. Some needles are torn off during a given time interval, and others have grown up without having already reached the maximum possible length.

2.3.4.4. *Temperature dependence of activation*

The rate of activation increases considerably with temperature up to a certain maximum temperature beyond which the ion current decreases again. This can be shown by passing currents of 1—5 mA through 2.5 μm platinum wires where the FI current starts to decrease. 1 mA corresponds to about 100 °C, 5 mA to about 450 °C wire temperature, depending on the length of the wire. The greater rate of activation can be attributed to the higher polymerization rate of the molecules forming the needles at higher temperature. The acceleration of the activation process proceeds up to about 400 °C.

The maximum temperature the needles could withstand can be tested by resistive heating of the wires to high temperatures in vacuo without high voltage being applied. After this heating, the temperature is reduced, high voltage applied and acetone at 5×10^{-3} Torr admitted to the ion source for testing of the FI properties of the wire.

Heating up to 950 °C does not change the total ion emission of 2.5 μm Pt wire activated with acetone. After heating to temperatures over 1000 °C the ion emission current is reduced. Electron micrographs prove that the needles disappear above 1000 °C.

2.3.4.5. The chemical nature of the micro needles

Theoretically, there exists a number of possibilities for the chemical nature of the micro needles. One could suppose that they consist of the material of the substrate wire. In the case of platinum or gold Wollaston wires, they could either consist of the core material (platinum, gold) or of silver from the cover, migrating to the etched platinum or gold core. Under these assumptions, the micro needles would consist of metal whiskers, the nature of which could be studied — in principle — by electron or X-ray diffraction. In practice, however, the size of the needles is too small for drawing safe conclusions from such experiments. A number of attempts have been made by the author's group to determine the nature of the needles grown in the presence of acetone using the electron diffraction method. However, no clear reflections were observed because the needles do not have crystalline structure, as will be shown at the end of this section. Hummel and Schweren [218] claim that they had observed platinum electron diffraction patterns from the needles grown on a platinum wire which was activated with acetone. They applied the dark field method to the study of the needles in order to exclude reflections obtained from the platinum wire itself. However, in view of the experiments which will be described now, doubt must be raised as to the origin of the reflections.

A large number of experiments have been made in the author's laboratory, comparing needle growth on platinum wires and carbon filaments (or whiskers) under the same conditions. Strong needle growth was observed on both substrates if acetone or — even more efficient — benzonitrile was used for activation. Clearly, in the case of a carbon substrate the needles cannot consist of metal, but only of carbon or a semiconducting organic material. Therefore, one is inclined to suppose that the needles on the metal consist also of carbon or organic material. However, in that case the possibility must be excluded that the needles on the carbon surfaces originate from the substrate material itself (the organic vapour would serve only for activation of the wire surface atoms, and the needles could then consist of carbon in the case of carbon surfaces, and of metal in the case of metal surfaces). This possibility can be excluded by measuring the wire diameters before and after the activation period. After complete and very efficient activation of a carbon whisker (3 μm diameter), an increase of its diameter by up to a factor of three is observed. Very long (up to several μm) and densely grown needles are observed on the surface

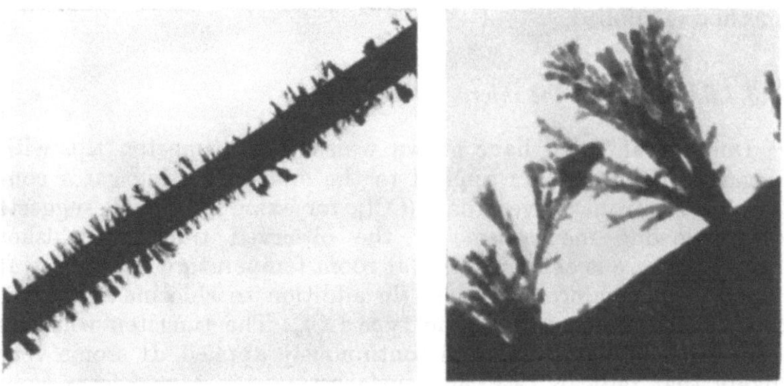

a b

FIG. 2.33. *a*) Electron micrograph of a 2.5 μm platinum wire activated in the presence of benzonitrile, showing micro needles. *b*) Part of fig. 2.33a) at larger magnification.

of the thickened whisker. Therefore, considering the material balance one must conclude that the material deposited on the whisker surface by the activation procedure cannot originate from the substrate, which otherwise had to decrease at least at certain sections in order to compensate for the increase at other sections (which, however, was not observed). For these reasons, the needles can only consist of organic, semiconducting polymers which are formed by a radical chain mechanism from the activating organic vapour (acetone, benzonitrile, for example) [260].

Systematic studies of the dependence of the activation efficiency on the organic substance by Migahed [224] showed that the efficiency increases with increasing polarity and decreasing ionization potential of the organic substance. Benzonitrile proved to be most efficient, as mentioned before (see fig. 2.33.).

The needles grown on platinum or carbon filaments in the presence of benzonitrile cannot be dissolved by either inorganic acids (HNO_3, H_2SO_4, HCl) or organic solvents like xylene or trichloroethylene.

On the other hand, the emission of a platinum wire activated with crotonaldehyde or acetone is reduced by a factor of 35—50 after immersion of the wire into butanoic acid. The emission current is only slightly reduced if an inactivated platinum wire is treated with butanoic acid.

From all these experiments it can be concluded that the needles on the metal wires do not consist of metal and the needles on carbon filaments do not consist of carbon. The needles very likely are of semiconducting organic polymers in both cases. This explains why the needles do not cause reflections if the electron diffraction method is applied.

2.3.4.6. Model for the micro needle growth

Duell et al [68, 247] have grown whiskers on tungsten tips with negative high voltages applied to the emitters, using gases containing electronegative atoms (CCl_4, for example). Duell suggests the following mechanism for the observed tungsten whisker growth. There is evidence that at room temperature carbon tetrachloride and chloroform yield (in addition to chlorine and possibly carbon) fragments of the type CCl_x. The tungsten whiskers grow if the high voltage is continuously applied. It seems very likely that radicals CCl_x and surface tungsten atoms form complexes which migrate to the ends of the whiskers where they dissociate and deposit tungsten atoms. Although this mechanism is well founded in the case of negative tip voltages and gases like CCl_4, it seems to play at least only a minor part in the case of positive tip voltages and gases like acetone or benzonitrile.

As discussed in the preceding section, there exists clear evidence that the micro needles grown under the latter conditions consist of semiconducting organic polymers. Very probably, the needles grow by a radical chain mechanism, including positively charged ions as suggested first by Röllgen [139]. Without giving the details of this mechanism, its first step may be indicated for the case of acetone. Two adsorbed acetone molecules are able to react in the following manner:

$$2\,(CH_3COCH_3)_{ads} \rightarrow (CH_3COCH_2)_{ads} + (CH_3COCH_3)H^+ + e^-$$

The adsorbed acetonyl radical is an active centre for growth of a chain. Dehydrogenation and connection between the polymer chains occurs by the heat treatment, thus transforming the initially loose needle structure to a much tighter one. Summarizing section 2.3.4. one obtains the optimum wire activation by using the benzonitrile activation technique as described in section

2.3.4.3. A 10 μm tungsten wire may be used preferentially. (But higher ion currents are obtained with 2.5 μm platinum-Wollaston wires at the expense of their smaller mechanical strength.) A voltage of about + 15 kV is applied to the wire. A quick increase of the ion current is observed. If a current of 1×10^{-5} A is obtained after a couple of minutes, the voltage is gradually reduced such that the FI current does not exceed 10^{-5} A. After about 3—4 hours one arrives at a stationary voltage of about 8—10 kV. Highly active and very stable emitters are obtained in this way.

2.3.4.7. Gas-induced whisker growth by FE emission of thin wires

Duell et al. [68, 247] reported — as mentioned above — that whiskers can be grown on tungsten tips if carbon tetrachloride is admitted into a FE microscope at a pressure of about 10^{-9} Torr and a negative high voltage is applied to the tip, sufficiently high to cause field electron emission. Robertson and Williams [69] have carried out experiments using a cylindrical field emission projection microscope in order to obtain information about the size of projections and their distribution on the wire surface after introduction of gases like CCl_4 and C_2H_2 and application of a negative high voltage to the thin wire.

Tungsten and platinum wires of 20 μm diameter and 20 cm length are used in these experiments. The growth of projections can be observed by image spots having elliptical shape on the luminescent screen of the FE microscope. The height of a projection can be estimated from the minor axis of its elliptical image. The calculation is similar to that given by Vibrans [70] for the spreading of an electron beam emitted by a cylindrical projection situated on an infinite plane surface.

There are many projections on the wire already before admission of the activating gases, as can be seen from the numerous spots on the luminescent screen of the FE microscope (fig. 2.34.). The lengths of the projections vary greatly.

The electron emission current of platinum or tungsten wires increases rapidly after admission of carbon tetrachloride at about 1×10^{-4} Torr. Many of the images grow larger, indicating increase in length of the emitting projections. Growth continues until the largest projections are destroyed due to the excessive heating by the high electron current.

This process is termed "blow up" by Brodie [71]. The voltage applied to the wire is lowered to keep the emission current constant and to preserve the projections. Quick growth occurs within a few seconds, until after 15—20 sec, growth becomes very slow.

The number of visible images after this growth depends on the size of the emission current during the growth. The number of projections grown is approximately 0.5 times the emission current (in μA). Fig. 2.35. shows the image of some protrusions grown

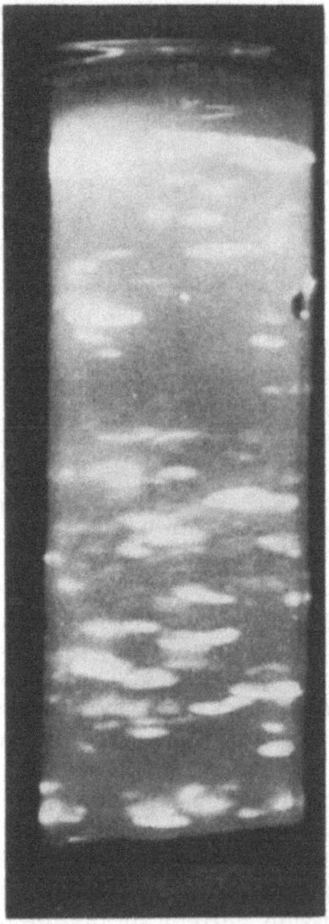

FIG. 2.34. Field electron image of projections on a 20 μm wire before beginning of the activation process, taken by Robertson and Williams [69]. ($I = 60$ μm A, $F = 6.9 \times 10^5$ V/cm)

at 5 μA. The field must be reduced to about 50–60% of its initial value in order to keep the current constant. Therefore, the original projections seen before gas admission (fig. 2.34.) become invisible after the field reduction, and only the few projections grown after gas admission at 5 μA emission current become visible. Growth does not occur in the absence of an electric field.

A growth of protrusions similar to that in CCl_4 also occurs in acetylene at about 6×10^{-4} Torr on a wire heated to 680 °C. On a cold wire with acetylene at the same pressure the projections grow singly, separated by induction periods ranging from a few

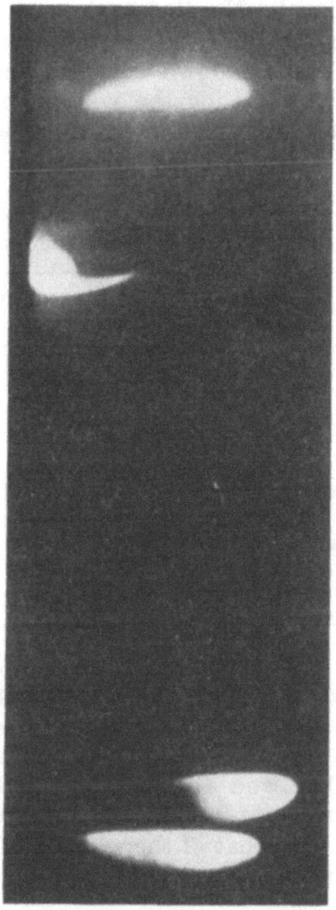

FIG. 2.35. Wire of fig. 2.34. after growth of projections at current of 5 μA. $F = 2.1 \times 10^5$ V/cm [69].

seconds to a few minutes. The inorganic gases (NH_3, O_2, I_2 and SF_6) used during the study of growth of projections by Robertson and Williams do not induce such a growth on platinum wires.

The field enhancement factors β and the work function Φ can be determined by measuring the current-voltage curve for single grown projections on platinum at low currents in CCl_4 or C_2H_2.

The measured total current is not identical with the electron current emitted by one projection. However, it can be concluded by inspection of the microscope screen that one projection is emitting predominantly. The wire must be heated to 700 °C in order to reduce fluctuations in the emission current. The date obtained by the Fowler-Nordheim plot are analysed using the method reported by Bradley and D'Asaro [72].

The maximum prebreakdown currents and the corresponding fields are obtained by raising the voltage until the projections blow up. The results obtained by Robertson and Williams are listed in table 2.3. It can be seen from the table that the

TABLE 2.3. PROPERTIES OF PROJECTIONS GROWN IN CCl_4 AND C_2H_2
(Φ = work function, β = field enhancement factor)

	CCl_4 — grown projection	C_2H_2 — grown projection
Φ (eV)	4.3	6.3
β	400	1350
height (Å)	$\sim 1 \times 10^4$	$\sim 1 \times 10^4$
tip radius (Å)	~ 25	~ 10
maximum prebreakdown current (μA)	160	18
applied field E at breakdown (V/cm)	4×10^5	4×10^5
true breakdown field βE (V/cm)	1.5×10^8	6×10^8

dimensions of the projections are rather different from those of the projections grown at positive wire potentials in acetone, as described in the previous sections. Electron micrographs show that those projections, termed needles there, have an average length of the order of 1000 Å and a radius about 100 Å. The β value of a whisker on the surface of a cylindrical emitter having a radius large as compared with that of the whisker can be approximated, according to Vibrans [70] by

$$\beta \simeq h/r \qquad (h \gg r)$$

where h is the height and r the radius of the whisker. Thus a β-factor of about ten can be derived from the electron micrographs of the projections grown at positive wire potentials.

The larger β values of the projections grown at negative wire potentials, as listed in table 2.3. ($\beta = 400$ for CCl_4, $\beta = 1350$ for C_2H_2) are due to the fact that their average length is about ten times larger and their radius about 4—10 times smaller than the dimensions of the needles produced by the acetone conditioning method. The maximum number N_i of protrusions grown at negative voltages is of the order of a hundred, with maximum prebreakdown currents of the order of 200 μA for CCl_4.

The number of needles grown at positive potentials in acetone is larger by a factor of about 10^4. This number (N) can be estimated approximately in the following way.

$$N = \frac{A_{macro} \times fr}{A_{micro}}$$

where A_{macro} is the surface area of the smooth wire, A_{micro} is the average area of the apex of a single needle, and fr is the fraction of the macroscopic area which is emitting ions. At 9 kV, $fr = 1/3500$ [51], $A_{micro} \simeq 3 \times 10^{-12}$ cm² (the radius of the needles is about 10^{-6} cm on average), $A_{macro} = 7.5 \times 10^{-3}$ cm² for a wire of 10 cm length and 2.5 μm diameter. Thus,

$$N \simeq \frac{7.5 \times 10^{-3}}{3 \times 10^{-12} \times 3500} \simeq 10^6.$$

The order of magnitude of this number is reconfirmed on average by electron micrographs.

In accordance with these considerations, Robertson states that the FI current from wires activated in CCl_4 is quite small. However, no quantitative data are available because the CCl_4 whiskers tend to be pulled off under FI conditions.

2.3.5. *FI mass spectra taken with heated wires*

FI mass spectra are dependent on temperature, as will be discussed later. Sometimes it is preferable to take the mass spectra at temperatures slightly above room temperature. This can be done either by heating the whole ion source, or by heating the wire only. The latter method has some advantages. Thermal decomposition of an unstable molecule may occur by multiple collisions with the wall of the ion source if the whole source is heated. This effect is reduced if the walls of the ion source are kept cool and only the wire is heated.

The temperature of a 2.5 μm platinum wire of about 3 mm length is raised to approximately 100 °C if a d.c. current of about 1 mA is passed through the wire. (This is an order of magnitude number only, because it depends critically on the length of the wire and on the geometry of the wire holder.)

The effect of heating the wire is twofold: adsorbed multilayers are removed, and the mass resolving power is increased.

The first effect may be illustrated by fig. 2.36. Fig. 2.36.a. shows the FI mass spectrum of acetone with an unheated wire. The parent peak is at $m/e = 58$. A parent plus one peak is observed which strongly exceeds the isotope contribution of ^{13}C

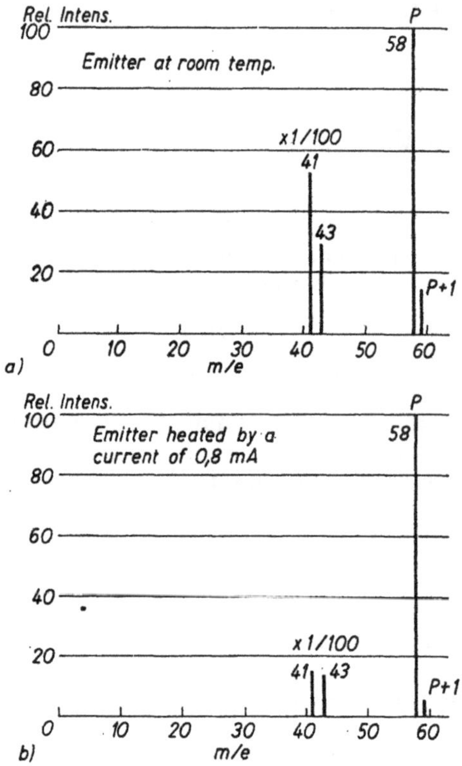

FIG. 2.36. FI mass spectrum of acetone on a 2.5 μm platinum wire.
a) FI wire at room temperature,
b) FI wire heated by a current of about 0.8 mA.

which should be 3.3% of the parent peak intensity. This $P + 1$ peak is due to a reaction in an adsorbed layer of acetone on the wire

$$2\,P_{ads} \rightarrow (P - 1)_{ads} + (P + 1)^+ + e^- .$$

Fig. 2.36.b. shows the acetone spectrum taken with a heated wire. The $P + 1$ intensity is reduced almost to the $^{12}C/^{13}C$-ratio. The m 43 peak and especially the m 41 peak which is due to a secondary reaction are reduced during heating of the wire.

The second effect of heating the wire consists in an increase of the resolving power if solid samples are measured. Solid organic substances are evaporated from a micro oven to the wire. A deposit of insulating or semiconducting organic material may be formed on the wire surface if the wire is not heated. A voltage drop across this deposit may occur which is irregular in magnitude, and this limits the resolving power of the mass spectrometer.

Fig. 2.37. shows the FI parent peaks of *d*-ribose which is evaporated from a small oven onto the wire. *a*) shows the wire at $t = 0$ where the solid deposit has not yet been built up, and the resolving power is reasonable. *b*) shows the parent peaks after

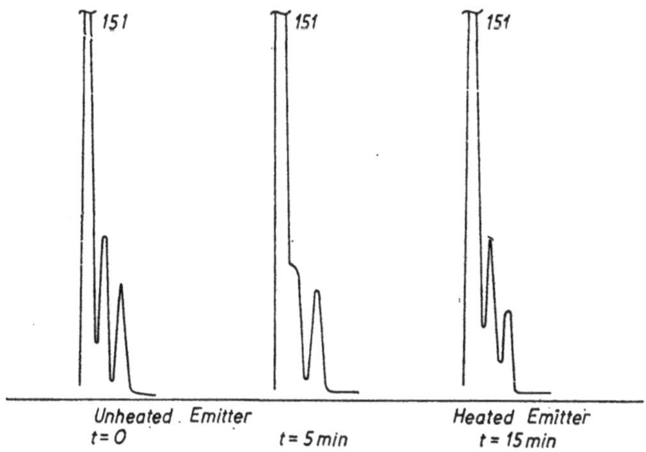

FIG. 2.37. FI mass spectrum of the parent ion group of d-ribose on a platinum wire.

a) spectrum at beginning of experiment,
b) spectrum after 5 min without heating of wire,
c) spectrum after heating the wire for 10 min by passing a current of 1 mA through it.

$t = 5$ min without heating. The resolving power is reduced. *c*) After heating of the wire by passing a current of 1 mA through it for 10 min, the resolving power of *a*) has been approximately re-established. The solid deposit has been reduced by the heating.

If an activated wire is heated (with a current of about 2.0—2.5 mA in the case of a 2.5 μ Pt-wire) and afterwards its temperature is reduced again to room temperature, irreversible changes of the FI mass spectra are sometimes observed, especially with alcohols. Fig. 2.38. shows a schematical plot of the FI mass spectra of n-butanol taken with an unheated wire (*A*) and with a heated wire (*B*) after temperature reduction [229]. The change of the spectra indicates an irreversible change of the microneedles grown on the surface of the activated wire. In spectrum (*A*) intense $(2M + 1)^+$, $(M + 1)^+$ and $(M + 1 - 18)^+$ ions are observed. In spectrum (*B*) these peaks are greatly reduced, and a very intense and broad fragment peak $(M - 18)^+$ appears which

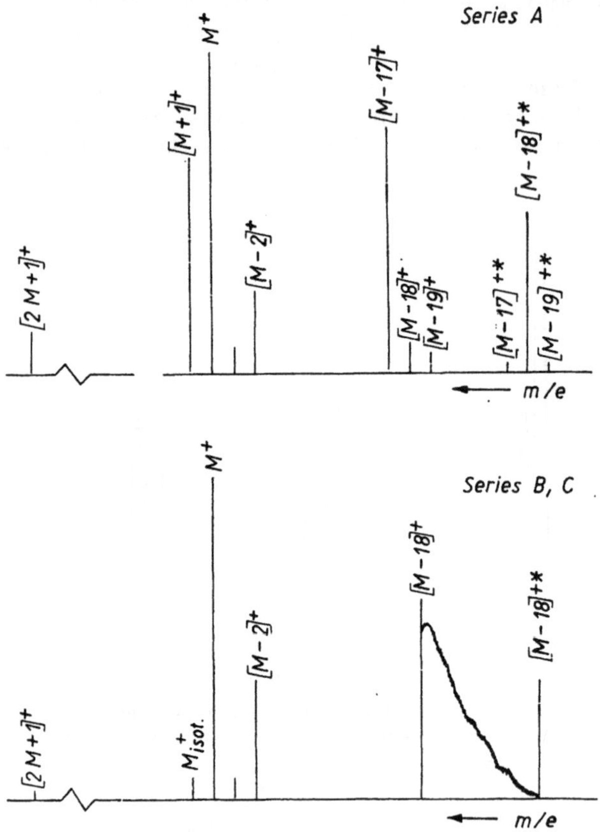

FIG. 2.38. Schematical representation of FI mass spectra of n-butanol.
(*A*) activated wire without heat treatment.
(*B*, *C*) After heating of the wire by a current of 2.5 mA and subsequent temperature reduction to room temperature.

extends down to the metastable $(M - 18)^+$ peak. The spectrum has changed in a manner as if the local field strengths on top of the micro needles had increased. Obviously, this occured by a change of the geometrical structure of the needles by the heat treatment.

According to Röllgen [139] a porous structure of micro needles of rather low mechanical strength is formed at low activating fields and temperatures. This structure favours multilayer formation of polar substances, similar to the phenomenon of "capillary condensation". This porous structure is transformed to a much tighter and stronger micro needle structure after heating of the needles to about 1000°C. The $(2M + 1)^+$ — and the $(M + 1)^+$ — ions are then greatly reduced.

2.4. SHARP METAL EDGES AS FI EMITTERS

The edges of sharp metal blades have been introduced as emitters for FI sources by Robertson et al. [10, 27] and by the author [73, 74]. The advantage of blades as compared to thin wires consists in avoiding rupture which easily occurs with thin wires. The disadvantages of blades will be discussed later.

The simplest way of preparing a blade for field ionization is to use a piece of a commercial razor blade without special treatment of the blade, except for dissolving grease on the blade by an organic solvent. Grinding of the blades, or chemical etching — according to Robertson — does not lead to improvement of the FI emission properties of the razor blades. The author has used Wollaston-platinum foils of 1 μm thickness and a silver cover of 100 μm. A part of the silver cover was etched away, and the foil was mounted into a FI source. The emission properties of these foils were found to be not essentially better than those of razor blades.

In contrast with this finding, Robertson could prepare thin platinum foils without silver cover which showed much better emission properties than razor blades. A platinum foil initially 8 μm thick was etched for 2 min in a 3% solution of KCN in water at a current density of about 1 A/cm² which is produced by an a.c. voltage of 10 V between the foil and another electrode 2—3 cm away from the foil. Experiments with gold foils seem to be less promising and will not be discussed here. Most of the commercial razor blades have similar emission properties with respect to field ionization. Only certain specially selected extra sharp stainless-steel blades (not covered with polytetrafluoroethylene) give a higher ion current than the average blades. Some of the results obtained by Cross and Robertson [75] are listed in table 2.4.

If a voltage of about 4 kV is applied to a new foil for the first time in the presence of a gas, a certain current i_0 is measured. After increasing the voltage to 10 kV and then reducing it again to 4 kV, the FI current decreases by a factor of about 100. This is due to tearing off of some field enhancing protrusions. The existence of these protrusions has to be considered if one would make a comparison of thin wires and sharp metal blades for field ionization. This comparison will be made in section 2.5.

TABLE 2.4. TOTAL FI CURRENTS AT VARIOUS VOLTAGES BETWEEN EDGE AND PLATE FOR OXYGEN AT 0.5 m Torr

Nature of edge	Voltage (kV)	Current (10^{-8} A)
Razor	11	< 0.04
blade	10	< 0.04
	9	< 0.04
Platinum	9	43
foil	8	25
	7	11.6

A current-voltage curve for ethylene at 5×10^{-4} Torr, taken with a blade 0.5 mm in front of a plane collecting electrode, and plotted in a double logarithmic scale is shown in fig. 2.39. The shape of the curve is rather different from the current-voltage curve of thin wires, as shown in fig. 2.25. Thus it can be concluded that the distribution of the emission areas of the protrusions is rather different in these two cases.

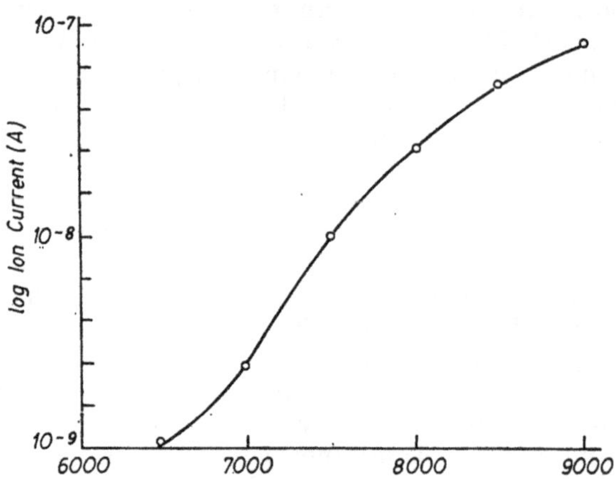

FIG. 2.39. FI current-voltage curve of ethylene on a sharp blade, distant 0.5 mm from a plane collecting electrode. (Robertson and Viney [10].)

2.5. COMPARISON OF TIPS, THIN WIRES AND SHARP METAL EDGES FOR FI

There are two distinct kinds of emitters for FI: Tips on the one hand and wires or sharp edges on the other hand. The latter two emitters are rather related and will be compared in more detail [76].

The properties and the range of application for tips are different from that of the other emitters. Much higher field strengths can be obtained with tips than with wires which may break and with edges from which the protrusions may be torn off. The highest field strength can be obtained with tungsten tips which are stable up to about 5×10^8 V/cm. At higher field strengths field evaporation of the tungsten atoms takes place. Tips are therefore preferably used in those cases where the physico-chemical properties of molecules should be studied at the highest possible field strengths. Further, tips are used if ions must be produced on clearly defined single crystal surfaces, or if a field ion image must be observed simultaneously with the mass spectrometric measurements.

On the other hand, the relative fragment ion intensity in FI mass spectra greatly increases with field strength. One of the advantages of FI mass spectrometry for chemical analyses is the high relative parent ion intensity. Therefore, in the normal analytical applications of FI mass spectrometers, the required field strength is only of the order 5×10^7 V/cm. This field strength can be easily realized with thin wires or sharp edges as FI emitters. The advantage of these emitters over a single tip lies in their much larger emission area and the correspondingly higher ion currents. An increase in stability of the emitted ion current is a consequence of this enlarged emission area, as discussed in 2.1.4.

The potential distribution near and the field strength at the surface of tips, wires and edges will be compared. The potential distribution is found by suitable transformations of the Laplace equation.

a) *Tips of parabolic shape*, surrounded by cathodes of parabolic shape at large distances.

$$V_{(r)} = V_0 \cdot \frac{2}{\ln (2R/r_0)} \cdot (\ln (2R)^{1/2} - \ln (2r)^{1/2}) \qquad (2.43)$$

$$F_0 = \frac{V_0}{r_0} \cdot \frac{2}{\ln (2R/r_0)} = \frac{V_0}{r_0} \cdot K_t \qquad (2.44)$$

where $V_{(r)}$ and V_0 are the potentials at distance r from the tip and at the tip surface, respectively, R is the distance between tip and cathode, r_0 the radius of curvature at the tip apex, and K_t is a geometry factor.

b) *Wire*, surrounded by a cylindrical cathode.

$$V_{(r)} = V_0 \cdot \frac{\ln R/r}{\ln R/r_0} \tag{2.45}$$

$$F_0 = \frac{V_0}{r_0} \cdot \frac{1}{\ln R/r_0} = \frac{V_0}{r_0} \cdot K_w \tag{2.46}$$

c) Edge, approximated by a thin parabolic cylinder surrounded by a parabolic cylinder as a cathode. (The cathode can be approximated by a plane at small angles counted from the edge.)

$$V_{(r)} = \frac{V_0}{r_0^{1/2} - (2R)^{1/2}} \cdot \{(2r)^{1/2} - (2R)^{1/2}\} \tag{2.47}$$

$$F_0 = \frac{V_0}{r_0\sqrt{2}} \cdot \frac{1}{\sqrt{R/r_0}} = \frac{V_0}{r_0} \cdot K_e \tag{2.48}$$

Robertson and Viney [27] reported on a method worked out by Gilliland and Viney [77] for calculation of the field strength at a sharp edge, in front of a plane cathode having a slit, using a conformal transformation. The resulting formula is virtually the same as that given above eqn. (2.48), if the slit is rather narrow as compared to the distance between the edge and the plane cathode.

A generalized formula can be given for the field strength at the surface of a tip, wire or edge:

$$F_0 = \frac{V_0}{r_0} \cdot K \tag{2.49}$$

where K is a geometry factor which has the form

$$K_{tip} = \frac{2}{\ln(2R/r_0)}, \quad K_{wire} = \frac{1}{\ln(R/r_0)}, \quad K_{edge} = \frac{1}{\sqrt{2}} \cdot \sqrt{R/r_0}$$

These formulae are applicable only to perfectly smooth emitter surfaces.

Inserting the values $V_0 = 10^4\ V$, $r_0 = 10^{-4}$ cm, $R = 0.5$ cm, one obtains:

$$F_{0\,tip} = 2.18 \times 10^7\ \text{V/cm}$$

$$F_{0\,wire} = 1.18 \times 10^7\ \text{V/cm}$$

$$F_{0\,edge} = 1.0 \times 10^6\ \text{V/cm}$$

The field strength at the wire is only about a factor of two smaller than that at the tip. However, the field strength at the edge is more than one power of ten smaller than that at the wire, which is due to the electrostatic field screening of the blade behind the edge. If the same field strength at a given voltage should be obtained with a sharp edge as with a thin wire, the radius of curvature of the edge has to be reduced by about two powers of ten, because

$$F_{0\ edge} \sim \frac{1}{\sqrt{r_0}}.$$

Resolving eqn. (2.48) for r_0 shows that the radius of an edge must be 72 Å in order to produce the same field strength as the wire, with the same values for V_0 and R as quoted above. This has to be compared with a wire having a radius $r_0 = 10000$ Å. If the same aperture for the emitted ions is considered in both cases, the emitting area of the edge will be 140 times smaller than that of the corresponding wire (both thought of as ideally smooth emitters of the same length). Thus, there will be a great loss in ion intensity if a sharp edge is used instead of a thin wire at the same applied voltage and anode — cathode distance and at the same field strength. Moreover, the statistical fluctuations of the ion intensity are expected to increase by a factor of $\sqrt{140}$, i.e. about twelve times, as pointed out in section 2.1.4.

If the blade-cathode distance is reduced to 0.035 cm the same field strength as quoted before is obtained with a blade having an edge of $r_0 = 1000$ Å. In this case, the emitting area and thus the ion current will be about ten times smaller and the fluctuations of the current about three times larger than the values given for a wire of $r_0 = 10^4$ Å and $R = 0.5$ cm.

All equations given so far, are derived for ideally smooth emitters. In reality, neither the wires nor the sharp edges represent perfectly smooth surfaces. There are micro needles on the surface of thin wires, and whiskers or other protrusions like sharp micro edges on the edge of metal blades or foils, as discussed before.

However, it can be shown that protrusions like whiskers change the field of ideally smooth wires or blades — termed "ground fields, F_0" — in the same way.

Using a method described by Gomer [16] for calculation of the field strength on top of a whisker, F_w, it can be shown [76] that the field enhancement factor for a whisker grown on FI emitters of different geometry is

$$\frac{F_w}{F_0} \equiv \beta \simeq \frac{h}{r_w} \quad (r_w \ll h \ll r_0) \tag{2.50}$$

as far as the whisker height h is small as compared to the radius of curvature of the substrate on which the whisker is grown. r_w is the radius of the shank of the whisker.

A more exact calculation of Vibrans [70] results in

$$\beta = 2 + \frac{h}{r_w} \tag{2.51}$$

which is approximated by eqn. (2.50) at $r_w \ll h$.

If projections of same β value are present on thin wires or sharp edges, the ratio of the radii of curvature, r_0, of these two emitter types remains the same as that given above. If the shape of the projections changes, the absolute value of the field enhancement factor will also change. But for any given protrusion, the field enhancement factor will be independent of the substrate as far as the condition $h \ll r_0$ is met.

Recently micro needles with very large field enhancement factors were produced on the edges of sharp blades by the benzonitrile activation method. The FI currents increased in that case to a value comparable to that of activated 5 μm wires.

2.6. MEANS FOR STABILIZING THIN WIRES DURING FI

In the preceding section it was shown that thin wires have theoretically some advantages over sharp edges for field ionization. In practice, however, Pt-wires of about 2.5 μm are easily broken if the applied voltage is raised over about 15 kV,*) unless certain steps are taken in order to reduce the probability of rupture.

These steps are:

1. It is not necessary to increase the voltage applied to 2.5 μm wires much above 10 kV if optimum micro needle growth had been achieved under conditions as described in section 2.3.4.
2. Wollaston wires of a few μm diameter as supplied from industry tend to have varying diameters and tensile strengths. By chance, one may get a piece from a specimen which has very good mechanical properties, but the next specimen may be much worse. If one has a coil containing many meters of Wollaston wire, the quality may vary from point to point. It is then advisable to cut the wire into one meter pieces and to test the tensile strength of the ends of each piece by a micro-tensile-strength apparatus [57]. Commercially available Wollaston wires for FI mass spectrometry should be pre-tested

*) 10 μm Wo-wires are very stable 15 kV.

by the supplier. Only the pieces of wire having the highest tensile strength should be used.

3. The mechanical force exerted by the high electric field on the wire can be largely compensated by an electrode behind the wire, which is at the same potential as the cathode in front of the wire.
4. Protective resistances of about 10^8 Ohms should be put into the admission lines for the field anode, the cathode and the lens electrode following the cathode. Thus the current produced during sparking in the ion source will be limited.
5. The ceramic spacers between the different electrodes should be designed to be as large as possible, and they should be cleaned carefully from time to time in order to avoid sparking in the ion source. Sparking has much more serious consequences for FI sources using thin wires than for EI sources.
6. The etched length of the wires should be as short as possible (about 2—3 mm), the supporting wires should be bent by about 30° as shown in fig. 2.21. and the FI wire should be slightly curved. (The reasons for this point were discussed under 2.3.2.2.).

2.7. ACTUAL DESIGN OF SOME FI SOURCES

An FI source as used by the authors group is shown in fig. 2.40. The upper sliding rod carries the FI wire which is at a positive potential of 6 kV. The two wires supporting the FI wire are separately fed through, so that a current can pass through the FI wire in order to heat it or to check whether it is in order or broken. Electrode (*1*) serves as support for a heating element capable of heating the source up to a maximum of 300 °C. A thermocouple measures the source temperature. The cathode (*1*) is on a variable negative potential of typically 5 kV and (*2*) is on ground potential. The lens electrode (*3*) is variable between 0 and + 3 kV. A pair of deflecting plates (*4a*) and (*4b*) can be supplied with voltage differences of max. 200 V. Electrode (*5*) is on ground potential and serves as a support for the whole source assembly. Samples in form of vapours are supplied through a glass tube. Solid samples are evaporated onto the FI wire from a micro-oven. The temperature of the oven is measured by a thermocouple contained in the horizontal sliding rod. The admission wires for the heater and the temperature sensor are fed through the back side of the sliding rod containing these elements. The sliding rod is part of a vacuum lock constructed similarly to that designed by Brunnée [79]. The surroundings of the oven are water cooled in order to shield the oven thermally from the ion source, as discussed by Brunnée. The glass tube for gas admission and the vacuum lock are not in one plane as in fig. 2.40. but perpendicular to each other.

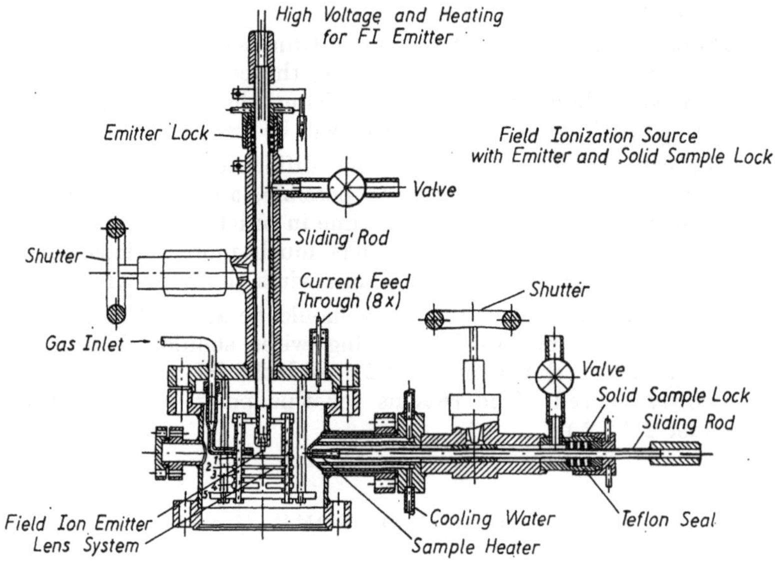

FIG. 2.40. Schematic drawing of an FI source for use with gas inlet or with evaporation oven for solid substances or for operation in the field desorption mode (see section 4.3.4.). The details of the source are described in the text.

An FI source using a sharp edge as FI emitter has been described in detail by Robertson and Viney [10, 27]. The first version of this source (fig. 2.41.a.) consisted just of a blade on positive high voltage opposite to a plane cathode with a slit followed by an electrode on ground potential.

An improved version of this source is shown in fig. 2.41.b. An additional electrode (C), is introduced acting as an electrostatic lens. This electrode is on a negative potential, but less negative than the cathode (P). The positive voltage on the FI emitter (B) is variable because the mass spectrometer is operated with a permanent magnet. Typical values of the potentials for ions of mass 86 are: $V_B = +418$ V, $V_P = -5850$ V, $V_C = -1000$ V (optimum focusing), 0 is on ground. The distance between the blade and the cathode is 0.25—0.5 mm. The slit is about 0.25—0.5 mm wide and parallel with the blade.

As mentioned earlier, Gilliland and Viney [77] calculated the electric field at the surface of a sharp edge of a thin blade using the method of conformal transformation.

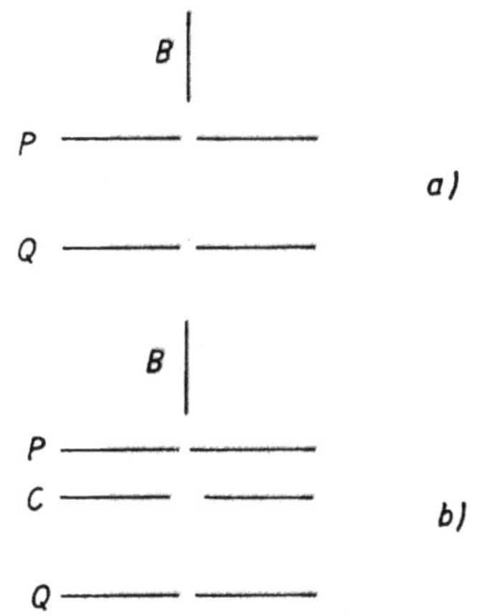

FIG. 2.41. *a*) Simple FI source using a razor blade *B* on positive high voltage. *P* is on negative high voltage and *Q* on ground potential.
b) Improved FI source with razor blade with additional slit plate *C* acting as focusing lens. Slit width of *P* and *Q*: 0.5 mm, of *C*: 1.6 mm. Distance P–$C = 2.3$ mm, C–$Q = 8$ mm. (Robertson and Viney [10].)

The radius of curvature of the blade is r_0; the sharp edge is distant l from the plane cathode having a slit of width $2b$. For an infinitely thin plate the field F is approximately

$$F = \frac{2V}{\pi r_0^{1/2}} \left\{ \frac{l + (l^2 + b^2)^{1/2}}{l^2 + l(l^2 + b^2)^{1/2} + b^2} \right\}^{1/2} \tag{2.60}$$

where V is the voltage between the blade and the cathode.

For an edge having a radius of curvature $r_0 \simeq 500$ Å and $l = 0.25$ mm, $2b = 0.5$ mm one obtains

$$F = 1.5 \times 10^3 \times V \text{ [V/cm]}.$$

Applying 10 kV, one obtains $F = 1.5 \times 10^7$ [V/cm]. Then a field enhancement factor of at least $\beta \simeq 5$ has to be assumed for obtaining a sufficiently large FI probability for most organic substances. An aged blade gives an ion current with most substances steady to within $\pm 2\%$ during about 1/2 h. The short time fluctuations (during a scan time of about 0.5 sec per peak) are much higher according to the authors experience with razor blades as FI emitters. With care a blade can be made to last a month [27]. (An average lifetime of a blade of a week is reported by Job [80].)

A spark in the FI source usually results in a loss of efficiency of the blade. Probably, sharp protrusions on the edge are torn off by sparks. It seems to be difficult to avoid sparks at voltages between blade and cathode exceeding about 10 kV.

By experiments with an electrolytic trough, Robertson and Viney found that field penetration through the cathode slit is appreciable with a very thin cathode, but it is considerably reduced with a plate thickness equal to that of the slit width. Altering of the slit width from 0.9 to 0.34 mm does not greatly change the collected ion current. Further reduction of slit width reduced the ion current.

The relatively low transmission of the mass spectrometer can be explained, according to Robertson and Viney [27] in the following manner. Consider the FI source shown in fig. 2.41.a. The ions accelerated between B and P are retarded between P and Q. The velocity parallel to the x — direction and parallel to the electrodes is not altered by the retardation (assuming parallel equipotentials between P and Q). But the velocity in the y-direction perpendicular to the electrodes is decreased. A loss in ion intensity may be caused by two effects. The half width (Δx) of the beam at plate Q may be larger than the slit width (S_2) in Q:

$$\Delta x = 2D\Theta_1 V_1^{1/2}/(V_1^{1/2} + V_2^{1/2}) \qquad (2.61)$$

where D is the distance between P and Q, V_1 is the total accelerating voltage, V_2 is the potential difference between B and Q, and Θ_1 is the angle with the normal to P at which the ions pass through the slit (S_1).

The ions pass through the slit (S_2) at an angle:

$$\Theta_2 = (\Delta x/2D)\{1 + (V_1/V_2^{1/2})\} \qquad (2.62)$$

Field penetration through the slits is neglected in these equations. Inserting the values $\Theta_1 = 14°$, $D = 6.5$ mm, $V_1 = 11$ kV, $V_2 = 1$ kV, one obtains $2\Delta x = 4.76$ mm which has to be compared with a slit width (S_2) of 0.5 mm. Thus, about 90% of the ions are lost for analysis by retardation. Another considerable fraction could be lost because the beam emerging from the slit in Q is diverging more strongly than is permissible for re-focusing by the magnetic field of the mass spectrometer.

In the general case, penetration of the field lines through the slits in the electrodes has to be considered, especially if a focusing electrode is placed between electrodes P and Q. The calculation of the change in transmission with retarding of the ions is then much more complicated. But the general result remains (which is confirmed experimentally) that stronger retardation of the ions leads to a decrease in transmission.

Combined Electron Impact and Field Ion Sources

FI and EI mass spectra are complementary, to some extent, as will be discussed in the chapter on organic structural analysis. Therefore, it is advantageous to combine both methods of ionization within one ion source. The prototype of such a combined

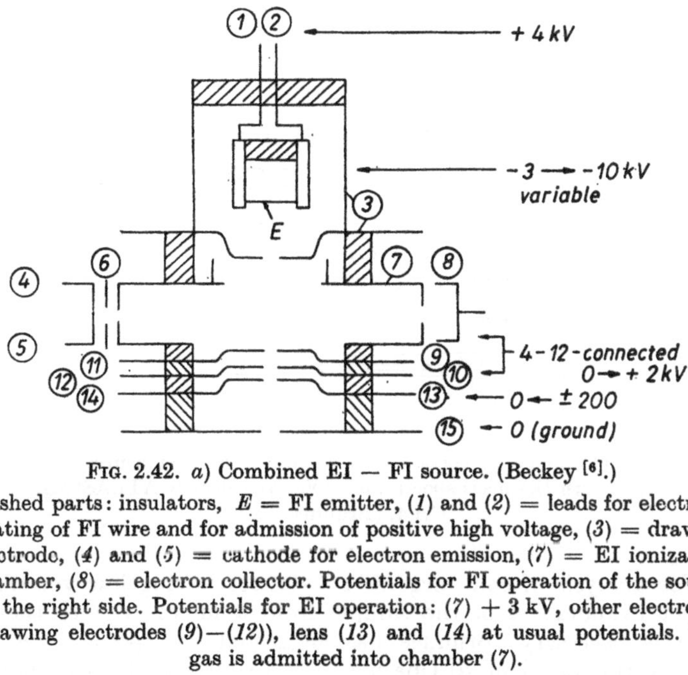

FIG. 2.42. *a*) Combined EI — FI source. (Beckey [6].)
Dashed parts: insulators, *E* = FI emitter, (*1*) and (*2*) = leads for electrical heating of FI wire and for admission of positive high voltage, (*3*) = drawing electrode, (*4*) and (*5*) = cathode for electron emission, (*7*) = EI ionization chamber, (*8*) = electron collector. Potentials for FI operation of the source on the right side. Potentials for EI operation: (*7*) + 3 kV, other electrodes (drawing electrodes (*9*)—(*12*)), lens (*13*) and (*14*) at usual potentials. The gas is admitted into chamber (*7*).

FI — EI source was constructed first by the author in 1964 and was described in the literature [6] (fig. 2.42.a.). Brunnée designed a refined commercial version of this source [81] which could be changed from one to the other mode of operation without changing parts of the source or opening of the envelope. All that has to be done is to change a multi-contact plug containing the connecting wires.

The construction of the source is shown in fig. 2.42.b. The "repeller" electrode on the back side of the EI source serves as a cathode (being on a potential of −3 to −10 kV) in the case of FI. Molecules emerging from the gas inlet diffuse through the slit in the repeller electrode to the FI wire. 2.5 μm platinum Wollaston wires are used for FI.

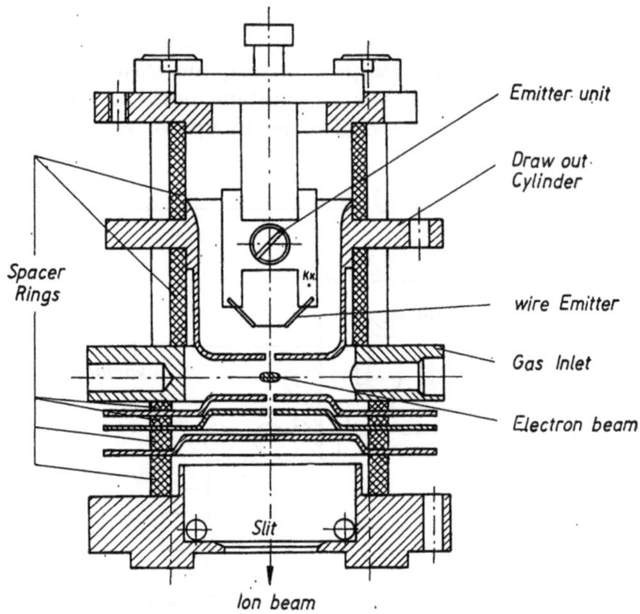

FIG. 2.42. *b*) Commercial version of a combined EI−FI source, as used in connection with the mass spectrometer Varian (formely ATLAS) MAT, type CH 4. (Brunnée [81].)

For analysis of solid organic substances having low volatility, direct evaporation of the sample from a micro-oven to the FI wire is necessary. Brunnée has suggested two possibilities for solving this problem. *a*) One may use a source with two separate ionizing regions as shown in fig. 2.42.b. Then it would be necessary to mount two separate direct inlet systems, or the vacuum lock containing the oven or the ion source must be movable. *b*) The second possibility is to make the ionization region for FI nearly coincident with the one for EI. Only one direct inlet is then necessary. The vapour beam leaving the oven has a certain spread so that molecules will cross the electron beam and hit the FI wire, even if these are slightly separated. The latter possibility has been realized and some EI/FI mass spectra have been measured with such a source [81].

An improvement of this principle has been achieved by drawing back the FI wire from the path of the electron beam by a few mm during EI operation of the source (fig. 2.43.).

FIG. 2.43. Combined EI — FI source used in connection with mass spectrometer type CH 5 or SM 1 (Varian-MAT). The lever (connected with two springs) driven by the rotatable feed-through (right side) moves the wire emitter either into the plane of the molecular beam, or a few mm backwards, respectively.

2.8. PULSED FI SOURCES

Pulsed FI sources may be used for the study of adsorption phenomena, as described by Inghram and Gomer [1] and by Block [9]. The applications of the source will be described in a later section. Here, the principle will be briefly outlined.

Molecules may be adsorbed onto the FI emitter. The number of molecules adsorbed in a monolayer may be n_m. If the high electric field is applied to the emitter only for very short time intervals (in the order of μsec), then the number of molecules arriving at the tip from the gas phase during that time interval can be made small as compared to the number of molecules n_m adsorbed in the monolayer. If the field is high enough to ionize the adsorbed molecules, these contribute much more to the ion beam than those ionized in the gas phase, provided that the time intervals between the electric field impulses are large enough to fill up the reservoir of adsorbed molecules. Thus, by changing the impulse frequency, one may study certain features of the kinetics of adsorption.

A high voltage generator for negative pulses of short rise time and high frequency has been described by the author and Röllgen [82]. Three conditions should be met simultaneously by the generator:

1 Amplitudes up to −20 kV
2 High repetition frequency of the pulses, up to 30 kHz
3 Very short rise times (a few nsec) and plateau times (a few 100 nsec.)

Condition 3 must be met because otherwise a large fraction of ions would be formed during a period when the energy of the ions is varying, which would lead to very diffuse mass lines. Some of the details of such a generator will be dealt with.

The negative high voltage square pulses are produced by a high voltage vacuum tube acting as an anode amplifier. This works only if the capacitive pulse load is small and a very small power of the pulse is required.

Fig. 2.44. shows a circuit diagram of the generator. Well known standard circuits are used for production of the primary pulses entering the generator. (This circuit is shown as a block diagram in fig. 2.44.)

The most important element of the generator is the high voltage tetrode (4 PR 60 B, Eimac). The maximum anode dc voltage is 20 kV, the maximum impulse current 20 A and the permissible anode plate dissipation 60 Watts.

A rise time of 30 n sec and a decay time of 220 n sec is calculated for typical operating conditions. These values are confirmed experimentally, as shown in fig. 2.45. The rise time of the grid impulses of the tetrode should not exceed 10 n sec by very much, and their amplitude should be 500−600 V. These grid impulses are produced by a hydrogen thyratron, PL 345. The length of these pulses can be varied by different lengths of the impulse-

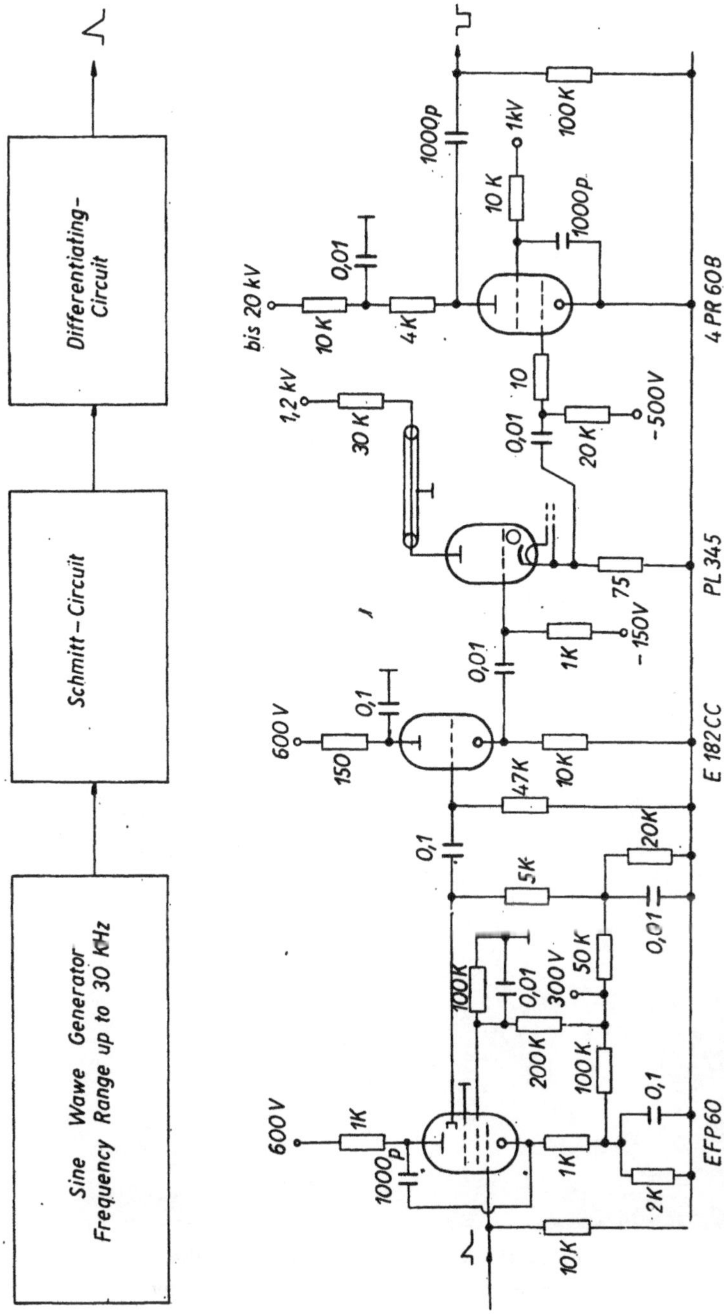

FIG 2.44. Circuit diagram of a high voltage impulse generator of high pulse repetition frequency and short rise times.

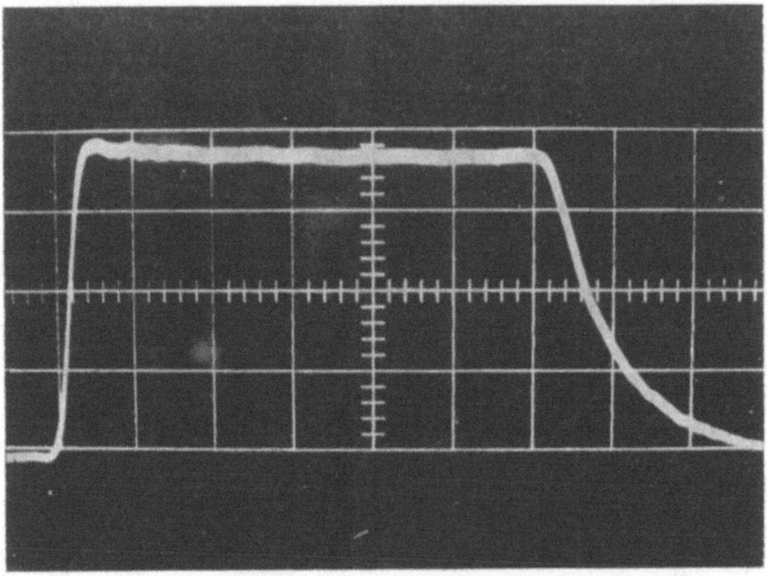

FIG. 2.45. Negative high voltage impulse.
a) amplitude -5.5 kV, time scale 50 ns/div

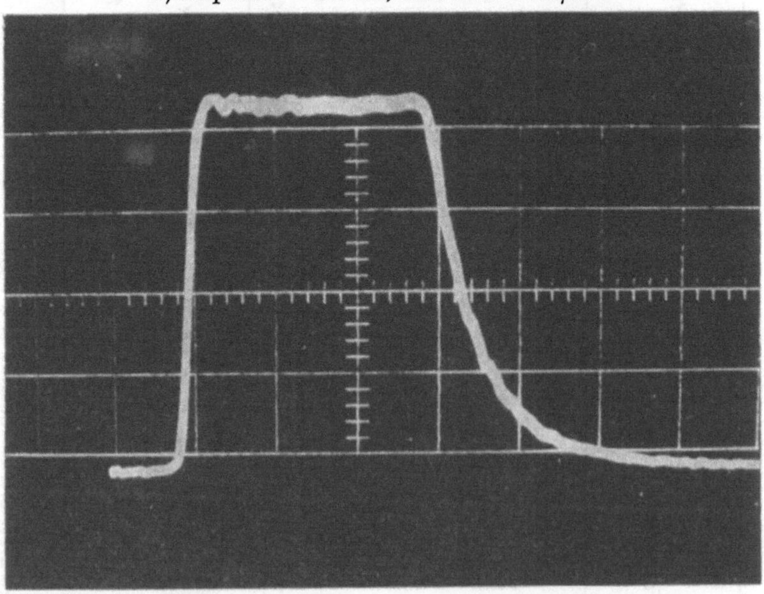

FIG. 2.45. Negative high voltage impulse.
b) amplitude -18 kV, time scale 100 ns/div

forming cables shown on top of the thyratron in fig. 2.44. (50 to 500 n sec.).

An univibrator using the secondary emission tube EFP 60 produces the pulses necessary for triggering the thyratron. A cathode follower is placed between the univibrator and the thyra-

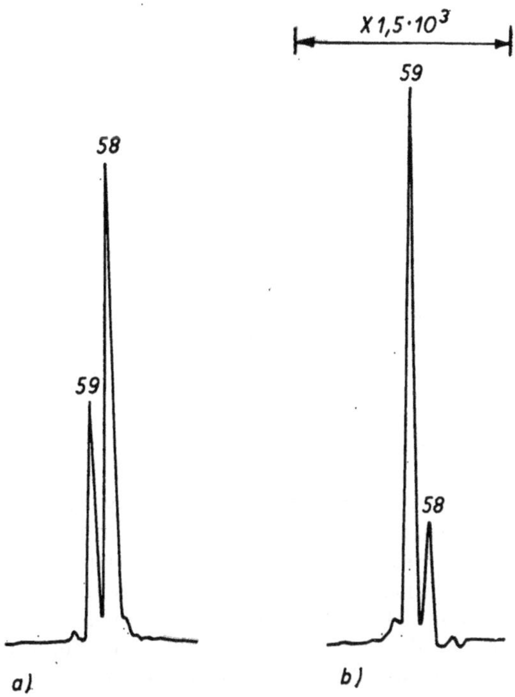

FIG. 2.46. FI mass spectrum of the acetone parent ions. (M = 58 and 59).
a) dc operation, $U_{cathode} = -10$ kV.
b) Impulse operation: $U_{cathode} = -4$ kV dc, superimposed impulse voltage = -6 kV, frequency 300 Hz, sampling ratio 10^{-4}.

tron. The trigger pulses for the univibrator are produced by a RC sine generator of variable frequency with subsequent pulse shaping by a Schmitt-trigger and differentiation of the pulse, as shown schematically in the block diagram.

The connection of the generator with the FI source must be very short and of small capacity. If a wire is used as FI emitter, resonance could occur between a mechanical vibration of the wire and a low pulse frequency. (The lowest frequency used is about 15 Hz.) This would lead to destruction of the wire. The probability of rupture of the wire can be greatly reduced, however, by using curved wires, as discussed earlier.

A typical change in the FI mass spectrum of acetone occurs if the dc operation is replaced by pulsed operation of the source as shown in fig. 2.46. The M 59 : M 58 ratio is much greater in the pulsed operation because the probability of the reaction $2\,C_3H_6O_{ads} \rightarrow C_3H_6OH^+ + C_3H_5O_{ads} + e^-$ is largely increased, as will be discussed in a later section.

2.9. FI MICROSCOPES AS SOURCES FOR MASS SPECTROMETERS

It would be outside the concept of this book to give a detailed treatment of FI microscopes. (For a review see [37].) However, some FI microscopes have been developed which enable one to make a mass analysis of the ions observed on the microscope screen by combination with a mass spectrometer. Therefore, the basic features of FI microscopes will be briefly discussed here.

2.9.1. FI microscopes

FI microscopes — as first designed by Müller [83, 84] — are similar in construction to FE microscopes. The basic features may be discussed by means of fig. 2.47. The description will be limited to a microscope with helium as imaging gas and tungsten tips for field production. The FI microscope essentially consists of an evacuated bulb filled with an imaging gas like helium or hydrogen and a cooled tip assembly. Cooling of the tip serves for reduction of the tangential velocity of the ions and hence for an increase in resolving power of the microscope. (A resolving power of 2—3 Å may be achieved.) Liquid hydrogen usually serves as a coolant, although liquid helium would be less dangerous. However, the average hopping height of helium atoms used as imaging gas at the tip surface is about 5 Å at 20 °K (liquid hydrogen), which is optimum for best resolving power of the microscope. Cooling of the tip and its surroundings is achieved by a cold finger. The metal leads supporting the tip are sealed in the bottom of the cold finger, which is conically ground from outside as seen in the figure. A cone of a metal like aluminium fits closely to the ground part of the glass finger, thus leading the low temperature to the surroundings of the tip.

The use of helium as imaging gas has double advantages. First, the resolving power of a liquid hydrogen cooled helium microscope is very high, and second, contamination of the tip during operation of the microscope is minimized. Helium is the gas having the highest ionization potential (24.5 eV). All other gases, which are potential contaminants, have ionization potentials below 15 eV. Therefore, with a field strength at the tip surface of about 500 MV/cm, all substances except helium will be field

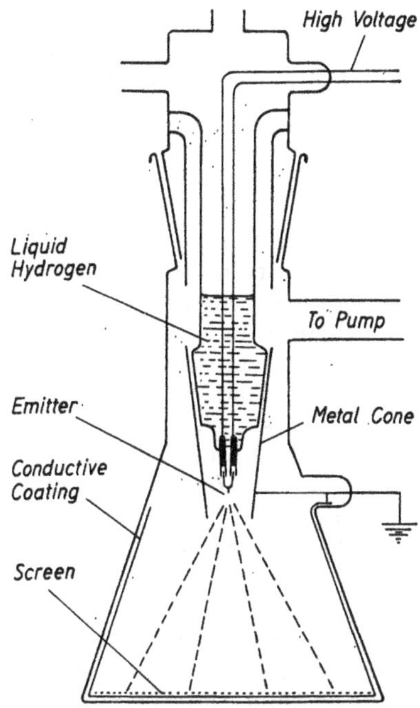

FIG. 2.47. Low temperature FI microscope. (Müller [37].)

ionized in space farther from the tip at a field strength of about 200—300 MV/cm. Thus the contaminant gases will not arrive at the tip as long as the high voltage is applied to the tip. The ions of the contaminant gases will produce a very faint background of light only on the microscope screen. A tip which has been cleaned by heating or field evaporation of adsorbed impurities will therefore remain extremely clean over a very long time under these conditions. It is not necessary to bake out the whole FI microscope before operation, and it is also possible to use ground joints with vacuum grease for changing the tip assembly, as shown in fig. 2.47. Thus a quick change of the tip is possible.

The helium FI microscope may be either filled statically with helium, or it may be operated with a dynamic gas supply [85].

In the latter case the metal cone surrounding the tip is closed from the bottom by a disc which has a small aperture. The gas is admitted by a side tube and is pumped off through the hole in the disc by a pump connected with the microscope tube. In this way, a higher pressure may be established in the vicinity of the tip without increasing the probability of scattering of the ions with the gas in the microscope bulb. The higher pressure leads to higher ion currents, and hence to brighter screen pictures.

The luminescent screen of the FI microscope must be made electrically conducting without appreciable loss in transparency. This is usually performed by coating the glass with tin oxide [20]. The conductive coating and the metal cone are on ground potential, whereas the tip is on a variable positive high voltage (max. about 25 kV).

The phosphorescent screen must be made so that it has a high efficiency at a total image current of $10^{-8}-10^{-10}$ A. The phosphor layer should be quite thin in order to avoid absorption of light which is produced by the ions in the upper layer of the phosphor. The screen may be made by dusting a fine powder of the phosphor on the glass treated with a binder like phosphoric acid [16]. Zinc sulfide or zinc silicate may be used as phosphor.

A flat glass screen should be used because cameras with very bright objectives must be used for photographing the very faint screen pictures. (F : 1 objectives, for example.) It would be difficult to obtain sharp photographs with these lenses if the screen were curved.

Very sensitive films must be used, like the Kodak Tri-X film, or Polaroid films of speed ASA 3000 or 10000, for example. The exposure time varies between a few seconds and an hour, depending on the tip radius. Increasing the tip radius increases the ion current and hence reduces the exposure time. Complete darkness of the laboratory is required for visual observation of the FI microscope images. Image intensifiers are now used in many cases, especially if gases other than helium are used for image production, which produce fainter images than helium.

The magnification of FI microscopes is approximately given by the ratio of the tip to screen distance (R) to the tip radius (r_0). Typical values are: $R = 10$ cm, $r_0 = 10^{-5}$ cm, magnification $M = 10^6$. The resolving power is about 2—3 Å.

2.9.2. Mass analyses in modified FI microscopes

Müller and Bahadur [3] reported on qualitative analysis of field ions in a simple mass spectroscope (Fig. 2.48.). This consisted of an FI microscope with a focusing lens in front of the FI tip. A normal FI image is observed on the microscope screen if the central electrode of the lens is on ground potential. If this electrode is on a high positive potential, the ion beam emitted into an angle of 0.2 radian is focused into a small spot on the screen. Then a

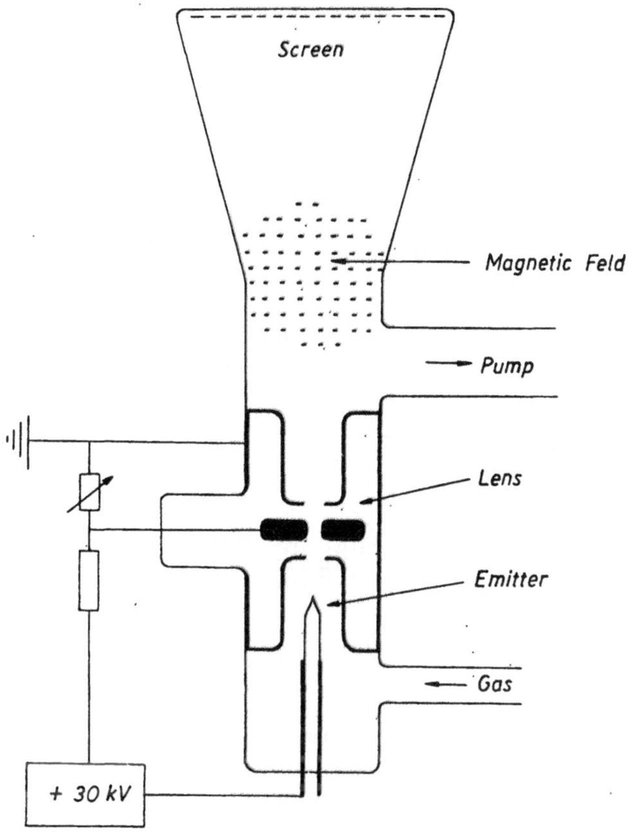

FIG. 2.48. FI mass spectroscope. (Müller and Bahadur [3].)

magnetic field as indicated is switched on, and a mass separation of the ions occurs. The resolving power of this simple mass spectrometer is limited to $M/\Delta M \simeq 80$, sufficient for resolving the krypton isotopes. The authors do not report on quantitative results obtained with this device. However, qualitative observation of the relative abundance of ions like H_2^+ and H^+ emitted from large fractions of the surface could be made.

More quantitative results were obtained with a device designed by Clements and Müller [23], fig. 2.49. The ions emitted by an FI tip can be either observed on the luminescent screen of an FI microscope, or deflected by a magnet into the analyser tube of a 180° mass spectrometer. The ions are measured after mass separation by a Faraday cup detector, the signal of which is recorded on the screen of an oscilloscope after amplification. The

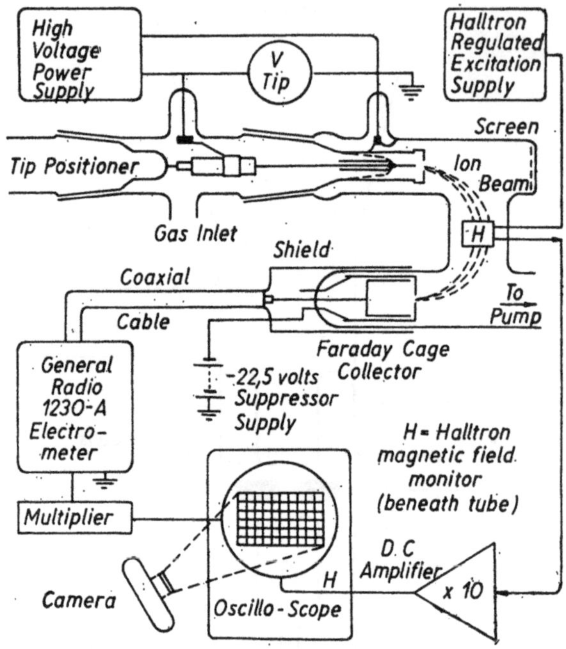

FIG. 2.49. Combined FI microscope mass spectrometer. (Clements and Müller [23].)

cathode ray of the oscilloscope is scanned in the direction of the mass scale, in parallel with the scan of the magnetic field. In that way the relative intensity of the H^+, H_2^+ and H_3^+ ions and of the corresponding deuterium ions was measured. The measurements with the 180° mass spectrometer average over the ion abundance emitted from a rather large, somewhat inhomogeneous tip area.

In contrast to this, two modified FI microscopes designed by Clements and Müller [23] served for mass spectrometric measurement of the ions emitted from single metal surface atoms of the tip. The two microscopes differed only in that the first one had a static gas supply operating up to 3×10^{-3} Torr, and the second one had a dynamic gas supply [85] working up to about 8×10^{-2}

Torr. A movable slit was placed about 20 mm from the emitter tip. The slit was formed by a gap about 0.05 mm wide between two 1 mm diameter tungsten wires. These wires appear on the 4-inch image screen as dark shadows with a bright line between them, extending over the whole screen (fig. 2.50.). The slit system

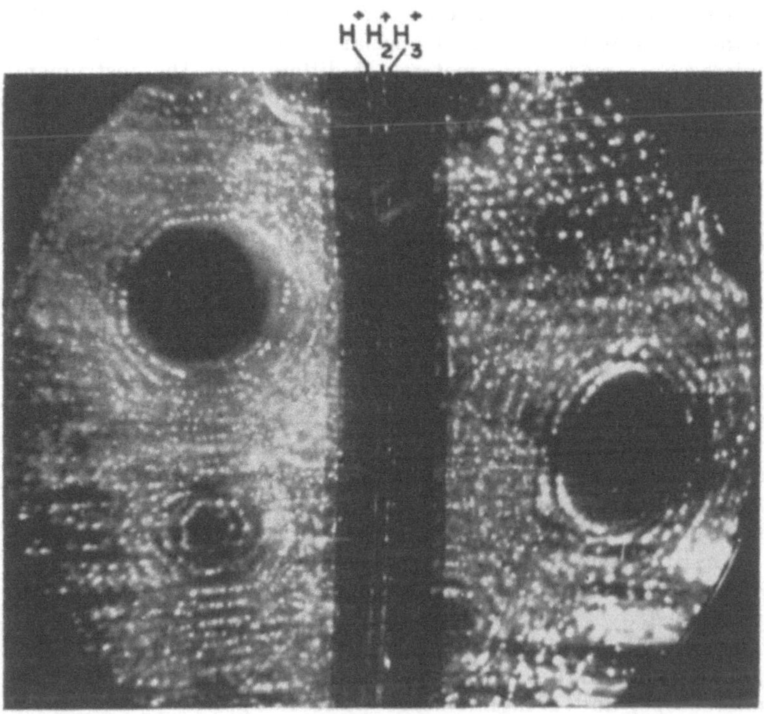

FIG. 2.50. Hydrogen FI mass spectrum in FI microscope near best image field. (Clements and Müller [23].)

may scan the whole image screen. An external electro magnet placed between the slit system and the screen deflects the slit image by about 10 mm. Mass separation of the ions occurs due to the action of the magnetic field. If hydrogen is introduced into the microscope, the image points in the slit image are split into 2—3 masses. In this way, the ions coming from an individual tip surface atom could be mass analysed. The relative intensities of the different ionic species were estimated, although they could be determined — in principle — more accurately by microphotometry.

The results of the investigations with these devices, especially the origin and relative abundance of the species H^+, H_2^+ and H_3^+ will be discussed in a later section.

2.9.3. The atom-probe FI microscope

The atom-probe FI microscope as designed first by Müller, Panitz and McLane [22] is a special version of the combined FI microscopes massspectrometers. It differs from the instruments as described in the preceding section in that mass analysis is not made by magnetic deflection of the ions but by time-of-flight measurements. Another important feature is that it is possible to make a mass analysis of single atoms of the FI tip, whereas atoms from more or less extended areas of the tip are collected by the other methods. Both types of techniques have their specific merits, depending on the problems to be solved.

The field ion microscope enables one to view the exact location of individual atoms on a metal surface. However, if the FI tip consists of an alloy or contains impurities or adsorbates it is not possible to identify unambiguously the different atomic species only by the microscopic technique. The fascinating feature of the atom-probe FI microscope is the possibility of making a correlation between one exactly localized atom on a metal surface and its mass, and hence in general with its chemical nature.

An FI image of the tip is produced on the fluorescent screen of the microscope in the conventional way. One image spot can be positioned over a probe hole in the screen by rotating the tip. The specimen tip is positioned at the center of rotation of a ball joint (see fig. 2.51.a.). The imaging gas is then pumped from the microscope, and a high voltage pulse is applied to the tip. If this pulse is adjusted to a proper value, the atomic species which had been selected desorbs from the tip and travels through the probe hole. This high voltage impulse also triggers the horizontal sweep of an oscilloscope. The ion — after passing through the probe hole — traverses a field free drift tube and is then captured by the first dynode of a secondary electron multiplier. The output of the multiplier is fed to an amplifier and then to the vertical deflection electrode of the oscilloscope. In this way the time-of-flight of the ion $t = d/v$ can be measured. d is the distance which the ion traverses, and v its velocity. v is correlated with the total desorption voltage V by

$$\frac{1}{2} mv^2 = ne V$$

where m is the mass of the ion and ne its charge. With $d = 82$ cm, one obtains

$$m/n = 0.288\, V t^2 \qquad (2.63)$$

A two-stage differential pumping system (see fig. 2.51.a.) provides for good vacuum conditions in the ion drift space. A view of the atom-probe FI microscope as described by Müller et. al. [22] is shown by fig. 2.51.b.

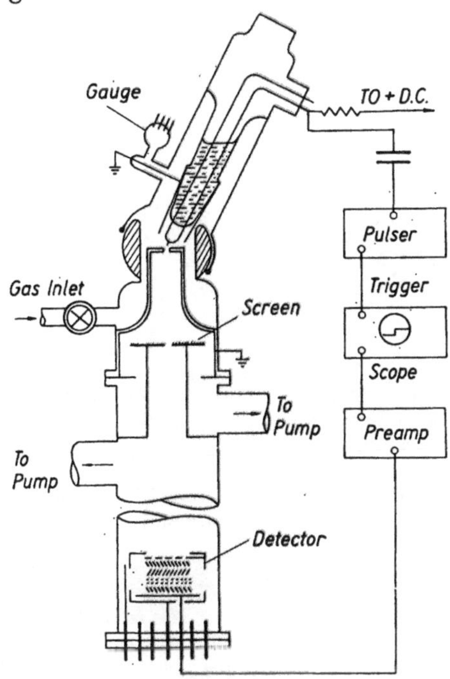

FIG. 2.51. *a*) The atom-probe FI microscope (schematically) [22].

Fig. 2.52. shows an example of the oscilloscope pattern arising from field desorption of single atoms from a tungsten tip at two different voltages.

From eqn. (2.63) one obtains $m/n = 71$ for the upper trace, and from left to right respectively and $m/n = 53.6$, $m/n = 71.2$ for the lower trace. Traces of oxygen were present in the FI microscope, and therefore it can be assumed that $m/n = 71.3$ is due to the tungsten-oxygen complex $^{182}W^{16}O_2^{+++}$, and $m/n = 54.0$ due to $^{184}W^{16}O_2^{++++}$. These species have been observed also by Vanselow and Schmidt [156] during field evaporation of oxidized tungsten tips in a magnetic FI mass spectrometer. It is possible,

FIG. 2.51. *b*) Actual design of the atom-probe FI microscope, as constructed by Müller, Panitz and McLane [22].

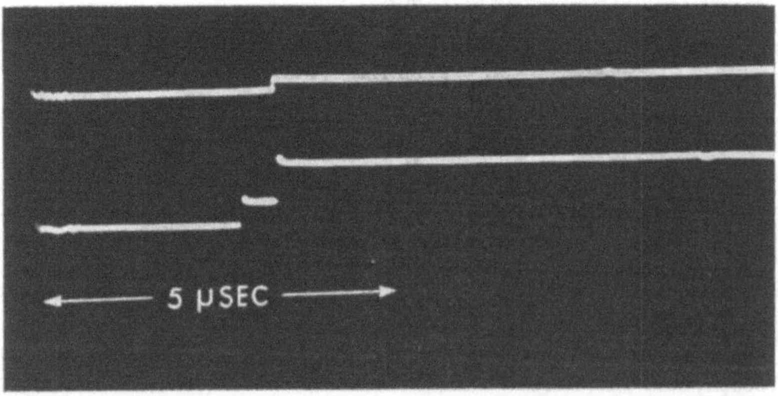

FIG. 2.52. Atom-probe examination of tungsten; upper trace: 2.5 kV/dc + 2.4 kV pulse; $m/n = 71$ ($t = 3.0$ μsec). Lower trace: 25 kV/dc + 2.5 kV pulse; (from left to right) $m/n = 53.6$ ($t = 2.6$ μsec), $m/n = 71.2$ ($t = 3.0$ μsec).

however, that the two species observed may also represent $^{182}W^{14}N_2^{+++}$ and $^{186}W^{14}N_2^{++++}$, respectively.

This brief account on the atom-probe FI microscope is rather preliminary because it was written after the main manuscript of this book was finished (Spring 1968). The first paper of Müller, Panitz and McLane on this subject had just appeared at that time. It is believed, however, that this new technique will play an important role in the future for the study of individual atoms at the surface of metals.

2.10. RESOLVING POWER OF FI MASS SPECTROMETERS

A number of different definitions of the resolving power of the mass spectrometer have been given in the literature. Here, mainly, only the "10% valley definition" will be used. This is given by

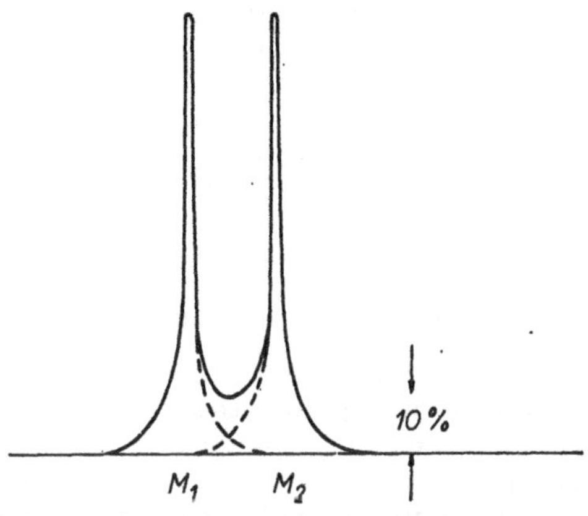

FIG. 2.53. Two mass lines illustrating the definition of resolving power.

$M/\Delta M$, where M is the maximum mass which can be resolved from the neighbouring mass $M \pm \Delta M$ (ΔM is unit mass difference) at the condition that the valley (measured from the base) between the two peaks of equal height is 10% of the peak height (fig. 2.53.). The author's experience with respect to the resolving power of FI mass spectrometers is as follows. Five FI mass spectrometers which resemble the former ATLAS CH 4 mass spectrometer are in use in his laboratories. (They are partially custom built, using parts of the components of the CH 4 mass spectrometer as available from the former Krupp-MAT*), like the m.s. and some electronic circuits.) The CH 4 mass spectrometer

*) Now Varian MAT, Brennen, germany.

consists of an analyser having a 60° magnetic field with a radius of curvature of the analyser tube of 20 cm. Using an entrance slit of 0.1 mm and an exit slit of 0.3 mm width, one obtains an optimum resolving power of $M/\Delta M \simeq 600$ with an FI tip. Using a platinum Wollaston wire at + 5 kV potential with a cathode of − 5 kV, one obtains a resolving power of maximum $M/\Delta M \approx 300$. The restriction in resolving power of FI mass spectrometers using tips or, more pronounced, using thin wires for field ionization is due to two effects. Adsorbed mono-or multilayers may be formed on tips or wires, the thickness of which depend on the nature of the adsorbed substances. A voltage drop across this layer may occur. If, for example, the layer would be 10 Å thick and the field strength would be 1 V/Å, the voltage drop across the layer would be 10 Volts. This has to be compared with the tip potential U_0 of 5000 Volts. If some of the ions are formed on tip potential, U_0, and others on the potential of the surface of the layer, $U_0 - 10$ Volts, then a spread in ion energy $\Delta E = 1/500$ is to be expected. The corresponding spread in the mass scale is $M/\Delta M = 500$. This effect of limitation in resolving power due to adsorbed layers is effective both with tips and with thin wires.

Another, more serious effect arises if wires are used instead of tips. Field ionization occurs on top of the whiskers*) formed on the wire surface. Although the spread of the field enhancement factors β for the predominantly emitting whiskers at a given voltage is not large, the factors $\beta = \frac{h}{r}$ may be realized either by longer and blunter, or shorter and sharper whiskers. It may be seen from the electron micrographs that there is a distribution of whisker length and diameter. Consequently, since β varies with r^{-1} and the electrical resistance of the whiskers with r^{-2}, there will be a distribution in the resistances of the whiskers. The specific resistance of thin whiskers having a radius r of a few Å only and being covered with an adsorbate is several orders of magnitude heigher than that of the compact metals. The distribution of these voltage differences causes again a spread in energy of the field ions, and hence a diffuseness in the mass lines. Fig. 2.37. shows how the deposit of solid organic material on the surface of a FI wire decreases the resolving power.

The resolving power should increase if the voltage applied to the wire, and hence the surface field strength, is raised such that the protrusions are torn off from the wire.

Brunnée [86] has reported that increasing the voltage between the wire and the cathode from the normal value of 10 kV to

*) The term "whiskers" is used here for organic micro needles.

18 kV results in an increase of resolving power from $M/\Delta M \simeq 300$ to $M/\Delta M \simeq 800$. However, wires of very good mechanical stability have to be used at these high field strength. The field enhancement factor must then decrease from about 7.4 to about 3.0. A β factor of three is realized, for example, by blunt protrusions of hemispherical shape on the wire surface. The voltage drop across these protrusions will be too small to cause an appreciable energy spread of the field ions. The field enhancement factor, on the other hand, is still large enough to allow for efficient field ionization.

A resolving power of 1225 [87] at mass 58 with the 1% neighbour peak contribution definition has been obtained with a single focusing mass spectrometer (type Varian-MAT CH 4), equipped with a new combined EI/FI source with improved ion optical design. The data were: +3 kV anode voltage, −10 kV cathode voltage, mass spectrometer entrance and exit slits 0.001 mm wide. The resolving power was about a factor of three higher at the EI mode of operation with the same slit widths.

A very high resolving power can be obtained not only with EI sources but also with FI sources if double focusing mass spectrometers are used. Beynon, Fontaine and Job [88] carried out a simple modification of the ion source of the A.E.I. double focusing mass spectrometer MS 7 such that it could be used as FI source. The source was a diode-system, consisting of a razor blade as FI emitter to which the full high voltage of the MS 7 mass spectrometer could be applied (+20 kV), and a slit on ground potential. The slit was 1 mm distant from the razor blade. The dimensions of this slit were 0.3 mm × 1 mm. A second slit, 0.12 mm wide acted as "beam defining slit" as in the unmodified MS 7 mass spectrometer. The resolving power was 4000 at $m/e = 100$ using the 50% peak height definition. The spectra were recorded on photoplates.

Beckey, Hilt and Krone [89] modified the electron impact ion source of the double focusing CEC mass spectrometer 21-110 B in such a way that it could be used as a combined EI/FI source (fig. 2.54.). The repeller plates of the EI source served as cathode opposite to the field anode. A 2.5 μm wire was used as anode. The anode admission leads were sealed into a glass plug which closely fits into a hole drilled into the plate closing the EI source.

A resolving power of $M/\Delta M = 10000$ could be obtained with this ion source, but it was not principally restricted to that value. A resolving power of $M/\Delta M$ 30000 was obtained with that source during the EI mode of operation. The sensitivity of the mass spectrometer is smaller by a factor of about 100 in the FI mode of operation. Therefore, increasing the resolving power above

10000 by further reduction of slit width would lead to extremely low ion intensities. The ion source could be operated as a combined EI/FI source only with gaseous, liquid or solid samples of high volatility introduced from a gas reservoir via a leak. This is because the point of sample admission to the source is different from the space of field ionization. Molecules of substances of high volatility could diffuse through the slit in the repeller plate to the field ionizing wire, whereas molecules of low volatility would condense at the repeller plate.

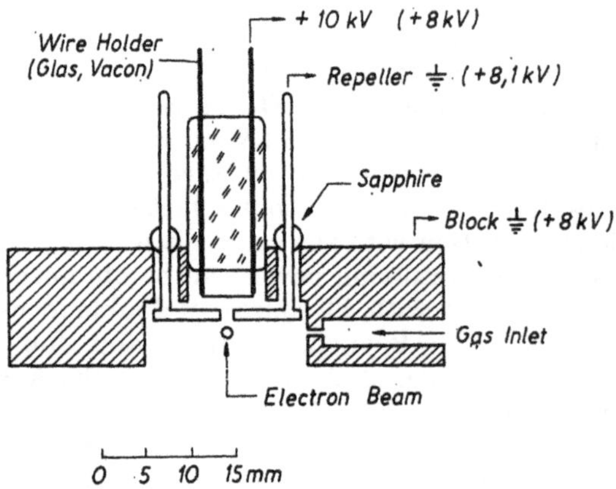

FIG. 2.54. Combined EI/FI source for the double focusing CEC mass spectrometer 21-110 B. (Non-commercial version [89].) The potentials in the brackets are given for EI work.

Another version of the ion source was used for direct introduction of solid samples of low volatility. The repeller plates were removed in this source, and the FI wire was placed to that line where the electron beam usually passes the ion source. The ion source cannot be used as EI source in this case, but only as FI source (see fig. 2.55.). The solid sample is evaporated directly onto the FI wire from a micro-oven admitted to the source through a vacuum lock.

Mead, West, Halliday, Riddoch and Howarth [255] have constructed a combined EI/FI source for the double focusing AEI mass spectrometer MS 9, which is similar in its basic construction with the first version of the combined source for the CEC mass spectrometer 21-110 B as just described. Here again, the repeller plate is used as the cathode for the FI source, and the sample diffuses through a slit in the repeller to the FI emitter. However, the FI emitter does not consist of a thin wire, but of a razor blade. The authors report on a resolving power of $M/\Delta M = 3000$ with 0.002 inch slits for the analysis of high molecular weight petroleum fractions (see section 4.4.2.).

Brunnée reported in a private communication on obtaining a resolving power of $M/\Delta M = 30000$ for FI with the new combined EI/FI source in connection with the double focusing mass spectrometer Varian-MAT-SM 1.

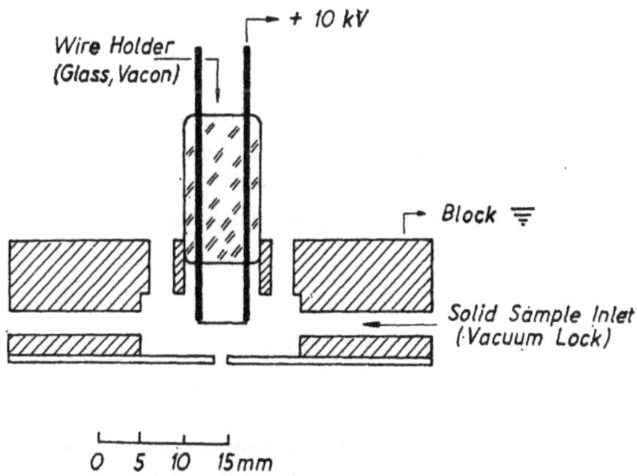

FIG. 2.55. FI source for the CEC mass spectrometer 21-110 B, for use with solid sample vacuum lock (Non-commerical version [89]).

Generally, one may conclude that with double focusing mass spectrometers a resolving power is obtainable in the FI mode which is comparable to the EI mode of operation. Recently Chait et al. [249] have modified the standard EI ion source of the CEC-21-110B mass spectrometer in such a way that FI mass spectra could be taken. The maximum resolving power $M/\Delta M$ obtained was 20000.

3. APPLICATION OF THE FI MASS SPECTROMETER TO PHYSICO-CHEMICAL PROBLEMS

As mentioned in the introduction of this book, there exists a number of problems which can be specifically solved by FI mass spectrometry. These problems may be summarized by "High Field Physico-Chemistry". Some of the more important problems of this kind will be discussed in detail in the next sections.

3.1. CHARGE DISTRIBUTION OF ORGANIC IONS IN HIGH ELECTRIC FIELDS

Neutral molecules and molecular ions are strongly polarized in high electrical fields of the order 10^7 to several 10^8 V/cm, i.e. the charge distribution is quite different for molecules in high fields from that in the absence of fields. For the sake of simplicity and in order to compare the experimental results with the few existing calculations of the charge density of molecular ions with and without electric fields, the homologous series of n-paraffins will be treated first. Before discussing this, it should be made clear whether one has to treat the n-paraffin molecules (or their ions) as being stretched to a regular zigzag-line, or as being irregularly bent. For a long n-paraffin chain, a statistically bent shape has to be assumed in the case of zero electrical field strength. One could consider a stretching of the chain in the presence of electrical fields of the order 10^7 to several 10^8 V/cm if there would be a considerable gain in polarization energy in the stretched position as compared with the statistically bent position of the molecule. Such a difference would result if the polarizability parallel and perpendicular to the axis of a C—C bond would be greatly different. Actually, however, this anisotropy of polarizability is quite small, and the corresponding difference in polarization energy is small as compared with kT [90], for molecules like n-heptane or higher n-paraffins. Therefore, the neutral n-paraffin molecule still remains statistically bent if it leaves the space of zero field and enters that of high field strength.

On the other hand, the FI mass spectra of n-paraffins can be interpreted consistently only on the assumption of a preferential orientation of the molecular *ions* parallel with the electrical field lines. This could occur in two different ways. First, the molecules may be adsorbed on the FI emitter in the form of a completed monolayer in a rather highly ordered state. The hydrocarbon chains may be ordered parallel with each other and perpendicular to the adsorbent metal surface. Some evidence for such a structure which is derived from the FI mass spectra of aliphatic alcohols will be given in section 3.6.5. An n-paraffin molecule in such a stretched position in the adsorbate is field ionized by tunneling of an electron from an atom at the outer end of the molecule. The stretched paraffin ion is then field desorbed from the metal surface.

The second possibility is the following one. A neutral n-paraffin molecule approaches the FI emitter in a statistically bent form. Tunneling of an electron may occur by chance from an atom near to one of the ends of the molecule. After this field ionization, a strong electrical force acts on the positive charge around the place where the electron has left the molecule. The force draws the charge in the direction of the cathode. Forces of inertia act in the opposite direction on the neutral remaining part of the molecule. In this way, a statistically bent neutral chain molecule may be straightened after field ionization. If an electron is removed from an orbital located more in the centre of the molecule, then a straight chain cannot be formed directly, or at least not in one step. Then the molecular ion might be accelerated in an only partially ordered shape. However, these ions do not contribute to the most prominent fragments in the FI mass spectra of n-paraffins which are produced by a splitting off of $C_2H_5^+$ ions, as will be discussed later. Moreover, at extremely high field strengths the $C_2H_5^+$-ion intensity is found to be an order of magnitude higher than the molecular ion intensity. Thus most of the molecular ions must be assumed to have the shape of a stretched chain with certain C—C bond angles. The charge distribution of n-paraffin ions in high electric fields will be discussed here only for the stretched shape of the molecules.

3.1.1. *Quantum Mechanical Treatment*

The charge distribution of neutral n-paraffin molecules and ions in the absence of an external electric field has been calculated by Lennard-Jones and Hall [91], Lorquet and Hall [92] and Lorquet [93]. These authors have carried out equivalent orbital calculations for stretched hydrocarbon chains. The charge distribution will be virtually the same for bent chains unless parts of the chain approach each other so closely that intermolecular forces between them become appreciable. These forces may lead to a chemical reaction of a positively charged n-paraffin molecule with specific parts of the chain, whereby elimination of smaller parts of the molecule may occur. These cases will not be considered.

Within these limitations, Lennard-Jones, Hall and Lorquet have calculated the charge distribution in n-paraffin molecules.

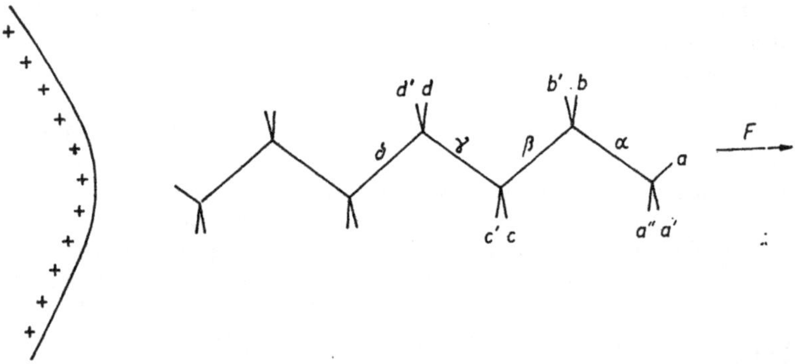

FIG. 3.1. Schematic representation of a linear paraffin adsorbed on an FI tip [92].

A number of different molecular orbitals of different energy and charge distribution result from these calculations. One of the most important results is that for the highest occupied orbital in a long n-paraffin molecule (like n-octane, for example) the maximum charge density is located in the centre of the molecule, if the external electric field is equal to zero. The charge density slowly drops to the ends of the molecule and has a specifically low value at the terminal C—C bonds (α bonds. For bond designation see fig. 3.1.).

The charge distribution of n-octane ions as a function of the field strength as calculated by Lorquet and Hall [92] is represented by tab. 3.1. for the different C—C and C—H bonds. It may be seen from the table that the charge density is maximum at the central δ-bond for zero field strength, and that it is virtually zero at the C—H bonds. The table lists the charge distribution for the highest occupied orbital of the n-octane molecule.

The calculations of Lorquet and Hall were made in the following manner. The Hamiltonian of an n-paraffin molecule oriented along the direction of the field lines as indicated in fig. 3.1. contains an electrostatic potential term whose value will vary linearly with the distance from the tip which produces the high electric field. A linear variation of the potential will be a sufficiently good approximation as long as variations over the length of the n-paraffin molecule are considered only (this length is a few Å as

TABLE 3.1. POSITIVE CHARGE DISTRIBUTION OF n-PARAFFIN IONS AS A FUNCTION OF THE FIELD STRENGTH ACCORDING TO LORQUET AND HALL [92]. (Bond designation as indicated in the figure)

	0 V/Å	0.3 V/Å	1.0 V/Å
CC_α	0.03	0.23	0.44
CC_β	0.12	0.34	0.11
CC_γ	0.21	0.19	0.02
CC_δ	0.24	0.07	0
CH_a	0	0.04	0.27
CH_a'	0	0.04	0.05
CH_b	0.01	0	0.03
CH_c	0	0.01	0.01

compared to the tip radius of the order of 1000 Å). The matrix elements e_{mn} of the equivalent orbitals (EO's) associated with each CC and CH bond then take the form:

$$\begin{aligned} e'_{mn} &= \langle \chi_m | \mathrm{H}^{scf} - eFz | \chi_n \rangle \\ &= \langle \chi_m | \mathrm{H}^{scf} | \chi_n \rangle - eF \langle \chi_m | z | \chi_n \rangle \\ &\simeq e_{mn} - eF\bar{z} \langle \chi_m | \chi_n \rangle \end{aligned} \tag{3.1}$$

where H^{scf} is the self-consistent Hamiltonian of the molecule in the absence of the electric field, χ_m and χ_n are the EO's associated to bonds m and n, e_{mn} are the matrix elements of H^{scf} [93] and $\bar{z}$ is the distance between the effective centre of the tip and the centroid of a particular EO.

It may be shown that the off-diagonal elements e'_{mn} are approximately equal to the e_{mn}, whereas the diagonal elements e'_{mn} differ from the e_{mn} by the term $-eF\bar{z}$. Instead of using $\bar{z}$, one may use $\Delta\bar{z}$, the differences between the coordinates of the EO's centroids along the direction of the field. Then one obtains:

$$e'_{mn} \simeq e_{mn} - eF\Delta\bar{z} \tag{3.2}$$

Lorquet and Hall made the calculations for all n-paraffins from propane to n-octane for two values of the field strength, $F = 0.3$ and $F = 1.0$ V/Å. It is interesting to note, that at a field strength of 0.3 V/Å the maximum of charge density has been shifted from the centre of the molecule at zero field to the $\beta - CC$ bond (see table 3.1.). This is in accordance with the predictions based on semi-classical considerations [95], and it is also consistent with experimental results concerning the different probabilities of bond rupture, which were found to reach a maximum at the $\beta - CC$ bond of n-paraffins [94].

At a higher field strength (1 V/Å, see table 3.1.) the positive charge density becomes much higher at the $\alpha - CC$ bond than at the $\beta - CC$ bond. Thus, because of the larger electron deficiency, the $\alpha - CC$ bond becomes much weaker than the $\beta - CC$ bond. This is reconfirmed by the experimental observation that the relative abundance of CH_3^+ ions may exceed that of the $C_2H_5^+$ ions at extremely high field strengths. The amount of positive charge at the CH_a bond is also rather high at a field strength of 1 V/Å. This is in accordance with the experimental observation that at relatively low field strengths (about 0.3 V/Å) there are virtually no parent-minus-one ions in the FI mass spectra of n-paraffins, but at about 1 V/Å strong parent-minus-one peaks appear.

An empirical scale for the field strength will be introduced now. It is rather difficult to determine the exact field strength at that part of the tip which emits ions collectible by the mass spectrometer detector. However, the relative intensity of the $C_2H_5^+$ ions from n-paraffins is a very sensitive measure for the field strength. As can be seen from table 2.2. the relative intensities of the $C_2H_5^+$ ions from n-heptane at different voltages between tip and cathode are

$$1\% \text{ at } 4.4 \text{ kV, and } 100\% \text{ at } 5.5 \text{ kV}$$

Those field strengths where the relative $C_2H_5^+$ ion intensity is $\leqq 1\%$ of the parent ion intensity are called "low field strengths", those where the relative $C_2H_5^+$ intensity is $> 1\%$ and $< 100\%$ are called "medium field strengths" and those where the relative $C_2H_5^+$ intensity is $\geqq 100\%$ of the parent ion intensity are called "high field strengths". Only very approximate absolute values of the field strengths can be given corresponding to this classification. The threshold field strength for n-heptane is about 6×10^7 V/cm as discussed in section 2.2.5. At this field strength the relative $C_2H_5^+$ intensity of n-heptane is about 1%. Thus the range of medium field strengths is rather narrow (between about 6×10^7 and 7.5×10^7 V/cm). The value of $F = 3 \times 10^7$ V/cm in the paper of Lorquet and Hall corresponds to the range of "low fields", and that of $F = 1 \times 10^8$ V/cm corresponds to the range of "high fields".

3.1.2. Semi-Classical Approximation

A semi-classical derivation of the charge distribution in stretched n-paraffin ions in high electric fields was given by the author [94, 95]. This was justified by the fact that the observed FI mass spectra are in accordance with conclusions drawn from these semi-classical derivations, and moreover by the fact that the charge distributions qualitatively agree with those derived more quantitatively by the quantum-mechanical treatment of Lorquet and Hall as discussed in the preceding section. The advantage of this

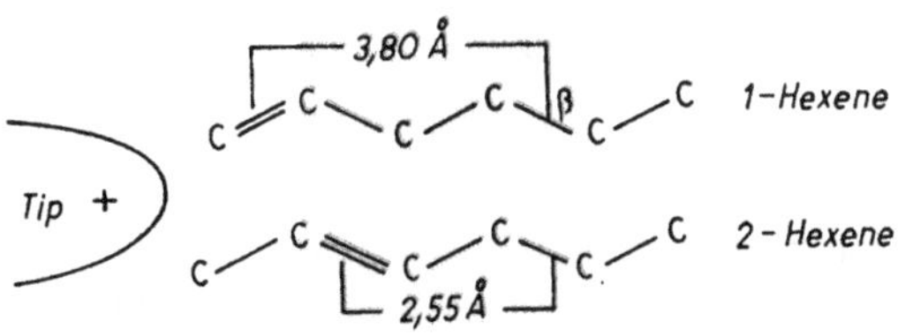

FIG. 3.2. Orientation and bond distances of a 1-hexene molecule in the high electric field produced by a tip.

semi-classical approximation lies in the possibility of a quick prediction of the most important field dissociation processes of chain molecules like n-paraffins or n-olefins. These predictions can be compared with the observed FI mass spectra.

The calculations are only of qualitative nature but may serve for a general judgement which factors govern the shift of the positive charge in an ion in a high electric field. This may be exemplified for the case of two olefin molecules.

Consider 1-hexene and 2-hexene as illustrated by fig. 3.2. [96]. In the absence of an electric field the π-orbital of the double bond is most easily transferred to the ionized state as compared to the other orbitals. The higher ionization potentials for removal of an electron from an orbital having a charge maximum at the β- or γ-bonds have been estimated from the ionization potentials of n-paraffins [96]. These ionization energies are listed in table 3.2.

TABLE 3.2. IONIZATION ENERGIES OF 1-HEXENE AND 2-HEXENE.

	$I_{double\ bond}$	$I_{\beta\text{-}bond}$	$I_{\gamma\text{-}bond}$ [eV]
1-hexene	9.59	10.80	10.43
2-hexene	9.16	10.80	—

Two factors influence the position of the positive charge in the two hexene molecules. In the absence of an electric field an energy amount $\Delta E_{(I)} = I_2 - I_1$ has to be transferred to the molecules if the positive charge would be shifted from the double bond (I_1) to one of the single bonds (I_2). In the presence of a high electric field, on the other hand, the shift of the positive charge in the direction of the field causes a lowering of the potential energy by $\Delta E_F = - e \cdot F(x_2 - x_1)$, where x_2 and x_1 are the distances of the bonds from a reference point.

Thus the net energy change for the charge shift in the electric field should be of the order

$$\Delta E_{(I)} + \Delta E_{(F)} = \Delta E = I_2 - I_1 - eF(x_2 - x_1) \qquad (3.3)$$

The distances between the double bond and the β-bond are shown in fig. 3.2. for 1- and 2-hexene. The values of $\Delta I_1 - eFx$ and ΔE are given for the different bonds in tab. 3.3. The results for E at $F = 0.5$ V/Å (column 5) indicate that the positive charge should be stabilized at the β-bond with 1-hexene (highest negative ΔE-value), but at the double bond with 2-hexene (positive ΔE-value). Column 7 indicates that at higher fields ($F =$ 1 V/cm) the positive charge should be stabilized at the β-bond of both 1- and 2-hexene. (It is assumed that the double bond is directed to the tip, as shown in fig. 3.2.)

These qualitative estimations of the charge location can be tested experimentally only by measurement of the abundance of the fragmentation processes of the different bonds by means of an FI mass spectrum. However, the fragmentation probability does not only depend on the distribution of the positive charge in the molecular ion, but also on the stability of the neutral and the ionized fragments formed by the ionization process.

In the case of the γ-bond rupture of 1-hexene the very stable $C_3H_5^{\cdot}$ radical is formed ("resonance stabilized"). Similary, the resonance stabilized $CH_3CH = CHCH_2^{\cdot}$ radical is formed by the β-bond rupture of 2-hexene. The resonance energy R is about 1.1 eV [97]. The two resonance structures of the radical can be formally described by two wave functions having equal weight factors. In the presence of a high electric field the two factors will be different due to polarization of the radicals. Therefore the resonance energy will be lower in that case and may be estimated to have roughly half of the normal value. R is assumed to have a value of about 0.6 eV for field strengths of 0.5—1.0 V/Å. The ($\Delta E - R$)-values are listed in column 9 of tab. 3.3. in order to indicate the most intense fragmentation processes to be expected.

TABLE 3.3. ENERGY CHANGE ΔE (IN eV) FOR THE SHIFT OF THE POSITIVE CHARGE FROM THE DOUBLE BOND TO THE β OR γ-BONDS AS SPECIFIED IN COLUMN 2, FOR TWO FIELD STRENGTHS [96].

Columns 10 and 12 give the intensity of the fragment ions relative to the most intense peak (normalized to 100) in the FI mass spectrum. The field strengths underlying columns 10 and 12 are not known exactly because the radius of the field-producing tip changes with time. They are classified as "medium" or "high".

R = Resonance energy.

1	2	3	4	5	6	7	8	9	10	11	12
Substance	Bond		$F = 0.5$ V/Å		$F = 1.0$ V/Å		Fragment	$F = 0.5$ V/Å	medium Field	$F = 1.0$ V/Å	High Field
		ΔI	$-eF\Delta x$	ΔE	$-eF\Delta x$	ΔE	Ion	$\Delta E - R$	rel. Intens.	$\Delta E - R$	rel. Intens.
1-Hexene	β	+1.21	−1.90	−0.69	−3.80	−2.59	C_2H_5 $R = 0$	−0.69	5	−2.59	100
	γ	+0.84	−1.28	−0.44	−2.56	−1.72	C_3H_7 $R = 0.6$	−1.04	50	−2.32	36
2-Hexene	β	+1.64	−1.28	+0.36	−2.56	−0.92	C_2H_5 $R = 0.6$	−0.24	3	−1.52	27

Two qualitative conclusions can be drawn from the results listed in columns 9—12 of tab. 3.3. First, the maximum of the fragmentation probability of 1-hexene is shifted from the γ- to the β-bond if the field strength is doubled. The ratios of the corresponding theoretical $\Delta E - R$-values change with the field strength in the right way.

Second, in the case of 2-hexene, the β-bond fission probability is low at medium field strengths because the double bond still tends to keep the positive charge nearby. However, at high field strength the positive charge is strongly shifted towards the β-bond, and hence the probability of fission of that bond is greatly enhanced.

It should be repeated that in view of the highly approximate nature of the estimation of the $\Delta E - R$-values, the absolute figures of the quantities have to be regarded sceptically. However, the experimental results confirm that the qualitative picture on which eqn. 3.3 is based is adequate. There are two opposing tendencies for the positive charge localization in a chain molecule:

In the field free case the positive charge is concentrated at a place where the highest orbital has the maximum value of the wave function. Under the influence of a high electric field the charge is shifted to the tip-distant side of the molecule as far as energetically possible.

The shift of the positive charge with increasing field from the δ- over the β- and γ- and then to the α-bond as calculated quantum mechanically by Lorquet and Hall [92] can be confirmed by the FI mass spectra of n-paraffins.

TABLE 3.4. ION INTENSITIES OF n-OCTANE FRAGMENTS AS A FUNCTION OF THE FIELD STRENGTH. (FIELD PRODUCED BY A TIP.)

Bond rupture	γ	β	α
fragment ion	$C_3H_7^+$	$C_2H_5^+$	CH_3^+
$F = 3 \times 10^7$ V/cm	6.6	100	0.1
$F = 4.1 \times 10^7$ V/cm	1.9	100	0.1

Table 3.4. shows the relative fragment ion intensities of the $C_3H_7^+$, $C_2H_5^+$ and CH_3^+ ions of n-octane at two different field strengths.

These experimental results are in good agreement with the calcualtions of Lorquet and Hall (see tab. 3.1.).

3.2. FIELD DISSOCIATION OF MOLECULAR IONS

It was shown in the preceding section that the positive charge of stretched chain hydrocarbon ions is more strongly localized at one bond near to one end of the molecule in a very high electric field than in the absence of a field. Although the larger deficiency of electronic charge at that bond leads to a more pronounced weakening of the bond, this does not yet give a sufficient explanation of why very abundant dissociation of the molecular ions does occur in the presence of electric fields of the order 10^8 V/cm. In contrast to ionization by 70 eV electrons, very little excitation energy is transferred to the molecular ion during field ionization. Thus, apart from the reduction of dissociation energy due to

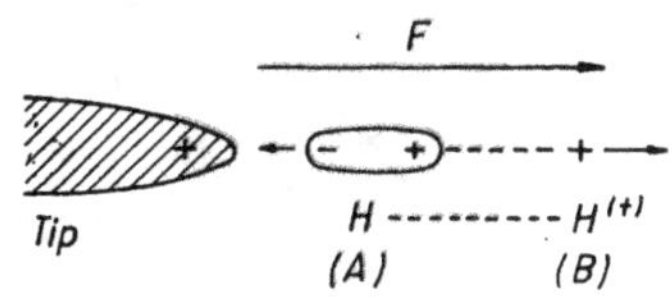

FIG. 3. 3. Schematic representation of a H_2^+ ion in a high electric field, illustrating the field dissociation effect.

the concentration of the positive charge at a specific bond, a further reduction of the dissociation energy of that bond by the high electric field must occur.

A model for this reduction of the dissociation energy, based on a classical approximation, was developed by the author [95], and the dissociation process caused by this reduction of dissociation was termed "Field Dissociation". Consider the simplest case for field dissociation, i.e. the dissociation of a H_2^+ ion into H^+ and H. The charge of the binding electron is distributed symmetrically around the two protons of the H_2^+ ion in the absence of an electric field. However, if the H_2^+ ion is located near to a positively charged tip producing a field strength of the order 10^8 V/cm, and if the H_2^+ ion is orientated parallel with the field lines as shown in fig. 3.3., the situation is different. The probability of finding the electron near the proton (*A*) adjacent to the tip will be much larger than the probability of finding it near the tip-distant proton (*B*). The H_2^+ ion can then be described approximately as consisting of a polarized H-atom and a bare proton. The polarized H_2^+ ion will vibrate with a frequency of the order 10^{13} sec^{-1}. During the phase of extension of the H_2^+ ion, as indicated by the arrows in fig. 3.3., work has to be done against the intramolecular forces, but the potential energy of the system is lowered due to the displacement Δr_B of the proton along the electric field lines. This lowering of energy is equal to $-eF\Delta r_B$. The energy change due to the interaction of the polarized H-atom with the external electric field during its displacement towards the tip is very small.

By superposition of the intramolecular potential and the potential term $-eF\Delta r_B$, the potential curve shown in fig. 3.4. results. Of the total change in internuclear distance $(r - r_0)$, 50% is due

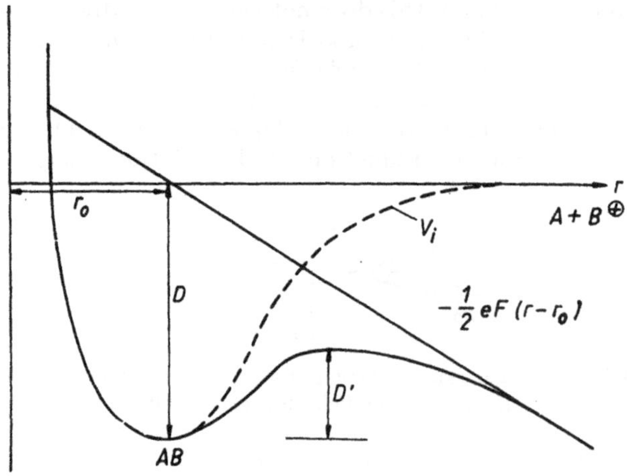

FIG. 3.4. Potential curves of a molecule with two atoms of equal mass in a high electric field in the "classical" approximation. (Special case: $AB = H_2$.) Superposition of the potential of interaction, V_i, of the proton with the polarized H-atom, and the potential of interaction of the proton with the external field, $V_F = -\,{}^1/_2\, eF\ (r - r_0)$.

to the displacement of (A) towards the tip and 50% to the displacement of (B) away from the tip. Therefore, the term $-eF\Delta r_B$ may be replaced by $-\frac{1}{2}\,eF\,(r - r_0)$, where $r = r_B - r_A$, and r_0 is the equilibrium distance of (B) and (A). It is seen from the figure that the dissociation energy D in the presence of the high field is much lower than the dissociation energy D in the absence of an external electric field. If the field strength becomes sufficiently high, the maximum in the potential curve vanishes, or in other words, field dissociation of the molecular ion could occur immediately after field ionization because the dissociation energy would be equal to zero.

It is an over-simplification to describe the potential curve for the change of the internuclear distance within the H_2^+ ion by a linear superposition of the potential curve in the absence of an external electric field and by the term $-\frac{1}{2}\, eF\,(r - r_0)$. This approximation will hold only for large internuclear distances where the intramolecular interaction is small and where the potential approaches the value $-\frac{1}{2}\, eF\,(r - r_0)$ asymptotically. A correct calculation of the potential curve for any distance has to be carried out by means of quantum mechanical methods. However, if one continues formally to describe the H_2^+ potential curve in a high external field by a linear superposition of the term $-\frac{1}{2}\, F\,(r - r_0)$ and the potential for the intramolecular forces, one has to keep in mind that the latter is not identical with that for the field free case because of the strong polarization of the ion. In other words, for high field strengths one has to draw a potential curve of the H_2^+ ion similar to that for the field free case, but with a different and not exactly known minimum of the curve (dissociation energy D'' instead of D), and one has then to superimpose the term $-\frac{1}{2}\, F\,(r - r_0)$ in order to obtain a qualitative description of the curve. The exact curve obtained quantum mechanically will be of very similar shape.

Since the exact depth of the minimum D'' of the curve will not be known exactly from the classical approximation, one will insert the normal value of the dissociation energy D, as an approximation.

The advantage of using this simple classical approximation lies with the fact that one can qualitatively describe the behaviour of a polyatomic molecular ion in the presence of a high field and that one can make certain predictions. For example, the field strength necessary for a spontaneous field dissociation can be estimated. This is obtained by postulating that the potential maximum D' should vanish. For H_2^+ the value of $r - r_0$ at the point of intersection between the intramolecular potential curve and the field potential (see fig. 3.4.) is of the order of 1 Å. The value of D is about 2 eV, and it can be seen from the figure that $eF\,(r - r_0)$ is about $1/2\ D$ at the point of intersection, i.e. about 1 eV. Thus, in order to obtain a vanishing small dissociation energy the field strength must be of the order 1 V/Å. This is in accordance with the experimental observations.

Quantum mechanical treatment

A quantum mechanical treatment of field dissociation of diatomic ions has been given by Hiskes [98, 99]. An outline of this treatment will be given here for the special case of the H_2^+ ion.

The Schrödinger equation for H_2^+ in the presence of an electric field F takes the form

$$\left\{\frac{-\hbar^2}{2m_e}\nabla_1^2 - \left[\frac{e^2}{|r_1 - {}^1/_2 r_n|} + \frac{e^2}{|r_1 + {}^1/_2 r_n|} - \varepsilon F z_1\right]\right\}\psi_k = E_{k(r_n)}\,\psi_k \tag{3.4}$$

Here r_1 is the distance of the electron from the centre of mass of the two protons, r_n is the internuclear distance, $\varepsilon = [1 + m/(2M + m)]e^*$); the direction of the z-axis is along the electric field. The primary effect of the term Fz_1 is to perturb the electronic eigenvalues E_k. z_1 is the coordinate of the electron along the z-axis.

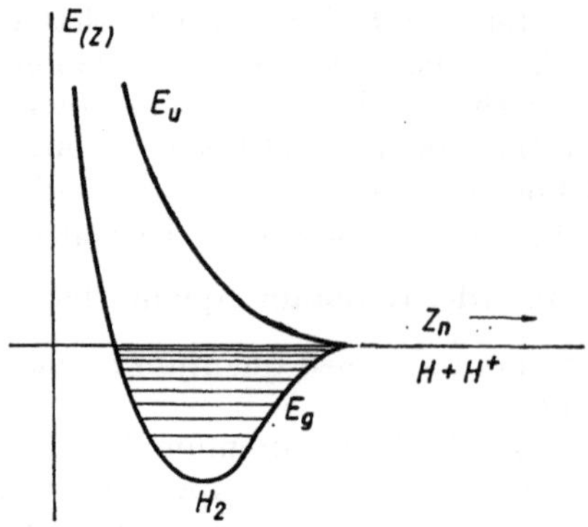

FIG. 3.5. Electronic ground state (E_g) and exited state (E_u) of the H_2^+ ion as a function of the nuclear distance z_n in the absence of an electric field.

Eqn. (3.4) is separable for $F = 0$ in confocal elliptic coordinates ξ, η, φ. These coordinates are defined by

$$\xi = (r_a + r_b)/r_n, \qquad \eta = (r_b - r_a)/r_n$$

where r_a and r_b are the distances of the electron from the proton a and proton b, respectively. Introduction of these coordinates leads to three separated equations. One of them is elementary, the others require numerical integration for their general solution [100].

The term $\varepsilon F z_1$ perturbs the resulting undisturbed electronic states. These new states, as derived by Hiskes, are

$$E_{g,u} = \frac{1}{2}(E_1 + E_2) \mp (E_2 - E_1)\,[1 + 4\,H_{12}^2/(E_2 - E_1)^2]^{1/2} \tag{3.5}$$

*) M and m are the masses of the proton and the electron, respectively.

Here E_1 and E_2 are the undisturbed ground state and the first exited state of the H_2^+ ion, which are shown as a function of the internuclear distance in fig. 3.5.).

H_{12} is a function of the electric field strength and of the internuclear distance, r_n. At large values of r_n, eqn. (3.5) reduces to

$$E_g = E_1 - \frac{1}{2}\varepsilon F r_n \quad \text{and}$$

$$E_u = E_2 + \frac{1}{2}\varepsilon F r_n \tag{3.6}$$

(The axis of the H_2^+-ion is assumed parallel with the direction of the field in both cases.)

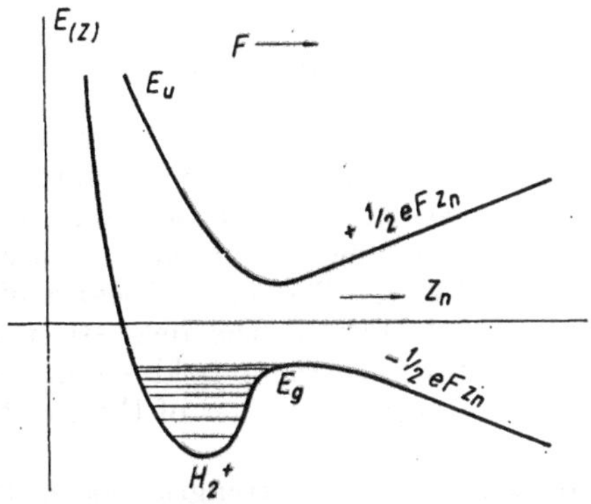

FIG. 3.6. The same curves as in fig. 3.5. in the presence of an electric field.

Fig. 3.6. shows the nuclear potential of the H_2^+ in the presence of a field for the ground state and the first exited state. It can be seen from the figure that the uppermost vibrational state of the E_g curve becomes instable, an effect which is termed "field dissociation".

Consider the field to be produced by a positively charged tip. In the ground state of the H_2^+ ion, the electron is associated with the proton (*b*) next to the tip. The proton (*a*) on the tip-distant side of the H_2^+ ion is then free, and its displacement away from the tip leads to a lowering of the potential energy if r_n is sufficiently large.

In the excited state, the electron is associated with the tip-distant proton (*a*), and the positively charged proton is moved against the field direction, thus raising the potential.

The effect of the Stark-shift on the ground state of the ion may be briefly discussed. If the Stark-shift is independent of the internuclear distance, only a uniform lowering of the potential curve would result, without changing the shape of the curve. However, the Stark-shift is stronger at larger internuclear distances than at smaller ones, and therefore a correction of the potential curve has to be made which is formally equivalent to a slight increase of the applied field strength. In the limit as r_n goes to zero, the molecular ion degenerates into a He^+ ion in its $1\,S$ state. The Stark-shift for this state is

$$E = -(9/64)\, a_0^3 F^2$$

where a_0 is the Bohr radius.

In the limit of large internuclear separation, the electronic state is that of a hydrogen atom in a $1\,S$ state. The Stark-shift for this state is

$$E = -(9/4)\, a_0^3 F^2 .$$

For field strengths of about 2×10^8 V/cm the correction of the potential due to this effect is relatively small. It is approximately equivalent to an increase of the electric field strength by some 5 to 10%. FI mass spectrometry of organic molecules is usually made at $F \sim 7 \times 10^7$ V/cm. Therefore, because of the quadratic dependence of the Stark-shift on the field strength, one should expect smaller corrections at these field strengths, even if one takes into account the different values of the constant before the F^2 factor.

Hiskes also compared the field strengths necessary to dissociate vibrationally excited ions in the non-rotating state with those in excited rotational states. For example, the field required for dissociation in 10^{-8} sec for an ion in the $v = 15$ vibrational state with the rotational state $J = 4$, $m = 0$ is about 30% less than for $J = 0$. Generally speaking, rotational excitation lowers the threshold field strength for dissociating a particular vibrational level.

After having derived the model for field dissociation of H_2^+ ions by the classical and the quantum-mechanical treatment, some generalizations of the model will be made. One can conclude from the simple classical model of field dissociation that this phenomenon is not restricted to H_2^+ but that, in principle, any bond of any molecular ion may be field dissociated at sufficiently high fields. For example, in the organic ion $R_1 - R_2^+$ (where R_1 and R_2 are polyatomic radicals) field dissociation of the bond between R_1 and R_2 may occur. It was mentioned in the preceding section that in the FI mass spectra of n-paraffins very strong $C_2H_5^+$ peaks appear at "high" fields. For example, in the n-pentane ion, rupture of the β-bond occurs with a high probability.

$$[\underbrace{CH_3-CH_2}_{R_1}-\underbrace{CH_2\overset{2}{-}CH_2-CH_3}_{R_2}]^+$$

The rupture of the β bond can be attributed to field dissociation because it is very sensitively dependent on the field strength. Thus, the n-pentane ion behaves like a diatomic molecular ion AB^+ with respect to field dissociation, atom (A) being replaced by the radical R_1 and atom (B) by the radical R_2. At present, no quantum mechanical calculations on field dissociation of polyatomic ions exist. Thus, the classical approximation has to be used to obtain a qualitative description of the phenomena, and a number of further simplifications have to be introduced.

First, the radicals R_1 and R_2 will be treated as rigid bodies vibrating with a certain critical frequency with respect to each other. In a more accurate treatment one has to regard all the normal vibrations extended over the whole molecule. Some justification for treating the radicals R_1 and R_2 as rigid bodies lies in the fact that the bond between R_1 and R_2 is much weaker than the other bonds. (The field dissociating bond will now be termed "critical" bond.) It was shown earlier that for n-paraffin ions at "medium" field strengths the β bond is the critical bond because the positive charge is concentrated there. Therefore, because of the large deficiency of bonding electronic charge, the critical bond is much weaker than the other bonds.

From the model of field dissociation the following rule can be derived [95].

a) Field dissociation of a specific bond of a molecular ion can occur only if the positive charge is located on the tip-distant side of the bond.

(It is assumed that the field is produced by a positively charged tip.)

In the case of an H_2^+ ion in a position relative to the tip as shown in fig. 3.3., the conditions imposed by rule a) are met. Now consider the case of an organic ion where the maximum of positive charge density is located between a specific bond and the positively charged tip. As an example, dipropyl ketone may be discussed (fig. 3.7.).

In the field free case, the positive charge is located at the oxygen atom because the ionization energy of the non-bonding electrons of oxygen is low. In the presence of an electric field of the order 7×10^7 V/cm the positive charge is shifted to a place between the carbonyl bond and the β bond. According to rule *a*), dissociation of the β bond should not occur in this case because the positive charge is located at the tip-neighboured side of the β bond. The situation is analogous to that underlying

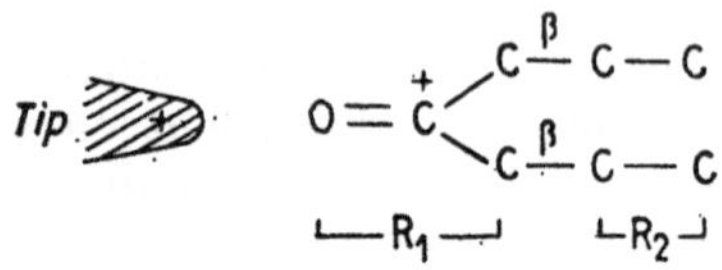

FIG. 3.7. Dipropyl ketone in the field of a positive tip. (Schematically.)

fig. 3.6. (upper curve). Consider the vibration of the radicals R_1 and R_2 (regarded as rigid bodies) along the β bond. During extension of the β bond the positively charged radical R_1 is moved against the repulsive force of the field at the positively charged tip. Therefore, the field potential term $eF\Delta r$ has a positive sign in contrast to the conditions underlying fig. 3.5. (lower curve) where the positive charge was on the tip-distant side of the critical bond, and where the sign of the field potential term was therefore negative.

It can be seen from fig. 3.6. (upper curve) that the bond is strengthened by the high electric field if the positive charge is between the critical bond and the tip.

This conclusion is reconfirmed by experiments. In the FI mass spectrum of dipropyl ketone one finds $C_2H_5^+$ fragments at a relative intensity of $<0.01\%$, at a field strength where $C_2H_5^+$ fragment ions are split off from n-pentane at a relative intensity of about 100% of the parent ion intensity.

Rule *a*) can be derived more quantitatively by calculation of the potential curve for the relative motion of two charged polyatomic radicals $A^{(a+)}$ and $B^{(b+)}$ in a high electric field, similar to the calculations of Hiskes [98, 99] for the motion of H and H^+ (*a* and *b* are the amounts of positive charge in units of the elementary charge *e*). Again the radicals *A* and *B* will be regarded as rigid bodies, an approximation which will be discussed later. The equations of motion of two charged, interacting particles $A^{(a)+} - B^{(b)+}$ in an electric field are

$$M_A \ddot{r}_A = -f(r_B - r_A) + eaF \tag{3.7a}$$

$$M_B \ddot{r}_B = +f(r_B - r_A) + ebF \tag{3.7b}$$

$f(r_B - r_A)$ is the force acting between *A* and *B*. Multiplying (3.7a) with M_B and (3.7b) with M_A, and subtracting (3.7a) from

(3.7 b), one obtains with $r_B - r_A = r$

$$\frac{M_B M_A}{M_A + M_B}\,\ddot{r} = f(r) + eF\,\frac{(b M_A - a M_B)}{M_A + M_B} \qquad (3.8\,a)$$

$f(r)$ approaches the value of zero at large values of r, and the relative potential is then

$$V_F = -\,eF\,\frac{M_A M_B}{M_A + M_B}\,(b/M_B - a/M_A)\,r \quad (r \to \infty) \quad (3.8\,b)$$

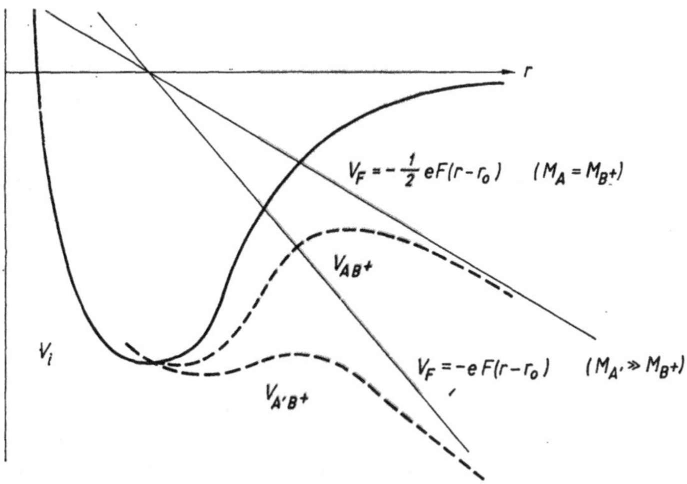

FIG. 3.8. Potential curves of two ions $A - B^+$ and $A' - B^+$ in a high electric field. $M_A = M_B^+$; $M_A' \gg M_B^+$. The potential V_i (solid curve) is assumed to be equal for both ions.

The zero point of the potential V_F can be fixed arbitrarily and will be chosen such that $V_F = 0$ at the equilibrium distance r_0 between A and B without external electric field. Thus, by adding this constant term, one obtains

$$V_F = -\,eF\,\frac{M_A M_B}{M_A + M_B}\,(b/M_B - a/M_A)\,(r - r_0) \quad (r \to \infty) \quad (3.9)$$

The radical nearer to the tip may always be termed A, and it is assumed that the axis of $A - B$ is orientated parallel with the electric field lines. It may be seen from fig. 3.6. and 3.8. that field dissociation of molecular ions only occurs if the ion takes energy from the electric field during vibration, i.e. if $d\,V_F/dr$ is negative. This requirement is met if the whole positive charge is concentrated at B (at the tip-distant side of the bond), and $a = 0$. If, however, the whole charge is located at A (at the tip-neighboured side of the bond), and consequently $b = 0$, dV/dF becomes positive and field dissociation cannot occur.

If a singly charged positive ion dissociates into a singly charged positive radical and a neutral radical, the whole unit charge will be located either on A or on B after dissociation. Thus, under the conditions underlying eqn. (3.9.), i.e. $r \to \infty$ either a or b must be zero. This holds even if the charge was originally more or less continuously distributed over the whole molecular ion before dissociation, or if it was concentrated in the centre of the bond to be broken, i.e. if a and b are fractions of the unit charge under the condition $r = r_0$. The absolute value of dV_F/dr will then be smaller than at large distances, where it approaches the limiting value

$$V_F = -eF \frac{M_A}{M_A + M_B} (r - r_0) \quad (r \to \infty,\ a = 0) \qquad (3.10\,\mathrm{a})$$

or

$$V_F = +eF \frac{M_B}{M_A + M_B} (r - r_0) \quad (r \to \infty,\ b = 0) \qquad (3.10\,\mathrm{b})$$

The first of these two equations is the more important one because it applies for the case of field dissociation.

The change of dV_F/dr with the change of the internuclear distance from r_0 to large r-values does not appreciably change the general shape of the potential curves like that shown in fig. 3.4., because the limiting value (3.10a) is approached very quickly. The shift of the positive charge (originally located at the centre of the dissociating bond) to the fragment B becomes pronounced even for small extensions of the bond.

The linear field potential of fig. 3.4. has only to be replaced by a potential which is curved near r_0 (with dV_F/dr smaller than that at $r \to \infty$), and which quickly approaches the limiting value (3.10a). It should be remembered, however, if one continues to apply eqn. (3.10a) to small r-values, although the equation was derived for $r \to \infty$, it serves only for a formal description of the potentials and for a qualitative understanding of how the change of the original charge distribution to the final distribution changes the potential diagram of fig. 3.4. Quantitative derivations have to be made quantum-mechanically, like the treatments of Hiskes [98, 99] and of Wind [101–103] for the case of the H_2^+ ion.

Now, field dissociation will be discussed for a series of molecules of different molecular weight. The equation

$V_F = -eF \frac{M_A}{M_A + M_B}(r - r_0)$ reduces further for two limiting cases:

$$a) \quad M_A \gg M_B \qquad V_F = -eF(r - r_0) \qquad (3.11)$$

$$b) \quad M_A = M_B \qquad V_F = -1/2\, eF(r - r_0) \qquad (3.12)$$

The first case applies for the dissociation of the β bond of a high molecular weight n-paraffin ion, the second for field dissociation of H_2^+ or of the β bond of n-butane. Although the applied external field is the same in both cases, only half the field potential has to be superimposed on the intramolecular interaction potential in the second case, as can be seen from fig. 3.8. The figure shows that at a certain field strength F' field dissociation can occur from the vibrational ground level in the case $M_A \gg M_B$. However, field dissociation occurs only at $F = 2F'$ in the case $M_A = M_B$. This can be formulated as a general rule [95]:

b) Field dissociation of a molecular ion $A - B^+$ occurs at the lower field strengths, the larger the mass of the neutral fragment A is as compared to that of the charged fragment B^+.

This effect may be termed "mass effect".

Qualitatively speaking, the reason for this effect is as follows. The centre of mass is located virtually at A if $M_A \gg M_B$. If a bond is extended by Δr, A is practically fixed, and B^+ moves the full distance Δr along the field lines. In the case $M_A = M_B$, B^+ moves only the distance $1/2\Delta r$ along the field lines, while A moves $1/2\,\Delta$ in the opposite direction. The change in potential energy of B^+ is correspondingly smaller in this case.

Rule *b*) is reconfirmed by the FI mass spectra of some n-paraffins. At the usual field strengths, the most intense fragment of these substances is the $C_2H_5^+$ ion formed by β bond rupture. Thus mass M_B is constant in this series whereas mass M_A increases with increasing molecular weight of the n-paraffins. Consequently,

$$dV/dF = -eF \frac{M_A}{M_A + M_B}$$

should increase with increasing molecular weight of the n-paraffins, and therefore field dissociation of the β bond should increase. Actually, one observes a strong increase of the $C_2H_5^+$ fragment ion intensity from n-butane to n-nonane at constant field strength, as can be seen from table 3.4.

A short justification for treating the radicals A and B^+ of the molecular ion $A - B^+$ as approximately rigid bodies has already been given. This will now be discussed further. As already stated, the "critical" bond $A - B^+$ is much weaker than the other bonds in the radicals because of the large deficiency of bonding electronic charge in the critical bond. Another reason is the different vibrational excitation of the critical bond and the other bonds, which can be concluded from the Franck-Condon principle. It may be

TABLE 3.4. RELATIVE INTENSITIES OF THE $C_2H_5^+$-IONS IN THE FI MASS SPECTRA OF N-PARAFFINS, RELATED TO THE PARENT ION INTENSITIES NORMALIZED TO 100.
(The applied field strength is constant.)

Substance	Propane	Butane	Pentane	Hexane	Heptane	Octane	Nonane
$C_2H_5^+$-rel. Intensity	7	18	90	155	240	300	440

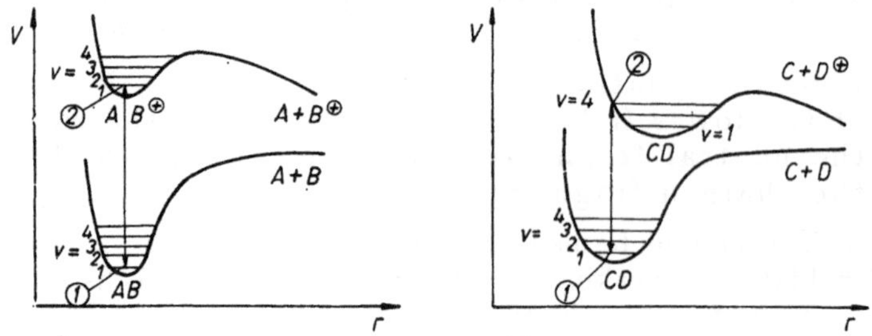

FIG. 3.9. Potential curves of a molecule AB in the neutral and the ionized state in a high electric field. Franck-Condon transition from state (*1*) to a state (*2*) by field ionization. (Actually, both curves intersect such that points (*1*) and (*2*) coincide because of lowering of the ionic curve by the field*). The curves are drawn separately in order to show the analogy to the case of Franck-Condon transitions during electron impact.)

a) Potential curves for a bond which is not, *b*) which is the "critical" bond.

assumed that the whole net positive charge of the molecular ion would be concentrated at the critical bond. Then after transition of the molecule from the neutral to the ionized state, the potential curves for the motion of the atoms along the normal coordinates will be changed only slightly, except for the motion along the critical bond. In the critical bond an increase of the equilibrium distance r_0 is associated in most cases with the decrease of the bond energy, after removing one electron from the critical bond. Thus, by the process of field ionization very little vibrational energy is expected to be transferred to all of the bonds except for the critical bond. This is shown schematically by fig. 3.9.

*) See fig. 3.10.b.

The neutral and the ionic curves are drawn separately for the sake of clearness in fig. 3.9. The actual position of the curves, taking into account the non-crossing rule for two different states of equal symmetry of a molecule, is shown in fig. 3.10.

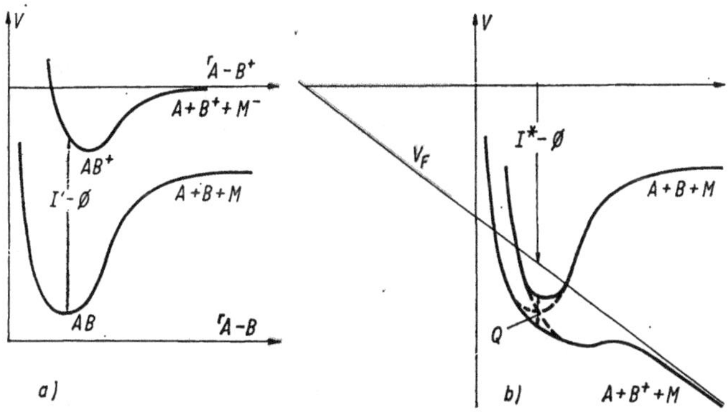

FIG. 3.10. Potential curves of the molecule AB and the ion AB^+. The distance between AB (or AB^+) and the metal surface is assumed to be so large that the corresponding interaction can be neglected. *a*) without external field, *b*) with field. An energy split Q arises by the effect of non-crossing of the neutral and the ionic curve.

Summarizing, the critical bond is much weaker and much more vibrational energy is transferred to it than to the other bonds. This is the reason why for a treatment of field dissociation of a polyatomic aliphatic ion $A - B^+$ only the motion of B^+ relative to A may be considered within certain limits of approximation.

The situation is more complicated with molecules where several bonds are field dissociated with comparable probabilities (alcohols, for example). Then the different parts of the molecular ion cannot be described at a good approximation as being rigid bodies.

3.3. BOND ENERGIES OF MOLECULAR IONS

In the previous section the potential curve for an ion $A - B^+$ was treated as a superposition of the intramolecular interaction potential and a field term $-e\,F\,\dfrac{M_A}{M_A + M_B}\,(r - r_0)$. The first term was approximated by the intramolecular interaction potential for the field free case. As a result of this one finds that the stronger a bond at zero field is, the larger will be the dissociation energy D' of that bond at a given field strength (see fig. 3.11.). Thus, the correlation between bond strength and dissociation probability existing for the case of zero field remains preserved for the case of field dissociation of a molecular ion.

The strength of the bond $A - B^+$ depends on the nature of the radicals A and B^+, and of the density of positive charge (deficiency of bonding electrons) at that bond in the presence of a high electric field. The dependence of the bond strengths and hence of the dissociation probability on the nature the radicals A and B^+

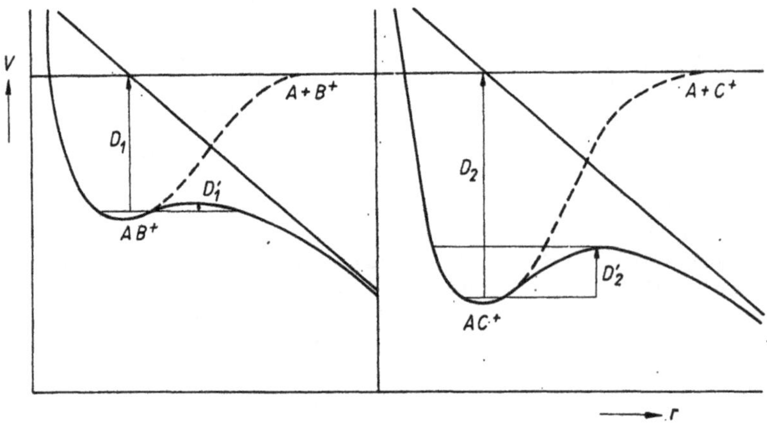

FIG. 3.11. Potential curves for two bonds of different strength without and with high external electrical field. D_i = dissociation energies.

which is observed in the electron impact mass spectra is qualitatively preserved in the field ionization mass spectra. Thus, in unbranched hydrocarbon ions the bond strength increases in the order C—C, C—H, C=C. Correspondingly, one finds a decreasing probability of direct bond split in the same order (rearrangement reactions will not be considered for the moment). For example, the relative abundance of ions originating from field dissociation of C—H bonds in n-paraffin ions is about two orders of magnitude smaller than that of C—C bonds, at field strengths where the $C_2H_5^+$ ion intensity is about equal to the parent ion intensity. Direct rupture of multiple carbon — carbon bonds is negligibly small at all field strengths. Similarily, *direct* field dissociation of C=O, C=S, C=N— and other multiple bonds is not observed. Dissociation of bonds after rearrangement of the molecular ion will be discussed in section 4.3.3.

A certain difference in the bond strengths of n-paraffin ions with and without high electric field (and hence a difference in the

FI and EI mass spectra) exists with respect to the distribution of the positive charge. Spiteller et al. [104] have measured the EI mass spectrum of n-triacontane at low temperature and low electron energy. All C—C bonds of the molecular ion (except for the terminal bonds) are broken with approximately the same probability. From this observation one can draw the conclusion that the strengths of the ionized C—C bonds are all approximately equal. However, in the case of FI the β C—C bond is field dissociated with very much larger probability than the other

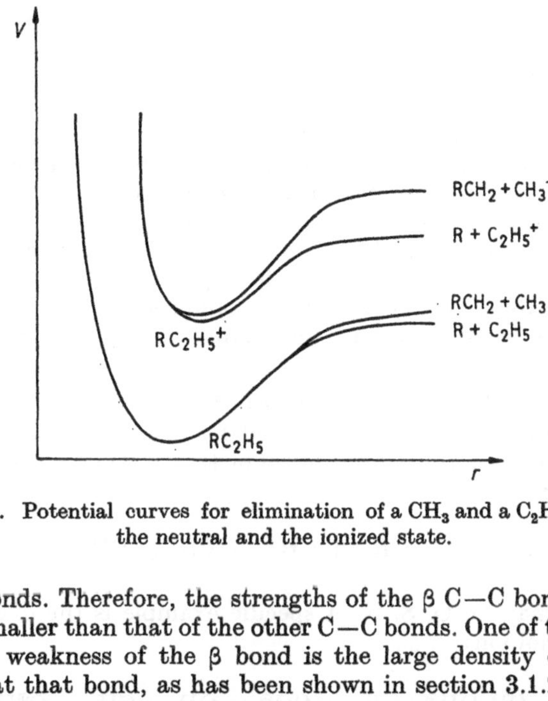

FIG. 3.12. Potential curves for elimination of a CH_3 and a C_2H_5 radical in the neutral and the ionized state.

C—C bonds. Therefore, the strengths of the β C—C bond must be much smaller than that of the other C—C bonds. One of the reasons for this weakness of the β bond is the large density of positive charge at that bond, as has been shown in section 3.1.2. Another reason for the weakness of the β bond relative to the terminal α bond at low or moderate field strengths is the following one. The heats of formation of a CH_3 or a C_2H_5 radical, respectively, from a neutral n-paraffin molecule are only slightly different. The ionization energy of the CH_3 radical is, however, much larger than that of the C_2H_5 radical ($\Delta I = 1.17$ eV). Therefore, the heat of formation of a CH_3^+ ion from a neutral n-paraffin parent molecule is larger by about 1 eV than that of the $C_2H_5^+$ ion (see fig. 3.12.).

Strictly speaking, the rate of dissociation of a bond is not a function of the heat of formation of the radicals produced, but it is dependent of the heat of activation, However, in general the heat of activation for the reverse reaction, the recombination of R_1 and R_2^+, is very small or zero in most cases. Therefore, the heats of activation may be approximated by the thermodynamic heats of formation of the radicals R_1 and R_2^+ from the molecular ion $R_1R_2^+$.

This result may be generalized. If very stable neutral or charged products are formed after dissociation of a molecular ion, the dissociation energy of the ruptured bond is in general smaller than that of a bond which can be dissociated into less stable products. This general rule is well known to the electron impact mass spectroscopist and has been discussed in detail in a number of original papers and monographs, for example by Beynon [109], McLafferty [110], Biemann [111], Budzikiewicz, Djerassi and Williams [112], Spiteller [113] and other authors. The rule is equally well applicable to FI mass spectra. The stabilization of the products may be described formally in many cases by resonance of several structures of one or both of the products. The corresponding rules elaborated mainly by organic chemists and applied to EI mass spectrometry may be equally well applied to FI mass spectrometry. Only a few examples will be given here. Others will be discussed in the chapter on qualitative organic analysis.

a) If one of the lone pair electrons of the hetero atom contained in an organic compound is removed by field ionization, a stable fragment ion is formed by rupture of the C—C bond next to the hetero atom:

$$R - \mathrm{CH_2}XH_n^{\oplus} \rightarrow R\cdot + \mathrm{CH_2^{\oplus}} - X\mathrm{H}_n \rightleftharpoons \mathrm{CH_2} = X^{\oplus}\mathrm{H}_n$$

where X may be O, S, N. A specific example is the dissociation of an alcohol ion:

$$R - \mathrm{CH_2OH^{\oplus}} \rightarrow R\cdot + \mathrm{CH_2^{\oplus}} - \mathrm{OH} \rightleftharpoons \mathrm{CH_2} = \mathrm{O^{\oplus}H}$$

It is found that the $CH_2OH^{\oplus}$ ion ($m/e = 31$) is very abundant in the FI mass spectra of primary alcohols.

b) If 1-olefins are field ionized, neutral or ionized fragments which are stabilized by resonance may be split off from the parent ion:

$$\mathrm{CH_2 = CH{-}CH_2\cdot} \rightleftharpoons \cdot\,\mathrm{CH_2{-}CH{=}CH_2}$$

or

$$\mathrm{CH_2{=}CH{-}CH_2^{\oplus}} \rightleftharpoons {}^{\oplus}\mathrm{CH_2{-}CH{=}CH_2}$$

In the FI mass spectrum of 1-hexene, for example, one finds strong $C_3H_5^+$ and $C_3H_7^+$ peaks ($m/e = 41$ and 43).

c) The stability of carbonium ions increases in the order: primary, secondary and tertiary carbonium ions. Correspondingly, the relative fragment ion intensities increase with increasing branching of hydrocarbon ions. For example, about 10% $C_2H_5^+$ ions are formed from n-pentane at low field strengths, 25% $C_3H_7^+$ ions from isopentane and 90% $C_4H_9^+$ ions from neopentane.

Rearrangement reactions involving cyclic intermediate structures — like the well know McLafferty reactions, for example — often lead to rather stable products which lead to very intense peaks in the EI mass spectra. In the FI mass spectra peaks resulting from these processes are of importance mainly in the form of "fast" or "normal metastable" peaks, in most cases, because metastable processes occur at a later time after FI than direct field dissociation processes. Strong FI rearrangement peaks on the normal mass position are found for the case of surface reactions on the FI emitter.

3.4. KINETICS OF UNIMOLECULAR DECOMPOSITION OF IONS IN HIGH ELECTRIC FIELDS

Dissociation of a molecular ion by a high electric field of the order 1 V/Å is a very fast process occurring within about several 10^{-14} to 10^{-12} sec. The decomposition time can be derived from the shape of the mass lines of a FI mass spectrum which are broadened in the case of field dissociation. Decomposition processes occurring in a region of low field strength within 10^{-10}—10^{-9} sec can also be resolved in the FI mass spectrum. Decompositions within about 10^{-9}—10^{-6} sec can be detected with a FI mass spectrometer as "metastable" processes.

Mass lines of different shape appear in the FI mass spectra if a spectrometer of the Nier type is used. Fig. 3.13. shows some mass lines of a molecular ion, metastable ions and fragment ions formed by field dissociation. The shape of the mass lines of the molecular ions and of the metastable ions is the same as that observed in an EI mass spectrometer. The shape of the lines of the molecular ions is sharp and symmetric at low and medium field strengths. The spectra will be considered only for these conditions in this section. At very high strengths (several V/Å), broadening of the molecular ion lines ("parent peaks") occurs due to ionization in space several Å apart from the minimum distance of field

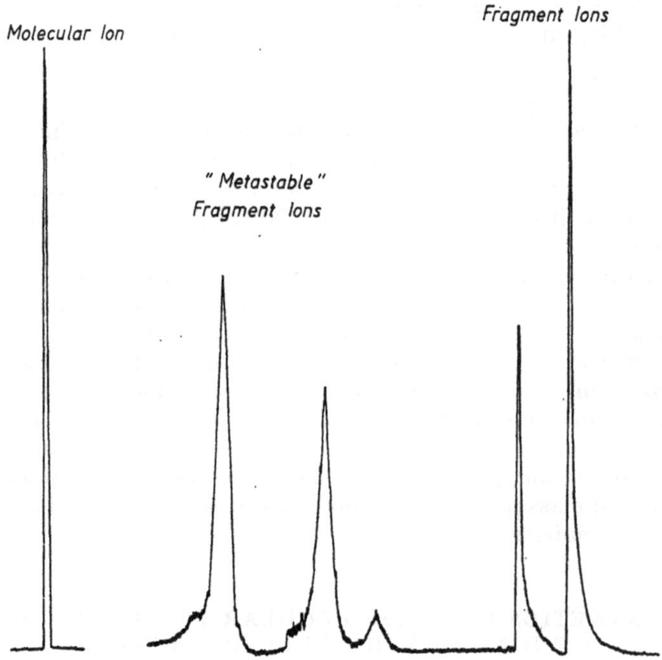

FIG. 3.13. Typical peak shapes of molecular ions, metastable and field induced fragment ions in FI mass spectra.

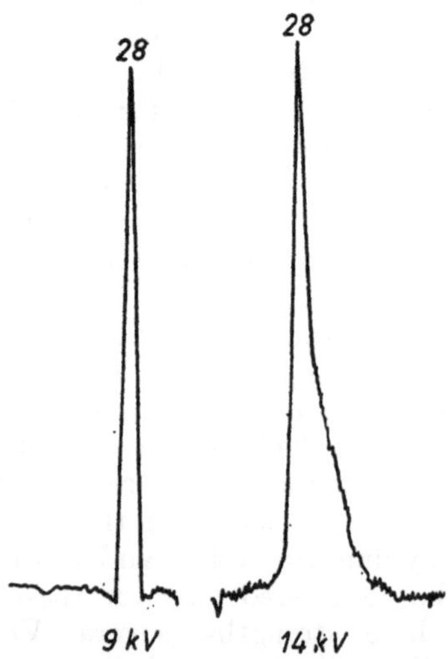

FIG. 3.14. Broadening of the N_2^+-peaks with increasing field strength (Tip-cathode voltages 9 and 14 kV).

ionization, d_{min}. Fig. 3.14. shows the peak shapes of field ionized nitrogen at medium and high field strengths.

The metastable peaks are symmetrically broadened from the centre of the line to lower and higher mass numbers. The mass lines of fragments formed in the FI source are broadened in the direction of smaller mass numbers, in contrast to fragments formed in EI sources, which produce sharp and symmetric mass lines.

3.4.1. Lifetime of Molecular Ions and Peak Shape of Fragment Ions

The broadening of fragment mass lines caused by field dissociation of molecular ions near the emitter tip was first described for the $C_2H_5^+$ ions originating from n-paraffins [105]. Later quantitative relations between the lifetimes of organic ions in high electric fields and the observed peak broadening were derived [106]. Fig. 3.15. shows schematically the equipotential lines between a tip at a positive potential of 10 kV and a cathode. (In reality, the density of equipotential lines near the tip increases much more strongly than shown in the figure.) A positive ion may be produced by field ionization very near to the tip. The ion is accelerated towards the cathode by the electric field. It may decompose at a certain distance Δr from the tip. The potential at the distance Δr is lower by ΔU than the tip potential U_0 (relative to the cathode). The fragment ions produced at Δr arrive at the cathode with smaller kinetic energy than those produced at the tip surface. Thus the fragments formed at Δx appear at smaller mass numbers in the FI mass spectrum. A relation between the observed shift in the mass scale Δm and the potential difference ΔU is obtained by means of the basic equation for mass spectrometers of the Nier type

$$R = \frac{\text{const}}{H} \sqrt{\frac{m \cdot U}{e}} \tag{3.13}$$

where R is the radius of curvature of the path of the ions in the magnetic field, H is the magnetic field strength, m the mass of the ion, U the potential of the ion corresponding to its kinetic energy in the magnetic field, and e the elementary charge.

The distance Δr from the tip may be calculated by means of the potential equation for a given electrode geometry. Finally, one obtains a correlation between Δx and the decomposition time t of the ion by integration of the equation of motion for the ion.

Schematically, the relations between the observed shift of a mass line Δm and the decomposition time t are

$$\Delta m \xrightarrow{\textit{m.s. equation}} \Delta U \xrightarrow[\textit{distribution}]{\textit{potential}} \Delta r \xrightarrow[\textit{motion}]{\textit{equation of}} t$$

These relations will now be derived in detail [107]. Fig. 3.16. shows schematically the potential distribution in a simple FI source, consisting of a FI tip on positive potential, a cathode on

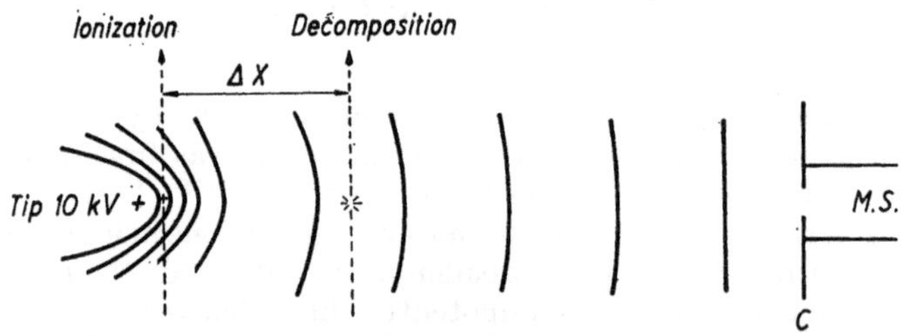

FIG. 3.15. Equipotential lines between a tip on positive potential and a cathode (schematically). C = cathode, $M.S.$ = mass analyser.

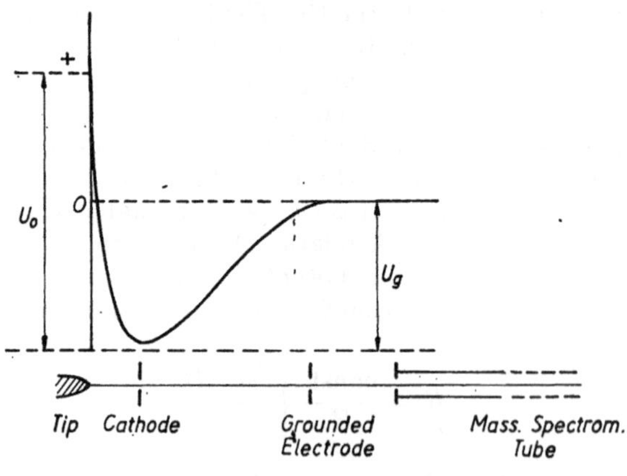

FIG. 3.16. Schematic representation of potential distribution in an FI source, indicating total voltage U_0 and retarding voltage U_g.

a negative potential Ug with respect to ground, and an electrode on ground potential. The whole mass analyser and the detector are also on ground potential. In practice, one or several electrostatic lenses will follow the cathode. They are not drawn in fig. 3.16. because they do not alter the principal considerations concerning the relation between shift of mass lines and lifetime of ions.

An organic molecule may be field ionized at the minimum distance from the tip, $d_{min} \simeq \frac{I - \Phi}{F}$. The potential at the place of ionization with respect to the cathode is termed U_0. The potential of the tip is only slightly higher, by the amount $F d_{min}$. The ion decomposes at a place Δx distant from the place of ionization, at a potential U. The fragment ion formed there does not enter the analyser tube of the mass spectrometer with energy $e(U_0 - Ug)$, like the parent ions, but with a lower energy eU^*, where U^* is given by

$$\begin{aligned} U^* &= (U_0 - U)\, m/M + U - Ug \\ &= (U_0 - Ug) - (U_0 - U)\ (1 - m/M) \end{aligned} \tag{3.14}$$

where m = mass of the fragment ion, M = mass of the molecular ion. $U_0 - U$ is the energy of the molecular ion at the moment of decomposition, $(U_0 - U)\, m/M$ the fraction of that energy which is transferred to the charged fragment according to the mass ratio.

The mass scale is calibrated with ions having the maximum possible energy $e(U_0 - Ug)$ (parent ions). The fragment ion of energy $eU^* < e(U_0 - Ug)$ is not recorded at its true mass in that mass scale, but at an "apparent mass" $m^* < m$, where

$$m^* = m\,U^*/(U_0 - Ug),$$

or

$$U^* = (U_0 - Ug)\, m^*/m \tag{3.15}$$

Substituting U^* of eqn. (3.14) into (3.15), one obtains the potential difference $\Delta U = U_0 - U$ between the place of formation and decomposition of a molecular ion as a function of m, M, m^*, U_0 and Ug:

$$U_0 - U = U_0 \cdot B \tag{3.16}$$

where

$$B = (1 - m^*/m)\,(1 - Ug/U_0)\ /\ (1 - m/M) \tag{3.17}$$

The lifetime t of the molecular ions is connected with the potential difference $U_0 - U_{(s)}$ by the equation

$$t = \int_{r_0'}^{r} \frac{ds}{v(s)} = \int_{r_0'}^{r} \frac{ds}{\left(2e(U_0 - U_{(s)})/M^{1/2}\right)} \tag{3.18}$$

Thus, the lifetime t can be calculated as a function of the measured apparent mass m^* at given values of m, M, U_0 and Ug.

r_0' is the distance of the ionization zone and r that of the dissociation zone from the centre of curvature of the tip apex. s is the variable for the distance along the orbit of the ion. $v_{(s)}$ is the velocity of the molecular ion. The minimum distance for FI from the tip surface is only a few Å as compared with the tip radius of several 1000 Å. Therefore, to a good approximation the distance r_0' may be replaced by the radius of curvature r_0 of the tip.

In an early paper the author evaluated eqn. (3.18) by replacing the tip and the cathode by two concentric spheres of radii r_0 and R [106]. This approximation leads to the right general results, however, the numerical values of the lifetimes are somewhat too small as compared to those obtained for the actual tip geometry. In a later paper the problem was recalculated by Knöppel to a better approximation [107]. The method of calculation was essentially the same as previously, but the tip and the cathode were approximated by co-axial rotational paraboloids with the radii of curvature r_0 and $2R + r_0$ at the apex. $U_{(s)}$ is the potential on the symmetry axis which is common to both paraboloids. The result is

$$U_0 - U_{(s)} = \frac{U_0}{\lambda} \ln (2\,s/r_0 - 1) \tag{3.19}$$

where

$$\lambda = \ln (2R/r_0) \tag{3.20}$$

Substituting $x = [\ln (2\,s/r_0 - 1)]^{1/2}$ into eqn. (3.18) one obtains

$$t = \frac{r\sqrt{\lambda}}{v_0} \int\limits_0^{\sqrt{\ln(2r_0/r-1)}} \exp (x^2)\, dx, \quad \text{where } v_0 = (2e\,U_0/M)^{1/2}$$

λ and B may be inserted into the upper limit of the integral, using the definitions (3.20) and (3.17). One obtains

$$t = \frac{r_0\sqrt{\lambda}}{v_0} \int\limits_0^{\sqrt{\lambda B}} \exp (x^2)\, dx \tag{3.21}$$

The integral of eqn. (3.21) is tabulated as a function of the upper limit [116].

In B a factor $(1 - m^*/m)$ is contained. Substituting $1 - m^*/m = 1/A$, where A is the resolving power of the mass spectrometer being used, one obtains approximately the time Δt_0 where fragment ions are formed from molecular ions which appear within the "natural line width" of the FI mass spectrum, i.e. without additional broadening of the mass lines. The resolving power of the mass spectrometer used for these investigation was 480, where A is defined as the ratio of the mass number of a peak to its half width expressed in mass units. With this value of A, the upper limit of the integral (3.21) becomes much smaller than one, and therefore, $\exp(x^2) \simeq 1$ within the limits of integration. As a result one obtains

$$\Delta t_0 = (\lambda r_0/v_0)\,[(1 - Ug/U_0)\,(1 - m/M)/A]^{1/2} \qquad (3.22)$$

For $M \sim 50-150$, $A \sim 500$, $m < M - 15$ one obtains with the usual values for tip radii and voltages

$$\Delta t_0 \simeq 5 \times 10^{-13} - 3 \times 10^{-12}\,\text{sec}$$

If H or H_2 are split off from the molecular ions, larger Δt_0 values up to about 1×10^{-11} sec result. Decomposition at $t > \Delta t_0$ leads to broadening of the mass lines.

Eqn. 3.18 was evaluated by Tenschert [130] for the case where an FI wire anode and a cathode are considered as two concentric cylinders. In reality, the field ions are formed at the tops of small whiskers of about 100 Å radius which are grown on the surface of an FI wire. However, the local deviations of the potential distribution from the cylindrical symmetry are pronounced only at distances of about 100 Å in front of the whiskers. At distances of more than about 1000 Å from the top of the whiskers, the potential distribution is again virtually the same as that between a smooth wire and a concentric cylindrical cathode. Ion lifetimes of more than 3×10^{-11} sec which are of interest here, correspond to displacements of more than 5000 Å from the wire surface. In this case,

$$U_0 - U_{(r)} = U_0 \frac{\ln \dfrac{r}{r_0}}{\ln \dfrac{R_0}{r_0}}$$

where R_0 is the field anode — cathode distance.

One obtains

$$t_{(r)} = \frac{2r_0}{v_0}\sqrt{\ln \frac{R_0}{r_0}} \int\limits_{y=0}^{y=\sqrt{\ln r/R_0}} e^{y^2}\,dy \qquad (3.18a)$$

The integral in (3.18a) is tabulated as a function of the upper limit in standard books. The upper limit of the integral can be replaced by

$$y = \left[\frac{1 - m^*/m}{1 - m/M} \ln R_0/r_0\right]^{1/2}$$

for the case of the retarding potential $U_g = 0$.

Thus a relation between the measured apparent mass m^* and the time-of-flight is established.

3.4.2. *Mechanisms for Ion Decomposition in High Electric Fields*

Several different mechanisms may apply if decomposition of molecular ions occurs at $t > \Delta t_0$. This will now be discussed.

3.4.2.1. *Tunneling of radicals through potential barriers*

It was shown in section 3.2. that a high electric field changes the potential curve of a molecular ion $A - B^+$ (where A and B are atoms or radicals) in a characteristic way (see fig. 3.6.). So far, dissociation has been considered to occur either from the vibrational ground state by vanishing of the dissociation energy D' at extremely high field strengths, or from vibrationally excited states which are above the maximum of the potential curve at lower field strengths (non-vanishing dissociation energy D'). The excitation of vibrational energy was explained by transitions from the neutral to the ionic state with increased equilibrium distance r_0, according to the Franck-Condon principle. In both cases dissociation of the molecular ion occurs in a classical sense by a jump of the system over the maximum of the potential barrier. Quantum mechanically, however, there exists a certain probability that the radicals will penetrate through the potential barrier some distance below the potential maximum (fig. 3.17.). This tunneling of heavy particles can be described by the same formulae as the tunneling of electrons, although the effect is much smaller for heavy particles because the inverse of the mass enters exponentially into the expression for the tunneling probability (eqn. 1.3.). Although tunneling of radicals through potential barriers has been considered in principle in ordinary reaction kinetics, the effect is usually very small. In the case of field dissociation, however, it cannot be neglected because the thickness of the potential barrier one or two vibrational levels below the potential maximum is only a few tenths of an angstrom. Narrow barriers lead to high tunneling rates. Tunneling may lead to a delayed decomposition of molecular ions at $t > \Delta t_0$. Fragment ions produced by the tunnel effect contribute only a slightly measurable amount to the mass line broadening if the tunneling probability, T, exceeds a certain lower limiting value, because of the low residence time of the molecular ions in the high electric field.

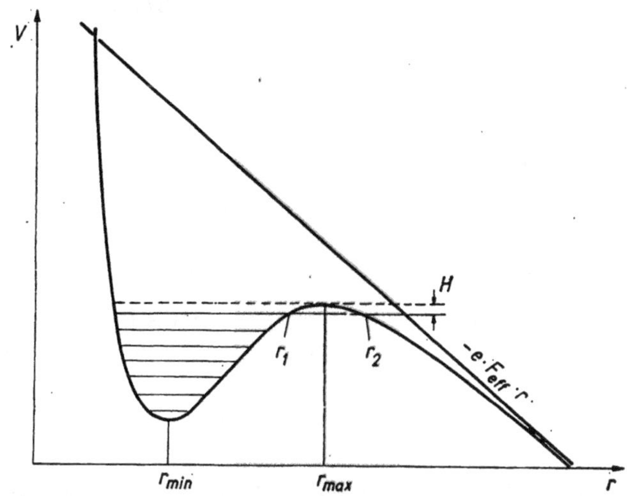

FIG. 3.17. Potential distribution for a bond in a high electrical field, indicating the maximum energy interval H below the potential maximum within which tunneling of radicals may occur with noticeable probability.

The contribution of the tunneling effect to the peak broadening of fragment ions as treated theoretically by Knöppel [90] will now be analysed in more detail. The lower limit of tunneling probability which must be imposed to make the line broadening visible will be derived. An energy quantity H will be calculated which defines the maximum amount of energy by which the highest vibrational level may be displaced below the maximum of the potential curve in order to result in a measurable tunneling effect of the molecular ion.

For this purpose, field dissociation of a certain bond in a molecular ion may be considered leading to a fragment ion of mass m. The fragment ions observed on the tail of the fragment ion peak at mass $m^* < m$ are ions of smaller kinetic energy which have the true mass m. They are formed from the molecular ions of lifetime $t(m^*)$ which can be calculated from eqn. (3.21).

This time can be calculated only with an uncertainty of Δt because of the limited resolving power $A = m^*/\Delta m^*$ of the mass spectrometer. Those ions contribute to the ion current at the apparent mass $m^* < m$ which are formed from the molecular ions within the interval of lifetimes

$$\Delta t = \frac{r_0 \cdot \sqrt{\lambda}}{v_0} \int\limits_{(\lambda B(m^*))^{1/2}}^{(\lambda B(m^* - \Delta m))^{1/2}} \exp(x^2)\, dx = \Delta t_0 \cdot K \tag{3.23}$$

The substitution

$$1 - m^*/m = n/A$$

is introduced. For the range of interest from

$n \simeq 2$ to $n \simeq 10$ it follows

$$K \simeq \sqrt{n+1} - \sqrt{n} \simeq 1/3.$$

The ion current at mass m^* which originates from tunneling of radicals from a specific vibrational level of a certain bond is given by

$$i = q\,(\dot{n}/p)\,\nu \cdot \frac{1}{3}\,\Delta t_0 T \tag{3.24}$$

where q is the elementary charge, $\dot{n}$ the number of molecular ions formed per unit time, $1/p$ the fraction of these ions which is not dissociated after $t(m^*)$ and which occupy the specific vibrational level under discussion, ν the vibrational frequency of the bond and T the tunneling probability of the specific vibrational level.

Only a certain fraction $1/s$ of all ions $i = q \cdot \dot{n}$ can be detected at the collector of the mass spectrometer because of the limited dynamic range of sensitivity of the FI mass spectrometer. (This is defined as the minimum fragment to parent ion ratio which can be measured at a signal to noise ratio of 3 : 1.) $1/s$ is typically in the order of 10^4, but up to a factor of about ten larger in favourable cases. Thus, the fragment to parent ion ratio,

$\frac{1}{p} \cdot T \frac{\Delta t_0}{3}$ must be larger or equal to $1/s$, or

$$T \geqq \frac{p}{s} \cdot \frac{3}{\nu t_0}$$

in order to make detectable the fragment ion current at m^* which originates from tunneling of radicals at the vibrational level under discussion. The vibrational frequencies are smaller than 10^{-14} sec^{-1}, and $\Delta t \geqq 3 \times 10^{-12}$ sec in FI mass spectrometers with tip emitters under typical conditions.

This imposes on T the condition

$$T \geqq p \times 10^{-2}/s$$

Moreover, at values of $T \geqq 10^{-2}$, the tunneling effect does not contribute to the broadening of the fragment ion peaks because practically all dissociation processes occur within Δt_0. Thus, for a specific vibrational level, T must meet the condition

$$10^{-2} \geqq T \geqq p/s \cdot 10^{-2} \tag{3.25}$$

in order to make measurable the peak broadening due to this process. The tunneling probability can be calculated approximately by using the WKB-method [14] if the radicals on both sides of the dissociating bond are regarded as rigid masses M_a and M_b. In analogy to eqn. (1.3) one obtains

$$T \simeq \exp.\left(-\frac{2\sqrt{2\mu}}{\hbar}\right)\int_{r_1}^{r_2}\sqrt{V_{(r)} - E}\,dr,$$ where μ is

$\dfrac{M_a M_b}{M_a + M_b}$, and E is the energy of the vibrational level which leads to dissociation by tunneling. $V_{(r)}$ may be roughly approximated by superposition of a Morse — potential [115]

$$V_{(r)} = D\left(1 - \exp\left[-a(r - r_{min})\right]^2\right.$$

with $a = (2\,\pi^2\,\mu\omega_0^2\,c^2/D)^{1/2}$, and the potential of the effective field, $V_{eff} = -e \cdot F \cdot \dfrac{M_a}{M_a + M_b} \cdot r$ (see eqn. 3.9), $\omega = 2\,\pi\gamma$.

The position of the relative maximum of the curve and the radius of curvature ρ at this maximum can be calculated as a function of the parameters D, a and F_{eff}.

$\int_{r_1}^{r_2} \sqrt{V_{(r)} - E)}\,dr$ may be replaced by a simpler expression by means of r_{max} and ρ if the potential near the maximum is replaced by the parabola

$$V_{(r)} = V_{max}\,(r - r_{max})^2/2\,\rho$$

This parabola coincides with the potential curve at its maximum and has the same radius of curvature at this place.

For an energy level at $E = V_{max} - H$ (see fig. 3.17.) one obtains

$$\int_{r_1}^{r_2} \sqrt{V_{(r)} - E} \cdot dr = \pi \cdot H \cdot \sqrt{\rho/2}$$

with

$$\rho = [a^2 D \alpha (1 - \alpha)]^{-1}, \; \alpha = \sqrt{1 - 2 F_{eff}/aD} \; (0 < \alpha \leqq 1)$$

For the tunneling probability one obtains

$$T = \exp\left(- H \frac{2\pi}{\hbar} \sqrt{\rho \cdot \mu}\right)$$

$$= \exp\left\{- \frac{H}{hc\,\omega_e} \cdot 4\,[2\alpha\,(1 - \alpha)]^{-1/2}\right\} \tag{3.26}$$

Condition (3.25) for T may be reformulated as a condition for H by means of (3.26)

$$10^{-2} \geqq \exp\left\{- \frac{H}{hc\,\omega_e} 4\pi\,[2\alpha\,(1 - \alpha)]^{-1/2}\right\} \geqq \frac{p}{s} \cdot 10^{-2}$$

or

$$2 < \frac{H}{hc\,\omega_e} \cdot \frac{4\pi}{2.3}\,[2\alpha\,(1 - \alpha)]^{-1/2} < 2 + \log \frac{s}{p}$$

Consequently, H must have a value in the interval

$$\Delta H = H_{max} - H_{min} = hc\,\omega_e \log (s/p)\,[2\alpha\,(1 - \alpha)]^{1/2} \frac{2.3}{4\pi}$$

The maximum value of $\alpha\,(1 - \alpha)$ is $1/4$ for $\alpha = 1/2$.
Therefore, the maximum of the interval ΔH is given by

$$\Delta H = 0.13\, hc\,\omega_e \log (s/p). \tag{3.27}$$

p may be put equal to one in order to obtain the maximum ΔH value, $s \simeq 10^4$ as stated before. With these values one obtains the result that H mustin be the order of half a vibrational ground quantum of the dissociating bond.

Hiskes [98] has shown that the distances of the vibrational levels in a high electric field near the maximum of the potential curve are not much different from those in the field free case.

Thus, one arrives at the conclusion:

Mass line broadening due to the tunneling effect of radicals from excited vibrational levels is not observable, except for tunneling from a level which lies in an energy interval of about half a vibrational ground quantum below the potential maximum.

Usually, molecules are transferred by field ionization to a vibrational state of the molecular ion which is not identical with the specific state which has just been considered. Thus in general, the contribution of the tunneling effect of radicals to line broadening will be negligible. Only in the specific cases where potential curves of the neutral and the ionic state (modified by the high electric field) are such that a vibrational level within ΔH is excited by a Franck-Condon transition, may the tunneling effect contribute considerably to line broadening (see fig. 3.18.).

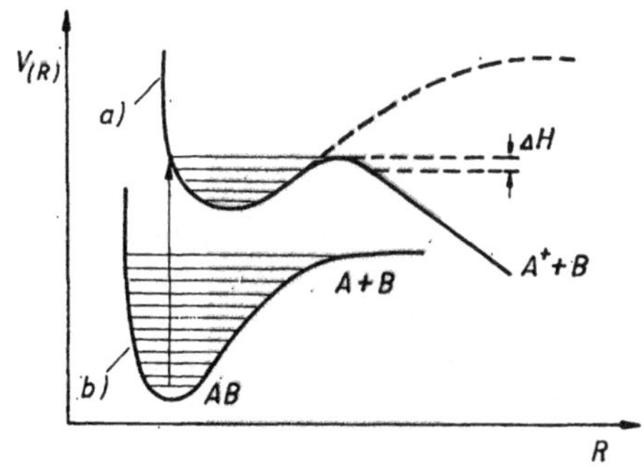

FIG. 3.18. Potential curves of a molecule AB in the neutral and the ionized state in a high electric field. Franck-Condon transition to a state which lies in the interval ΔH, capable to dissociation by tunneling (for separate drawing of curves see subscript to fig. 3.9.).

Assume that the very specific condition for dissociation due to tunneling of radicals may be met by a molecule. Then dissociation can occur only during a limited time interval. At a certain time after field ionization of the molecule, the ion has moved to a distance far from the tip as compared with the tip radius. The field strength is much lower at this place, and the potential maximum is increased (see dashed curve of fig. 3.18.). Dissociation by tunneling can no longer occur if the potential maximum is raised by at least $hc\omega_e/2$. This quantity has been estimated to be about 0.03 eV [107].

At the usual effective field strengths of 0.2—0.7 V/Å, the field strength must be reduced by about 20% in order to increase the maximum of the potential curve of an ionized bond by about 0.03 eV.

Table 3.5. shows for three characteristic examples the changes of field strength with the time-of-flight of the ions.

The table shows that field dissociation cannot occur due to tunneling of radicals after a lifetime of the parent ions of maximum 6×10^{-12} sec.

TABLE 3.5. FIELD STRENGTH F AFTER t sec RELATIVE TO F_0 AT TIP APEX AT DIFFERENT CONDITIONS.

Time of flight t sec	F/F_0 for $M = 80$ $r_0 = 5000$ Å $U_0 = 12$ kV	 $M = 100$ $r_0 = 8000$ Å $U_0 = 16$ kV	 $M = 100$ $r_0 = 3000$ Å $U_0 = 6$ kV
1×10^{-12}	0.99	0.99	0.99
2×10^{-12}	0.95	0.98	0.95
4×10^{-12}	0.83	0.92	0.83
6×10^{-12}	0.70	0.83	0.69
8×10^{-12}	0.57	0.74	0.56

The fragment ion mass lines of many organic substances show peak broadening — if a tip is used for field ionization — which correspond to lifetimes of molecular ions $< 6 \times 10^{-12}$ sec. In these cases tunneling of radicals may contribute to the line broadening. However, other effects like rotation of the molecular ion or statistical fluctuation of excitation energy may cause the same peak broadening. Presently, it is not possible to differentiate experimentally between these processes. Generally speaking, the probability of line broadening due to the tunneling effect will be smaller than that due to the effects to be discussed in the following sections, because of the very specific requirements for excitation of the molecular ion.

3.4.2.2. Broadening of mass lines due to rotation of molecular ions

The charge distribution of a molecular ion and the shape of a potential curve for a specific bond in a high electric field depend on the orientation of the molecular ion. Therefore, changes in orientation of the molecular ion or parts of it lead to changes of field dissociation probability of specific bonds. If the molecule or parts of it are turned from an unfavourable position after field ionization to a favourable one with respect to field dissociation, a delayed dissociation may occur which leads to line broadening.

A rotation of a fully stretched n-heptane molecule by 360° requires a time of 1.1×10^{-11} sec at 300 °K. Turning a stretched n-heptane molecule from a position perpendicular to a position parallel with the field lines by 90° requires about 3×10^{-12} sec.

However, neither the n-heptane nor other chain molecules are stretched in the neutral state. The molecule approaches the FI emitter in a statistically bent shape. In the vicinity of the FI emitter a slight preferential order of the molecule parallel to the field lines occurs due to the very high electric field. This effect is caused, as mentioned in section 3.1., by the anisotropy of polarizability of a C—C bond parallel and perpendicular to the field lines. However, this effect is very small, and there is only a very small time interval available for transition of the molecule from the nearly field free region to arrival at the tip.

On the other hand, the FI mass spectra of chain molecules can be interpreted only in a consistent manner if one assumes a stretched position of the ions at the moment of dissociation. This may be arrived at in the following way. The molecules which are not ionized in the gas phase while approaching the tip (which are the magority under typical field conditions) are adsorbed for a while at the tip. The hydrocarbon chains of these molecules may be adsorbed either parallel or perpendicular to the tip surface (see section 3.6.5. Butanol is adsorbed with the hydrocarbon chain parallel, pentanol perpendicular, to the tip surface). Those molecules which are desorbed with the position of the chain parallel to the surface do not field dissociate because the field lines are not parallel with the C—C axes. However, if the positive charge is positioned more to one end of the molecule, the electrostatic force will move it away from the tip whereas the forces of inertia are directing the opposite neutral end of the molecule into the opposite direction. Therefore, the molecular ion is turned into a position parallel with the field line within a time interval in the order of a quarter of a rotational period. Thus a delayed dissociation may occur after a max. time of about 3×10^{-12} sec. The time necessary for turning a stretched primary alcohol molecule from a position where either an OH group or a CH_3 group is nearest to the tip, respectively, i.e. turning by 180°, requires a time of the order of 6×10^{-12} sec. These are order of magnitude estimations only, but they show that line broadening due to the rotational effect must be small, but just observable. The observed, much larger line broadening of some fragment lines of polar molecules must be due to other effects which will be discussed in the next section.

3.4.2.3. *Line broadening due to fluctuation of excitation energy. (Fast metastable decomposition)*

A molecule may dissociate due to fluctuation of vibrational excitation energy and accumulation of this energy in a specific bond (preferentially a weak bond). This phenomenon which is well known from general reaction kinetics of neutral molecules has been treated theoretically for the case of unimolecular decomposition of ions by Rosenstock, Wallenstein, Wahrhaftig and Eyring [117]. The theory has been extended and refined mainly by Rosenstock [118] who termed it "Quasi Equilibrium Theory" of unimolecular decomposition of ions (ΘET). It is outside the scope of this book to discuss on the one hand the success of this theory as applied to simple small molecules or to specific problems like isotopic substitution or metastable decomposition, and on the other hand to comment on the criticism of some authors on this theory. However, it is without any question that one possibility for a delayed decomposition of a polyatomic excited ion is — among others which have been summarized by Lorquet [119] — the statistical fluctuation of vibrational energy. Delayed decomposition means here a dissociation process which occurs many vibrational periods after primary ion formation.

Such delayed decomposition processes are observed in the case of field ionization of polar organic molecules. Fig. 3.19. shows part of the FI mass spectrum of diethyl ether as measured by Knöppel. The mass line of the fragment $m/e = 59$ is broadened over a very large mass range. The mass scale may be converted into a time scale by means of eqn. (3.21). The result is that delayed decompositions are observed for this process extending over a period of several 10^{-12} to several 10^{-8} sec. After this period a "normal metastable" decomposition occurs within the field free region of the mass spectrometer between the ion entrance slit and the first boundary of the magnet. These ions are recorded as a "normal metastable peak" at mass number 47.04. The upper limit of lifetimes of these ions is of the order 10^{-6} sec. Thus, there is a spectrum of lifetimes of the molecular ions which extends over six orders of magnitude. The broad intermediate peak between the normal fragment peak and the normal metastable peak is termed "fast metastable peak".

These fast processes cannot be due to field dissociation without fluctuation of excitation energy because the field strength has already droped by about 20% after a time-of-flight of the ions of several 10^{-12} sec, as shown earlier. By this reduction of field strength, the originally field-reduced dissociation energy increases to such an extent that neither direct dissociation from the highest vibrational level (as excited by FI), nor dissociation by quantum mechanical barrier penetration can occur to a measurable extent. Thus, the delayed dissociation can be explained only by transfer of vibrational energy quanta from the neighbouring bonds to the bond under discussion. These decomposition processes may be termed "statistical" decompositions.

Fluctuation of vibrational energy may also play some role in the case of dissociation within the time interval between one vibrational period (about 5×10^{-14} sec for a C—C bond) and 20 vibrational periods (about 10^{-12} sec, i.e. the time corresponding to the line width of unbroadened peaks). In this case, the activation energy for dissociation is greatly reduced by the extremely

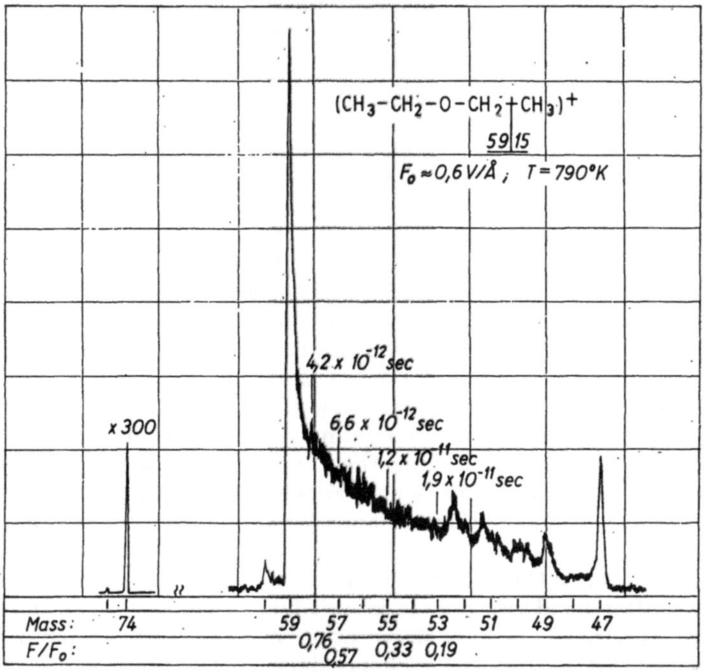

FIG. 3.19. Parts of FI mass spectrum of diethylether, on a heated platinum tip [107].

high electric field. Transfer of only one or a few vibrational quanta from the bonds neighbouring to the bond to be dissociated may suffice for rupture of this bond if a vibrational level which is only one or a few quanta below the reduced dissociation limit were occupied as a result of FI. This process may be termed "field induced statistical decomposition". The time required on the average for transfer of a vibrational quantum from one to the other degree of freedom may possibly be involved, but for ions no quantitative statements on these times have become known to the author.

The transfer of vibrational energy in the case of field ionization may be due to Franck-Condon transitions from the neutral to the ionic ground states. (Shift of interatomic equilibrium distances in the neutral and the ionic states.) Another source of the vibrational energy is the thermal energy of kT on the average per vibrational mode, which may result in a total amount of several tenths of an eV for a molecule with many vibrational modes at room temperature. Tunneling of electrons from lower lying electronic states of the neutral molecule may also occur. This electronic excitation energy may be converted into vibrational energy by radiationless transitions. However, the contribution of these processes will be small because the tunneling probability falls off rapidly with the height of the potential barrier. Finally, part of the kinetic energy of the neutral molecule approaching the FI tip (enhanced over the thermal value by polarization effects) may be partially converted into internal energy of the molecule because of incomplete accommodation to the tip during impact. No quantitative statements on this effect can be made at present for organic molecules. However, it is assumed that this kind of energy conversion plays only a minor role, because from the FI mass spectra as a function of tip temperature (see section 3.5.) it may be concluded that organic molecules impinging onto the tip are largely accommodated to the tip temperature.

Very broad mass lines extending over the whole mass range between the "normal" and the "metastable" peak correlated with the same decomposition process are observed in the case of polar molecules like ethers or amines. Fragments of this type are always formed by rupture of a C—C bond which is next to the hetero atom to C bond. The positive charge remains at the hetero atom after bond rupture. It is well known from electron impact mass spectra that these dissociation processes are favoured because of stabilization of the charge due to resonance

$$RXCH_2R' \xrightarrow{-e} R' + RX^+ = CH_2 \rightleftharpoons RX - \overset{+}{C}H_2 \qquad (3.28)$$

where X stands for O, NH etc.

Reaction (3.28) is also observed as a metastable process with ethers and amines, whereas metastable elimination of a neutral paraffin is predominant with ketones, where a hydrogen atom is rearranged during the dissociation process.

The shape of the fragment peak at mass 59 produced by field ionization of diethylether (fig. 3.19.) will be further discussed.

The peak extends over 12 mass units, down to mass 47.04 which is due to metastable formation of $C_2H_5OCH_2^+$ ions. The parent ion peak at mass 74 is also shown in the figure. It is very narrow as compared with the fragment peak beginning with mass 59. The conversion of the mass scale to a time scale has been carried out for several mass values, as seen from fig. 3.19. The range of decomposition times extends from several 10^{-12} to several 10^{-8} sec. The latter value characterizes the beginning of the "normal" metastable decomposition, which covers a range of decomposition times up to several 10^{-6} sec. Some intermediate peaks are observed between mass 53 and 47. These peaks are due to retardation of the ions by electrostatic lenses. They disappear if the tip is at full positive potential and all other electrodes at zero potential. The ratio of the field strength F at the place of decomposition to the field strength F_0 at the place of production of molecular ions has been calculated by means of $F/F_0 = e^{-\lambda B}$ where λ and B are defined by eqn. (3.20) and (3.17). Some of these values are also shown on the abscissa of fig. 3.19. From curves of the type shown in fig. 3.19., average rate constants for unimolecular decomposition can be derived.

The equation for unimolecular decomposition with a continuous distribution of rate constants is

$$\frac{dN_{(t)}}{dt} = -\int_{k=0}^{\infty} k \frac{\delta N_{(k,t)}}{\delta k} dk \equiv -N_{(t)}\bar{k}_{(t)} \tag{3.29}$$

where $N_{(t)}$ is the total number of undecomposed parent ions at the time t, $N_{(k,t)}$ is the fraction of the parent ions which decompose with the rate constant k at t, $\bar{k}_{(t)}$ has the meaning of an average rate constant:

$$\bar{k}_{(t)} \equiv \frac{\int_{k=0}^{\infty} k \frac{\delta N_{(k,t)}}{\delta k} dk}{N_{(t)}} \tag{3.29a}$$

$\frac{dN^P_{(t)}/dt}{N^P_{(t)}}$ may be replaced by the measured ratio of the ion currents:

$$\frac{-dN^P_{(t)}/dt}{N^P_{(t)}} = \frac{di^{m^*}_{(t)}/dt}{i^P_{(t)}} = \bar{k}_{(t)} \tag{3.29b}$$

where the indices P and m^* refer to the parent and the fast metastable fragment ion current, respectively. Only one decomposition reaction of the parent ions will be discussed here. If the relative fragment ion intensity is small, i^P may be taken as constant for the time interval of a few μ sec, and hence the parent ion intensity measured at the detector of the mass spectrometer may be inserted into eqn. (3.29b) for $i^P_{(t)}$. (The decomposition of the parent ions in the magnetic sector field is neglected here.)

For evaluation of the average rate constants $\bar{k}_{(t)}$ in the range of ion residence times between 3×10^{-11} and 1×10^{-8} sec one has first to transform a plot of the measured fragment ion

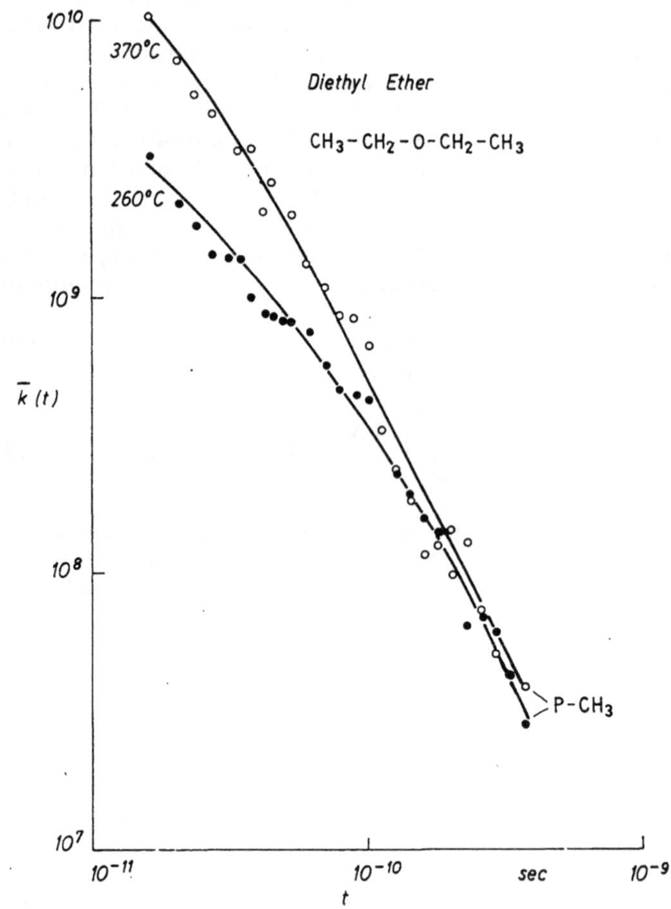

FIG. 3.20. Average constants $\bar{k}$ for the elimination of CH_3 from diethyl ether as a function of the ion residence time [130].

intensity in the mass scale (see fig. 3.19.) into the time scale by means of eqn. (3.18). Then the differential quotient $di_{(t)}^{m^*}/dt$ is derived from the curve, by dividing the measured ion current i_m^* of the fast metastable ions at mass m^* by Δt_m^*. $\Delta t_m^* = \left|\frac{dt}{dm^*}\right| \Delta m^*$, and $\Delta m^* = \frac{m^*}{A}$ where A is the resolving power of the mass spectrometer. The quantity $i^P_{(t)}$ of eqn. (3.29b) is replaced by $\frac{1}{\Delta t^P} \int i^P_{(t)}\, dt$, where the integration extends over the peak area, and Δt^P has the meaning analogous to Δt^*_m. An average curve has to be formed from several individually measured curves in order to reduce the scatter of the measured values of the ion current as shown in fig. 3.19. Curves $\log \bar{k}$ versus $\log t$ for diethyl ether as derived from measurements by means of the equations given here have been drawn by Tenschert [130] (see fig. 3.20.). A detailed discussion of these and other curves (for n-heptane, for example) will be given in section 3.4.2.6.

3.4.2.4. *Different shapes of fast metastable peaks*

The shape of the fast metastable peaks is similar to that shown in fig. 3.19. in many cases; i.e. the peak extends from the normal fragment peak to the normal metastable peak with continuously decreasing intensity. On the other hand, the $C_2H_5^+$ fragment peak in the FI mass spectra of n-paraffins, for example, is only slightly broadened at $F = 7 \times 10^7$ V/cm and room temperature. Other molecules exhibit fragment peaks which are intermediate in shape between these two limits. Aliphatic amines, for example, show strong $CH_2NH_2^+$ fragment peaks ($m/e = 30$) which have an intermediate degree of broadening.

Still other peak shapes are observed in the FI mass spectra of many *rearrangement reactions*. Figs. 3.21.a. and 3.22.a. show the fast metastable elimination of water from n-pentanol and n-hexanol. There exists a broad maximum of the $P - H_2O$-fragment peak intensity between the position of the normal fragment peak and the metastable fragment peak, indicating a maximum rate at ion residence times of several 10^{-11} to 10^{-10} sec. Such more or less broadened peaks with maxima shifted from the normal mass position are characteristic for most of the rearrangement reactions, as will be shown in section 4.3.3.

Fig. 3.21.b. shows a plot of the logarithm of the decomposition rate versus the ion residence time, as measured by Knöppel. The rate can be shown to be equal to the fragment ion current, i, divided by the time interval Δt which is given by eqn. 3.23. A maximum rate is found at 2×10^{-11} sec. The rate constant, k, is given by the slope of the curve. Fig. 3.21.b. seems to indicate that the rate constant does not change between 2×10^{11} and 1.3×10^{10} sec. However, the range of ion residence times is too small for such a statement. It will be shown in section 3.4.2.6. that the rate average constant for water elimination from n-hexanol drops from about 2×10^9 to 1.5×10^7 sec^{-1} within ion residence times ranging from about 3×10^{-11} to 6×10^{-9} sec.

It cannot be decided from fig. 3.21. wheter the broad fast metastable peak originates from elimination of wather from the parent peak (rearrangement reaction) or from the protonated parent

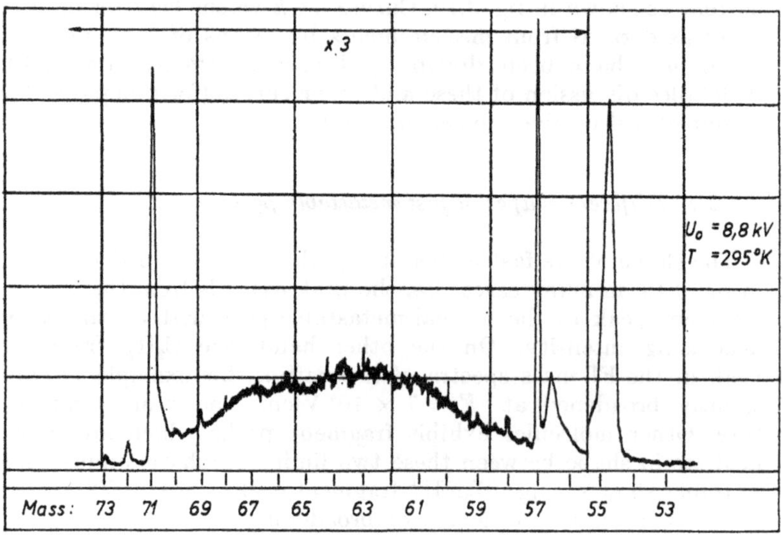

FIG. 3.21.a. Part of FI mass spectrum of 1 — pentanol [107].

ion peak (direct bond split). Normal metastable peaks are found for both processes ($(P + 1) - H_2O$, $m^* = 55.7$; $P - H_2O$, $m^* = 56.6$). The latter one is much more intense in fig. 3.21.a. A relatively strong $(P + 1)^+$ peak is observed in the FI mass spectrum of n-hexanol, with a corresponding normal fragment peak $[(P + 1) - 18]^+$, according to the reaction

$$R - CH - \overset{+}{O}\langle^{H}_{H} \; R - CH_2 - {}^{+}O\langle^{H}_{H} \rightarrow R - CH_2^+ + H_2O .$$

This is a fast direct bond split, leading to a sharp $[(P + 1) - H_2O]^+$-peak. It was proved experimentally that this process makes only a small contribution to the broad fast metastable peak [212]. This was done by using a double focusing mass spectrometer and applying the technique of measuring the metastable ions which decompose before reaching the electric sector of the mass spectrometer [213-216] (see section 3.4.3.5.). The energy deficit of ions of a given mass can be measured by leaving the voltage of the electric sector constant and by matching the ion accelerating voltage with the energy deficit of the fragment ions. Fig. 3.22.b. shows the energy distribution of ions of mass 84 ($P - 18$) and of mass 85 of n-hexanol. It becomes evident from this figure that the broad fast metastable peak of n-hexanol is due mainly to elimination of water from the parent ion. There are two more very broad metastable peaks at lower mass numbers in the FI mass spectrum of 1-hexanol (see fig. 3.22.a.). They are due to elimination of CH_3, CH_4 and C_2H_4, respectively, from the parent-minus-water ion

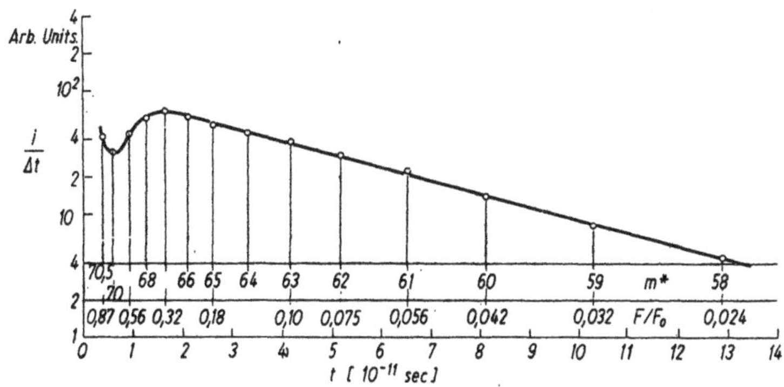

FIG. 3.21.b. Change of the decomposition rate $r = i/\Delta t$ with time for formation of fragments $m = 71$ from parent ions of 1-pentanol. (Re-calculated from the curve shown in fig. 3.21.a.)

($m/e = 84$). The elimination of these neutral particles from the ion $m/e = 84$ occurs as a normal metastable process in the field free space of the mass spectrometer. However, since the precursor ion $m/e = 84$ shows a large energy spread, this spread is transferred also to the metastable daugther ion.

For a fragment ion m_2 which arises from a precursor fragment ion m_1 originating on potentials between U_0 and zero, one derives an apparent mass number from eqns. 3.14 (with $U_g = 0$) and 3.33

$$m_2^* = [(1 - U/U_0)\, m_1/M + U/U_0]\, m_2^2/m_1 .$$

Two limiting cases may be considered:

1. $U = U_0$, $m_2^* = m_2^2/m_1$
2. $U = 0$, $m_2^* = (m_1/M)\,(m_2^2/m_1) = m_2^2/M$

For example, the $[(P - H_2O) - C_2H_4]^*$ – peak should extend from $m^* = 37.33$ to 30.74. The spectrum supports this view. These results clearly show that those consecutive reactions occur

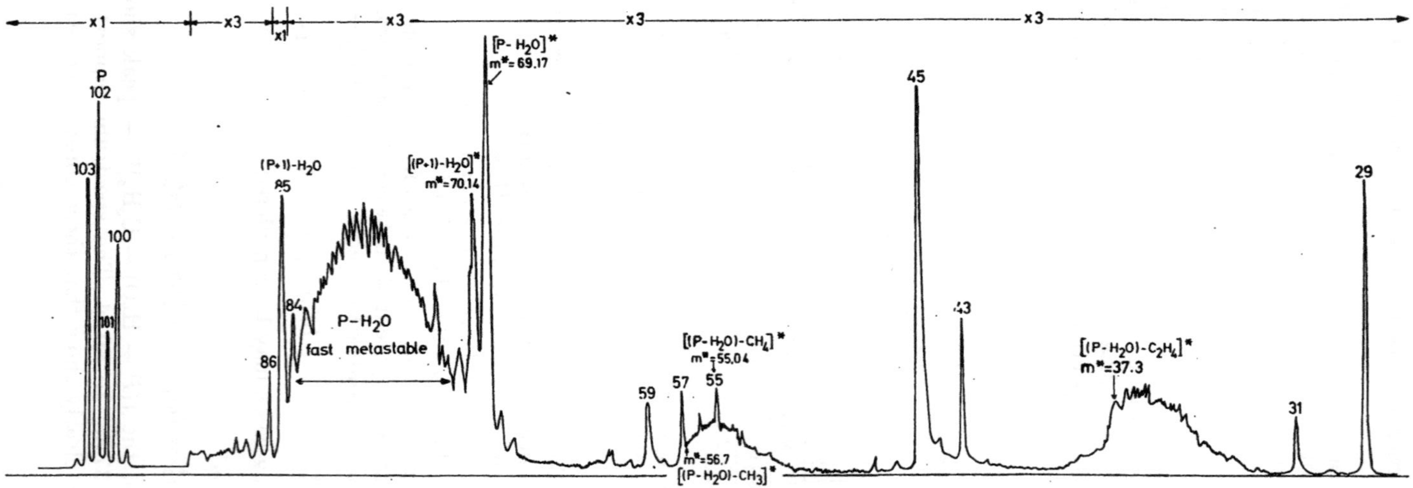

FIG. 3.22.a. FI mass spectrum of n-hexanol on a 2.5 μm platinum wire.

with two greatly different rate constants. The water elimination from 1-hexanol occurs very quickly (about 10^{-10} sec on the average), and the subsequent elimination of CH_3, CH_4 or C_2H_6 much more slowly (about 2×10^{-6} sec, on the average).

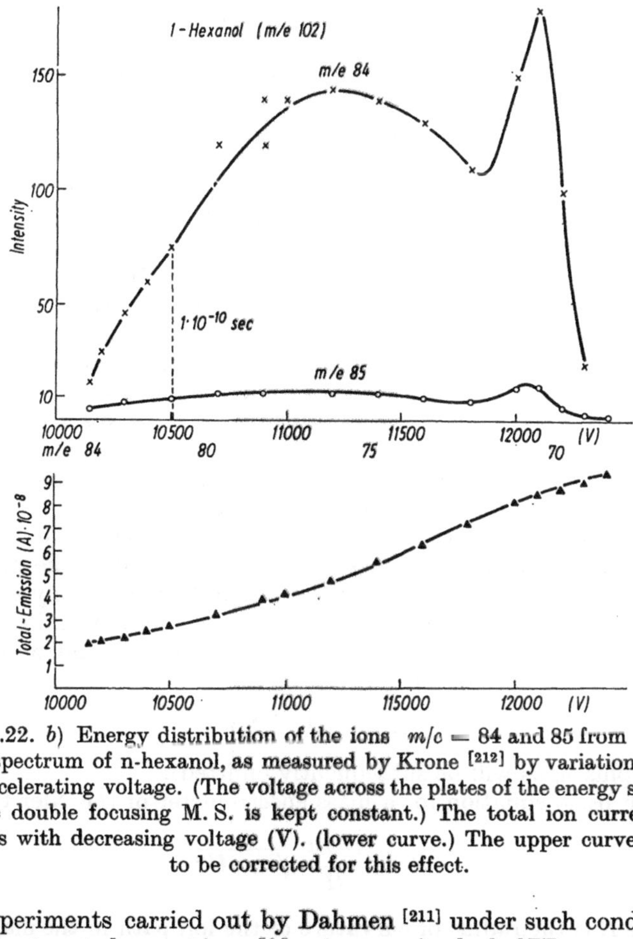

FIG. 3.22. *b*) Energy distribution of the ions $m/e = 84$ and 85 from the FI mass spectrum of n-hexanol, as measured by Krone [212] by variation of the ion accelerating voltage. (The voltage across the plates of the energy selector of the double focusing M. S. is kept constant.) The total ion current decreases with decreasing voltage (V). (lower curve.) The upper curves have to be corrected for this effect.

Experiments carried out by Dahmen [211] under such conditions that protonated parent ions did not appear in alcohol FI mass spectra also indicated that the broad fast metastable peak originates from water elimination of the parent ion by a rearrangement process.

Further, Dahmen was able to show by means of deuterated alcohols that the elimination of water from the field ionized parent molecules is a 1.4 and a 1.3 elimination via a cyclic transition configuration, similar to the conditions under electron impact (where the 1.4 elimination is more preferred, however).

3.4.2.5. Normal metastable decomposition of ions

In electron impact mass spectrometry, metastable processes are defined as decomposition processes occurring after full acceleration of the parent ions. The metastable fragment ions do not appear at their true mass number in the mass spectrum. It depends on the geometry of the mass spectrometer where these ions appear on the mass scale. Usually — apart from random coincidence — these ions appear at non-integral mass numbers. In most cases, the metastable mass lines are broader than the "normal" fragment mass lines. "Normal" mass line means a position of the fragment line in the mass spectrum which corresponds to the true mass of the fragment. (The normal fragments have the same energy as the parent ions.) All considerations made so far for EI mass spectra are equally well applicable also for FI mass spectra. However, these peaks should be termed "normal metastable peaks" in the case of FI, in order to differentiate between them and the "fast metastable peaks" as discussed in the preceding section.

First, normal metastable processes in a mass spectrometer of the Nier type may be discussed. This mass spectrometer consists of an ion source, a field-free drift tube, a magnetic sector field of 60°, a second field-free drift tube and a detector.

The parent ions formed by EI or FI are accelerated to the full potential U_0 in the ion source. U_0 is the potential at the place of origin of the ions with respect to the grounde dmass analyser tube. In the first field-free drift tube a metastable decomposition of the parent ion M^+ into a charged fragment ion m^+ and a neutral ragment Δm may occur:

$$M^+ \rightarrow m^+ + \Delta m$$

(The plus signs will be subsequently omitted).

The basic equation for a mass spectrometer of the Nier type is

$$R = \frac{const}{H} \sqrt{m \cdot U^*} \tag{3.30}$$

where R is the radius of curvature of the ion path in the magnetic sector, H is the magnetic field strength, m the actual mass and $e \cdot U^*$ the actual kinetic energy of the fragment ion. The energy of the parent ions in the drift tube is $e \cdot U_0$. If a parent ion decomposes in the drift tube, its energy is distributed over the two fragments according to the ratio of masses

$$U^*_{(m)} = \frac{m}{M} \cdot U_0 \tag{3.31}$$

The position of the mass line of the metastable ion m at energy $e \cdot U^*$ is the same as that of a parent ion of mass m^* with energy $e \cdot U_0$.

Therefore
$$m^* = m \cdot \frac{U^*}{U_0} \qquad (3.32)$$

Substituting (3.31) into (3.32), one obtains

$$m^* = \frac{m^2}{M} \qquad (3.33)$$

m^* is the "apparent" mass of the metastable ions of actual mass m which characterizes the position of the metastable mass line in the "normal" mass scale.

The ions decomposing in the magnetic sector field are not focused to a sharp mass line. They contribute to the background of the mass spectrum over a wide mass range.

Fragment ions produced in the second field-free drift tube cannot be differentiated from the ions from which they originate.

The broadened shape of the metastable mass lines can be explained by the excess kinetic energy of the two fragment ions which are formed by dissociation of the precursor ion. Although this energy is usually small (of the order of several tenths of an eV), the effect on the shape of the mass line is "amplified" by the high original velocity of the decomposing ions. Details of the calculation of metastable peak shapes have been given by different authors [120–126].

It was pointed out in the preceding section that the metastable peaks produced by an FI source in connection with a 60° magnetic sector mass spectrometer result from a spectrum of residence times in the first drift tube between several 10^{-8} and 10^{-6} sec. Therefore, a differential analysis of the actual distribution of lifetimes of metastable ions produced after field ionization would be of interest. A knowledge of this distribution could serve as a test of the quasi-equilibrium theory of mass spectra [117, 118]. From the measured "decay function" which will be defined next, one could either derive the distribution of the abundance of the rate constants, provided one knows the energy distribution of the ions, or vice versa, with a known distribution of rate constants one could calculate the energy distribution [124]. The principle of a retarding potential technique for differential analysis of metastable ion lifetimes was given earlier [127]. In a more recent investigation by Tenschert and the author, this retarding potential technique was improved [128, 129]. This method will now be described.

Fig. 3.23. shows the potential distribution within a special FI — source suitable for lifetime measurements of metastable ions. The abscissa represents the distance x from the field anode. The symbols t_c, t_A, t_E and t_R represent the ion time-of-flights corresponding to X_c, X_A, X_E and X_R. The field anode at $x = 0$ consists of a thin wire of 2.5×10^{-4} cm diameter which is at a con-

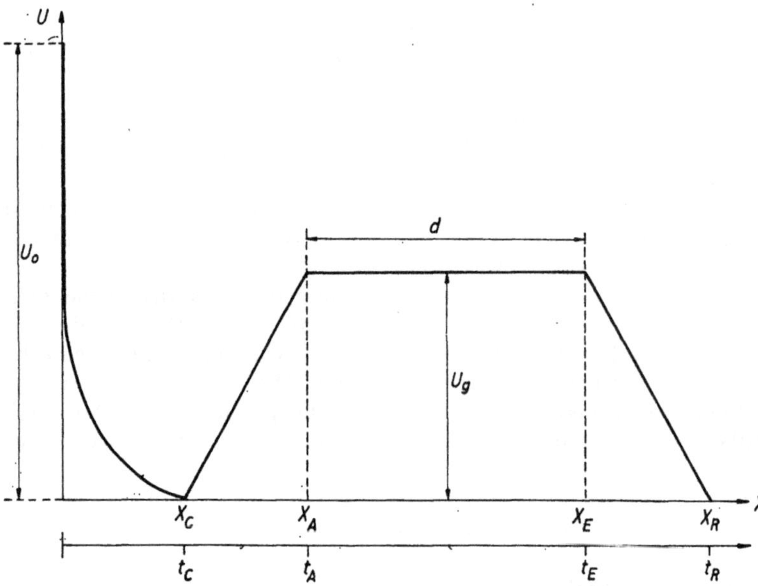

FIG. 3.23. Potential distribution in a FI source with retarding electrodes. X-scale: distances of electrodes from the field anode. t-scale: Time-of-flight of the ions from the field anode.

stant potential of + 10 kV with respect to ground. The field ionized molecules are accelerated by a cathode at distance $x_c = 2.5$ mm which is at ground potential. After a distance of 2 mm from the cathode, at x_A, there is the first of five retarding electrodes, some or all of which are on the same, variable between $+2$ and $+9.5$ kV, retarding potential U_g. The rest of the electrodes are on ground potential. In this way, the "plateau length d" of the retarding potential and thus the residence time of the metastable ions on the plateau and the time for their arrival at the field free space of the mass spectrometer can be varied. The last of these retarding electrodes is at x_E. The residence time can also be varied — in addition to altering the plateau length — by regulating the retarding potential U_g.

Fig. 3.24. shows the most important parts of this special FI source schematically [129]. The plateau length d can be varied from 0.25 to 3.37 cm by suitable electrical connection of the electrodes. For reduction of the penetration of field lines through the slits, fine meshes of high transparency are mounted across the electric slits.

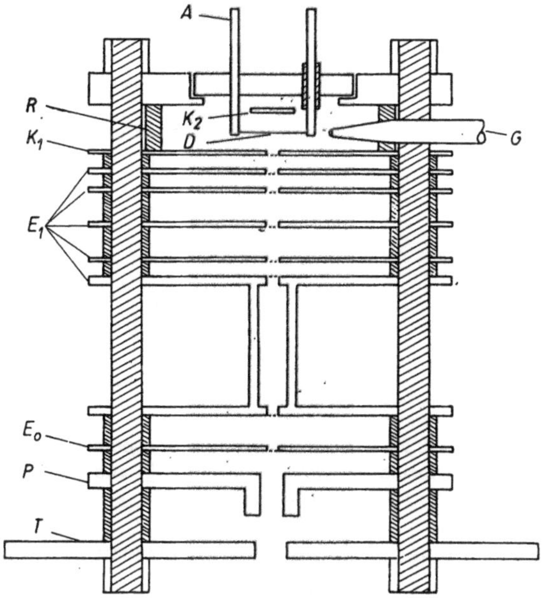

FIG. 3.24. FI source with retarding electrodes for measurement of lifetimes of metastable ions.

A = Field anode; K_1 = cathode; R = ceramic ring; K_2 = electrode for compensation of mechanical force; G = gas admission; D = Wollaston wire, $Ei_{(i=1-5)}$ = retarding electrodes; E_0 = ground potential, P = deflecting plates; T = mounting plate on ground potential.

The distribution of rate constants of field ionized molecules is measured in the following way. First, all retarding electrodes are put on ground potential. The ions decomposing in the first field free drift tube after full acceleration are recorded as normal metastable ions at the apparent mass number $m_n^* = m^2/M$. Fig. 3.25. shows a group of normal metastable peaks in the FI mass spectrum of n-octane at $m = P-29$ and $P-30$. Then one or several of the retarding electrodes E_{1-5} are brought onto a potential between $+2$ and $+9.5$ kV with respect to ground. A new group of mestastable peaks, which is similar to the normal metastable peak group, appears in the FI mass spectrum. These new peaks are termed "metastable plateau peaks" [129]. The apparent mass m^*_{pl} of these peaks is between the normal metastable mass m^*_n and the actual mass of the fragments m. The position of the metastable plateau peaks in the mass spectrum is variable (in contrast to the fixed position of the normal metastable peaks), depending on the retarding potential (see fig. 3.25.).

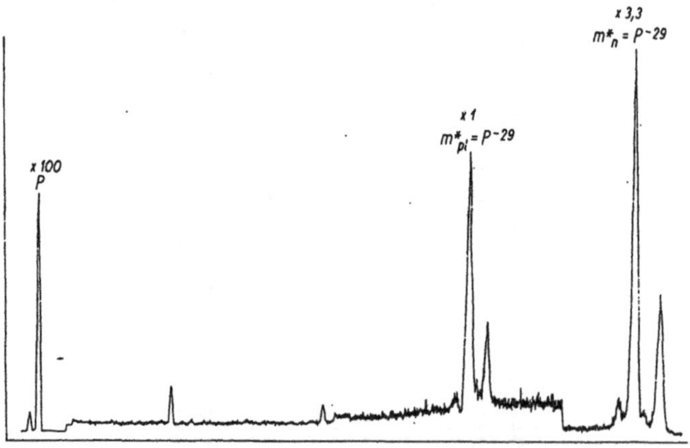

FIG. 3.25. FI mass spectrum of n-octane.
Molecular peak (P), normal metastable peaks (m^*_n) and metastable plateau peaks m^*_{pl}.

The apparent mass of the metastable plateau peaks is given by

$$m^*_{pl} = m\left[(1 - \alpha)\frac{m}{M} + \alpha\right] \tag{3.34}$$

where $\alpha = U_g/U_0$, m and M are as defined before.

There are two limiting cases:

1. The retarding potential $U_g = 0$. Then

$$m^*_{pl} = m^*_n = \frac{m^2}{M}$$

2. $U_g \simeq U_0$:

$$m^*_{pl} \simeq m$$

Now the measurement and calculation of lifetimes of normal metastable ions will be discussed.

If there would be only one decomposition process of the molecular ions with one decay constant k, the ratio of the normal metastable ion current, i_{mn}, to the parent ion current, i_p, would be given by

$$i_{mn}/i_p = A \cdot e^{-kt_R}\,(1 - e^{-k \cdot \tau}) \tag{3.35}$$

where t_R is the time-of-flight of the ions from the field anode to the entrance of the first-field free drift tube, and τ the residence time in the drift tube. (The first term of the right side represents the number of undecomposed molecular ions entering, the second term the number of undecomposed molecular ions leaving the drift tube.) If t_R is variable, but τ is kept constant, eqn. (3.35) may be written as

$$\ln i_{mn}/i_p = const. - k \cdot t_R \tag{3.36}$$

Thus, a plot of $\ln i_{mn}/i_p$ against t_R should result in a straight line, the slope of which represents the decay constant.

However, a plot of $\ln i_{mn}/i_p$ against t_R for normal metastable ions at variable retarding potential (and hence t_R-values) does not lead to a straight line. Thus it can be concluded that decomposition of the molecular ions occurs with a distribution of decay constants k_i.

For this case, the rate equation of unimolecular decomposition with a certain number of discrete rate constants, is:

$$\frac{dN_{(t)}}{dt} = -\sum_i Ni_{(t)} k_i = -N_{(t)} \cdot \bar{k}_t \tag{3.37}$$

where $N_{(t)}$ is the total number of undecomposed parent ions at t, $Ni_{(t)}$ the number of parent ions decomposing with decay constant k_i at t. ($\Sigma\, Ni_{(t)} = N_{(t)}$).

From eqn. (3.37) one obtains

$$\frac{\Sigma\, Ni_{(t)}\, k_i}{\Sigma\, Ni_{(t)}} = \bar{k}_{(t)} \tag{3.38}$$

Thus $\bar{k}_{(t)}$ may be interpreted as the average rate constant at t, which varies with time. If one assumes a continuous distribution of rate constants, one obtains the integrated form of eqn. (3.37)

$$N_{(t_2)} = N_{(t_1)}\, exp\left(-\int_{t_1}^{t_2} \bar{k}_{(t)}\, dt\right) \tag{3.39}$$

With $-(N_{(t_2)} - N_{(t_1)}) = N_m$, N_m = number of metastable ions, one obtains

$$\int_{t_1}^{t_2} \bar{k}_{(t)}\, dt = \ln\left\{1 + \frac{N_m(t_1, t_2)}{N_{(t_2)}}\right\}, \tag{3.40}$$

Now the metastable ions decomposing on the plateau and the normal metastable ions will be treated separately.

a) metastable plateau peaks

The time for acceleration of the ions, t_c is constant. The variation of the retarding time $t_A - t_c$ with varying retarding potential is small as compared with the average length of the plateau time.

Therefore

$$t_1 = t_A = const.$$

Furthermore: $t_2 = t_E \simeq t_R$ (see fig. 3.23.).

Under these conditions, (3.40) is written as

$$\int_{const.}^{t_R} \bar{k}_{(t)}\, dt = \ln \left\{1 + \frac{N_{mpl}}{N_{tR}}\right\} \tag{3.41}$$

N_{tR} is the number of parent ions at the end of the retarding potential at x_R (see fig. 3.23.).

$N_{tR} = N_P + N_{mn}$, which is approximately equal to the number of parent ions at the mass spectrometer detector, plus the number of normal metastable ions decomposing in the field-free region.

$$\bar{k}_{(tR)} = \frac{1}{1 + N_{mpl}/N_{tR}} \left\{ \frac{N_{tR} \dfrac{dN_{mpl}}{dt_R} - N_{mpl} \dfrac{dN_{tR}}{dt_R}}{N_{tR}^2} \right\}$$

Now, $\dfrac{dN_{mpl}}{dt_R} = - \dfrac{dN_{tR}}{dt_R}$ and $N_{tR} \gg N_{mpl}$, hence

$$\bar{k}_{(tR)} = \frac{1}{1 + M_{mpl}/N_{tR}} \cdot \frac{1}{N_{tR}} \cdot \frac{dN_{mpl}}{dt_R} \tag{3.42}$$

N_{mpl} and N_{tR} can be measured as a function of $t_R \cdot N_{mpl}$ is replaced by the measured ion current of the metastable plateau peak i_{mpl}, and N_{tR} is replaced by the measured parent ion current i_p plus the normal metastable ion current $i_{mn} \cdot t_R$ can be calculated for a given retarding potential and plateau length d:

$$t_R = t_c + (t_A - t_c) + (t_E - t_A) + (t_R - t_E) \quad \text{(see fig. 3.23.)} \tag{3.43}$$

t_c is given by eqn. (3.18a), $t_c = t_{(R_0)}$, and

$$t_A - t_c = \frac{2d'}{v'^2}(v_0 - v) \tag{3.44}$$

where $d' = x_A - x_c$, $v_0^2 = \frac{2}{M} e U_0$ where v_0 is the initial velocity of the ions when entering the retarding potential, $v'^2 = \frac{2eU_g}{M}$, and $v^2 = \frac{2e(U_0 - U_g)}{M}$. Also

$$t_E - t_A = \frac{d}{v} \tag{3.45}$$

and

$$t_R - t_E = t_A - t_c \tag{3.46}$$

From the measured i_{mpl} values at different retarding potentials one can draw a plot i_{mpl} versus t_R. By differentiating this curve with respect to t_R one obtains $\frac{d}{dt_R} i_{mpl}$. This is inserted into eqn. 3.42, and with $N_{tR} = N_p + N_{mn}$ one obtains the average rate constants:

$$\bar{k}_{(tR)} = \frac{1}{i_p + i_{mpl} + i_{mn}} \cdot \frac{d}{dt_R} i_{mpl} \tag{3.48}$$

Results for n-heptane are discussed in section 3.4.2.5.

b) Normal metastable peaks

From eqn. (3.40) one obtains for the normal metastable ions

$$\int_{t_R}^{t_M} \bar{k}_{(t)}\, dt = \ln \left\{1 + \frac{N_{mn}}{N_{(tM)}}\right\} \tag{3.49}$$

where t_M is the time when the ions enter the magnetic sector field. The difficulty in evaluating this equation lies in the fact that both the upper and the lower limit of the integral vary with varying retarding potentials. This difficulty can be overcome by a little experimental trick as pointed out by Tenschert [130]. The total voltage between the field anode and the cathode is always kept constant at 10 kV as before, but with increasing retarding potential both the anode and the cathode potentials are increasingly lowered with respect to ground as shown in fig. (3.26.).

By this lowering of potentials — which amounts at the maximum to a few percent of the total voltage — the shortening of residence time $t_M - t_R$ with increasing retardation of the normal metastable ions is compensated by the somewhat smaller velocity of the ions.

Thus $t_M = t_R + \Delta t$, where $\Delta t = const.$
Eqn. (3.49) may now be written as

$$\int_{t_R}^{t_R+\Delta t} \bar{k}_{(t)}\, dt = \ln \left\{1 + \frac{N_{mn}}{N_{(tR+\Delta t)}}\right\} \tag{3.50}$$

$$\bar{k}_{(tR+\Delta t)} - \bar{k}_{(tR)} = \left\{\frac{1}{1 + \frac{N_{mn}}{N_{(tR+\Delta t)}}}\right\}$$

$$\left\{\frac{N_{(tR+\Delta t)} \cdot \frac{dN_{mn}}{dt_R} - N_{mn} \cdot \frac{dN_{(tR+\Delta t)}}{dt_R}}{N^2_{(tR+\Delta t)}}\right\}$$

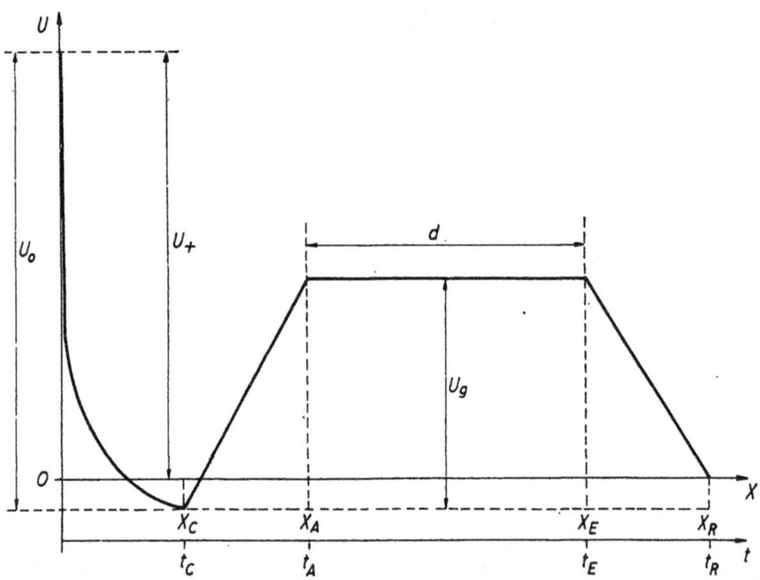

FIG. 3.26. Potential distribution in a FI source with retarding electrodes. Anode and cathode potentials are lowered with respect to ground in order to keep $t_M - t_R$ constant.

With $\frac{dN_{mn}}{dt_R} = -\frac{dN_{(tR+\Delta t)}}{dt_R}$ and $N_{(tR+\Delta t)} \gg N_{mn}$, one obtains

$$\bar{k}_{(tM)} = \frac{1}{N_p + N_{mn}} \cdot \frac{dN_{mn}}{dt_R} + \bar{k}_{(tR)} \tag{3.51}$$

As before, the numbers of parent ions, N_p and metastable ions N_{mn} may be replaced by the measured parent and normal metastable ion currents:

$$\bar{k}_{(tM)} = \frac{1}{i_p + i_{mn}} \cdot \frac{di_{mn}}{dt_R} + \bar{k}_{(tR)} \qquad (3.52)$$

Thus, the average rate constant for the time when the ions enter the magnetic field can be calculated as the sum of the previously calculated rate constant $\bar{k}_{(tR)}$ for the metastable plateau decay at tR, and the derivative of the plot i_{mn} versus tR.

3.4.2.6. *Rate constants in the range between* 10^{-11} *and* 10^{-6} *sec*

The average rate constants have been measured by Tenschert [128, 130] for a number of polar and non-polar substances with a variation of the ion residence times by about five orders of

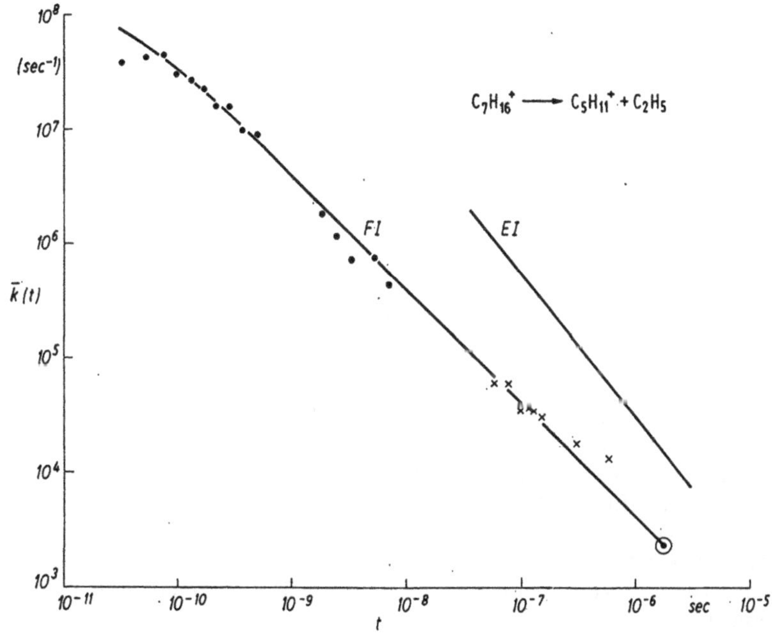

FIG. 3.27. Average rate constants $\bar{k}_{(t)}$ for the decomposition of n-heptane as a function of the ion residence time. $\bar{k}$ values derived from: ··· broad fast metastable peak, **** metastable plateau peaks, ⊙ normal metastable peak. For comparison: $\bar{k}_{(t)}$ values for the case of electron impact as derived by Ottinger [123].

magnitude. The results for n-heptane will be discussed first because measurements with the electron impact method made by Ottinger [123] were available for comparison.

Fig. 3.27. shows the average rate constants $\bar{k}$ for n-heptane as a function of the ion residence time. The points for the curve were obtained from the three different techniques described above, i.e. from the fast metastable peak, the metastable plateau peak

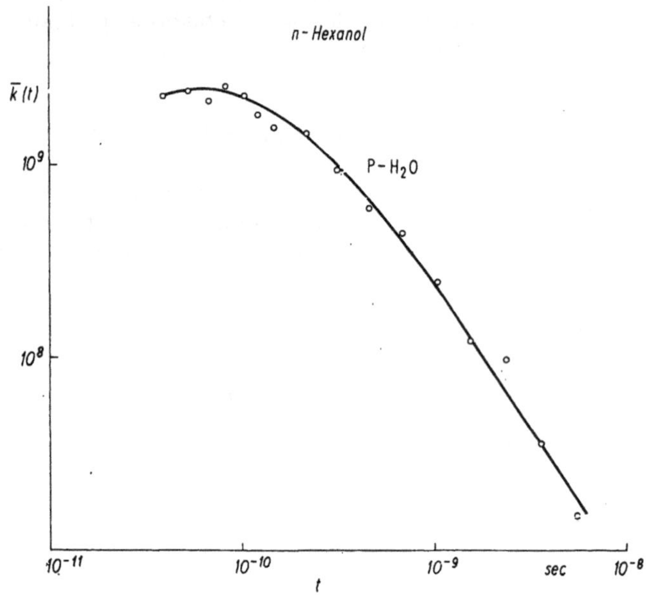

FIG. 3.28. Average rate constants $\bar{k}_{(t)}$ for the water elimination from n-hexanol as a function of the ion residence time.

and the normal metastable peak. It is remarkable that the points originating from all of these different techniques fit well together, resulting in a smooth curve. This gives one some confidence that effects of different transmission of fragment ions formed at different places are of minor importance. This might be partially due to the fact that the absolute fragment ion intensities do not enter into the equations, but only the ratio of the fragment and parent ion intensities.

The time is plotted logarithmically in fig. 3.27. for convenience of further theoretical treatment.

The most striking result consists in the finding of a continuous variation of the average rate constants $\bar{k}$ over about four orders of magnitude within the time interval of 5×10^{-11} to 5×10^{-6} sec.

From this a quasi-continuous distribution of the individual rate constants can be concluded [124], in qualitative agreement with the statistical theory of mass spectra.

The curve for the average rate constants under electron impact conditions [123], which is also shown by fig. 3.27., is largely parallel

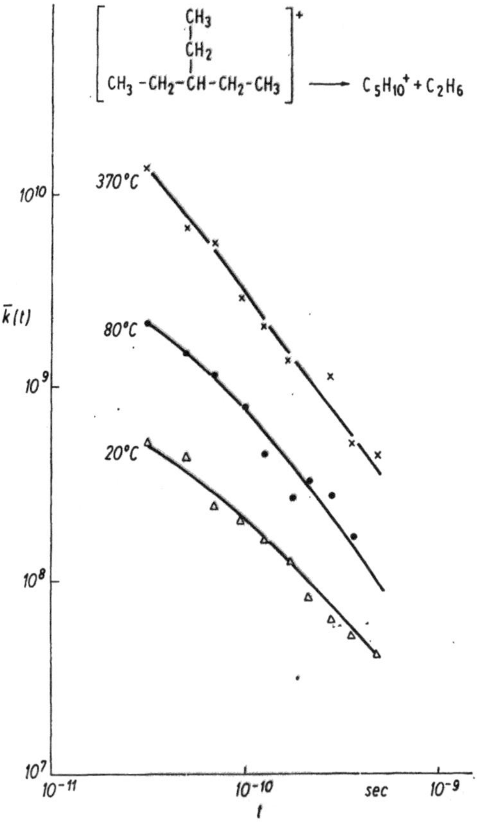

FIG. 3.29. Temperature dependence of the average rate constants $\bar{k}$ for the decomposition of 3-ethyl pentane.

with the curve derived under field ionization conditions. The EI curve is shifted towards higher rate constants, as compared with the FI curve. This is again in qualitative agreement with the statistical theory because the rate constants are strongly dependent on the internal energy transferred to the molecules. This energy is much larger in the case of impact of 30 eV electrons than in the case of field ionization.

Fig. 3.28. shows the distribution of the average rate constanst for water elimination from n-hexanol. Only the fast metastable processes have been evaluated here. It is seen from the figure that the rates are about 50 times larger than those measured for n-heptane. Moreover, the figure gives an indication that there seems to exist a maximum rate constant for the elimination of a water molecule from n-hexanol at a residence time of about 7×10^{-11} sec. This is in accordance with the observed maximum of the fast metastable peak in the analogous n-pentanol mass spectrum as shown in fig. 3.21.a. and the n-hexanol spectrum (fig. 3.22.a.).

The strong dependence of the average rate constants on the internal energy can also be seen from the strong dependence of the rate curves for field ionization on the temperature (see fig. 3.29.)

3.5. TEMPERATURE DEPENDENCE OF FI MASS SPECTRA

The temperature dependence of FI mass spectra has been investigated in detail by Knöppel and the author [90, 131]. The temperature of the FI tip was determined by measuring the voltage drop across the electrically heated wire loop carrying the tip.

The intensities of the fragment ions show a pronounced temperature dependence, see fig. 3.30. The temperature dependence is represented as plots of $\log i_{fr}/\Sigma i$ versus $1/T$, where i_{fr} are the fragment ion intensities, Σi is the total ion intensity and T the absolute temperature. It was pointed out earlier that one can differentiate between the very fast decomposition processes occurring in the region of extremely high electric fields within a time interval Δt_0, and delayed decomposition processes at moderate fields. The first type of decomposition processes occurring within a few times 10^{-14} to 10^{-12} sec is characterized by the height of the fragment lines. (The interval of lifetimes Δt_0 corresponds to the "natural line width".) The delayed processes lead to broadened fragment peaks; the fragment peak area is characteristic for these processes. Only the temperature dependence of processes occurring within Δt_0 will be discussed here, and hence the peak heights are plotted in the figures (if not noted otherwise). Fig. 3.30. shows for the case of n-hexane that the fragment ion intensity increases approximately exponentially between 100 and 560 °C, but that the average fragment ion intensity is approximately constant or slightly decreasing with temperature between room temperature and 100 °C. This behaviour near room temperature is due to adsorbed layers on the FI tip, as will be discussed later.

At temperatures above 100 °C the adsorbed multilayers are largely removed, and therefore dissociation within Δt_0 will be treated

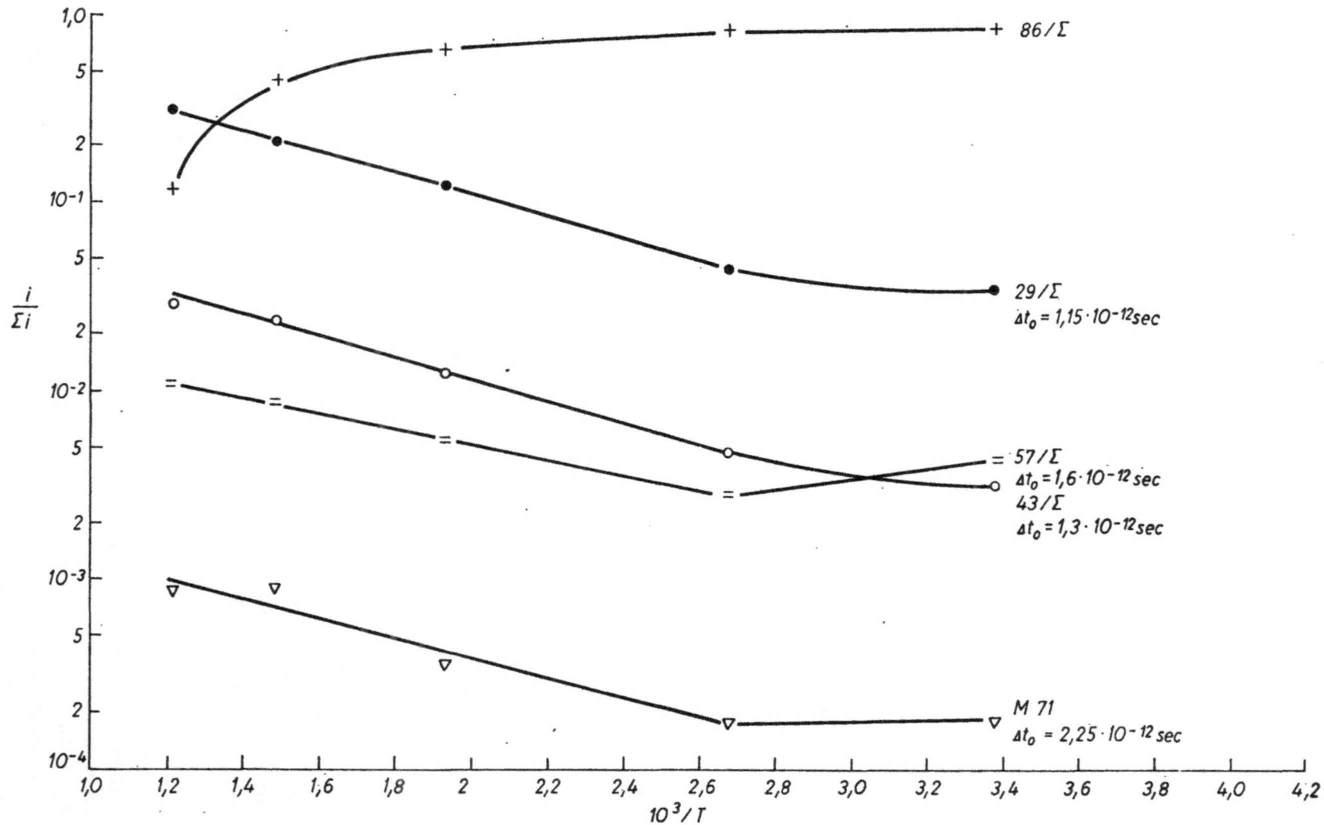

Fig. 3.30. Temperature dependence of the relative ion intensities of n-hexane.

as gas phase dissociation. The effect of dissociation within adsorbed layers at room temperature will be discussed afterwards. The temperature dependence of field dissociation of organic ions in the gas phase will be discussed in terms of the semi-classical model treated in section 3.1.2.

First it will be assumed that field dissociation occurs within one vibrational period after field ionization, i.e. without fluctua-

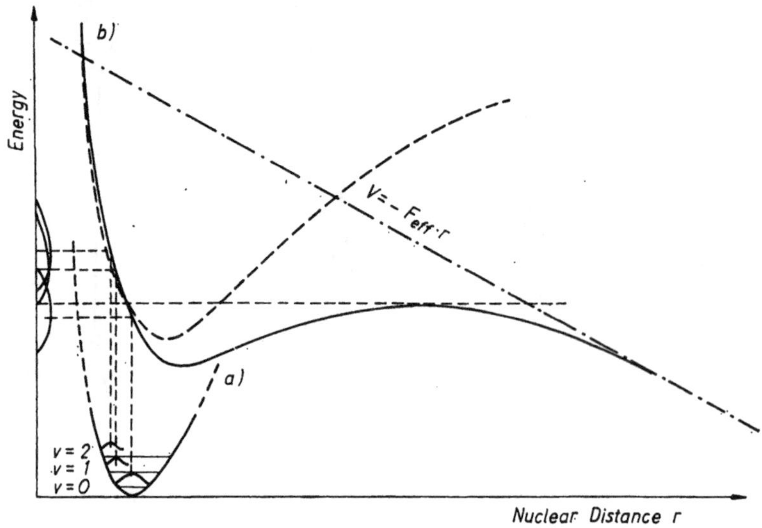

FIG. 3.31. Potential curve of a bond in a molecule.

a) prior to, *b*) after field ionization. (Dashed curve: ionic state without field.) The Franck-Condon factors for transition from the different vibrational levels of the neutral to the ionic state are indicated schematically on the energy axis.

tion of vibrational excitation energy. Fig. 3.31. shows the potential curves of the neutral and the ionized ground states of a molecule in a high electric field. The figure represents an intersection through the multi-dimensional potential hypersurface along the reaction coordinate. Ionization occurs according to the Franck-Condon principle. The energy difference for a vertical transition is overcome by the electrical field. (The neutral and ionic curves are drawn separately, as they are for the case of electron impact ionization. The drawing of crossed ionic and neutral curves for the case of field ionization is avoided, as discussed in section 3.4.2.1., fig. 3.18.)

Several vibrational quanta are transferred to the molecule by the vertical transition from the neutral to the ionic state if the ionic curve is shifted to larger bond lengths as compared with the neutral curve. Ions excited to a state above the dissociation limit can decompose within one vibrational period. Molecules excited to a state below the dissociation limit cannot dissociate at all as long as the assumption made above is valid, which excludes fluctuation of energy across the different vibrational modes. (The very small contribution of tunneling of radicals will be neglected here.)

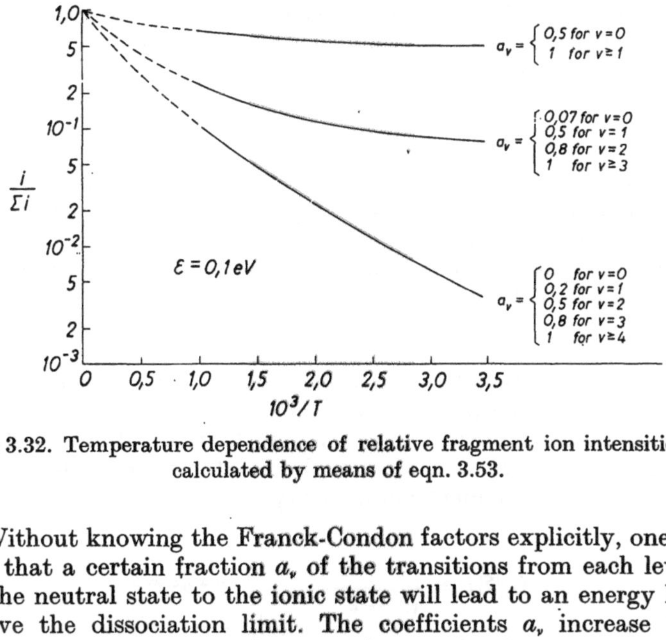

FIG. 3.32. Temperature dependence of relative fragment ion intensities as calculated by means of eqn. 3.53.

Without knowing the Franck-Condon factors explicitly, one can say that a certain fraction a_ν of the transitions from each level ν of the neutral state to the ionic state will lead to an energy level above the dissociation limit. The coefficients a_ν increase with increasing excitation of the neutral state if only transitions between the repulsive part of the curve are considered first (which is a simplifying condition). Then

$$i_{fr}/\Sigma i = (1 - e^{-\varepsilon/kT}) \sum_{\nu=0}^{\infty} a_\nu \, e^{-\nu\varepsilon/kT} \tag{3.53}$$

where ε is the vibrational energy quantum of the neutral ground state.

Fig. 3.32. shows some curves log i_{fr}/Σ_i versus $1/T$ for different arbitrarily assumed distributions of the a_ν — factors according to eqn. (3.53). All these curves are "curved downwards". The tangents of these curves at $1/T > 0$ intersect the ordinate at $i_{fr}/\Sigma_i < 1$. This is in contrast to the experimental curves like that for n-hexane as shown in fig. 3.30. For the fragment $m = 29$

a tangent can be drawn to the curve at $1/T > 0$ which intersects the ordinate (drawn through $1/T = 0$) at $i_{fr}/\Sigma_i = 1$. This discrepancy between eqn. (3.53) and the experiment can be explained by the idea that — in contrast to the assumption underlying (3.53) — fluctuation of vibrational energy through the molecule occurs. This will now be discussed.

The molecule is regarded as a system of loosely coupled harmonic oscillators. The rate constant for a given dissociation with activation energy E_0 and excitation energy E is given by the first approximation of the quasi-equilibrium-theory (QET) by

$$k = \nu\,[(E - E_0)/E]^{s-1} \qquad (3.54)$$

where ν is a frequency factor of the order 10^{13}—10^{14} sec^{-1}, and s is the effective number of oscillators. (The improved rate expression obtained by Marcus [132] by more exact counting of the states of the molecules will not be discussed here.)

The distribution of energies transferred to the molecules during field ionization is not known at present. Therefore an assumption must be made: The energy E_c is transferred to some of the coupled vibrational modes the number of which is k. The rest of the modes, $s - k$, are vibrationally excited before and after ionization, corresponding to the tip temperature.

With the wellknown expression for the thermal energy distribution of a $(s - k)$ — dimensional oscillator system, one obtains for the relative fragment ion intensity

$$i_{fr}/\Sigma i = \frac{1}{(s-k-1)!} \int\limits_{Max\,(x_0 - x_c,0)}^{\infty} e^{-x}\, x^{s-k-1} \times \left\{1 - exp\left[-\nu t_0 \left(\frac{x + x_c - x_0}{x + x_c}\right)^{s-1}\right]\right\} dx \qquad (3.55)$$

where $x_c = E_c/kT$, $x_0 = D/kT$ and $D =$ dissociation energy. $Max\,(x_0 - x_c, 0)$ means the greater of the numbers $x_0 - x_c$ or zero.

Fig. 3.33. shows the calculated curves $\log\,(i_{fr}/\Sigma i)$ versus $1/T$ according to eqn. (3.55) for $s = 15$ and $k = 3$, and for some values of D and E_c. $k = 3$ was chosen because the positive charge of n-paraffins at field strengths of several 10^7 V/cm is distributed mainly over three bonds, according to Lorquet and Hall [92]. Tangents can be drawn to all curves at $1/T > 0$ which intersect the ordinate at $i_{fr}/\Sigma_i = 1$. This is in accordance with the experimental curves for the most intense fragment ions (see fig. 3.30.). This agreement indicates that — at least at high temperatures — field dissociation occurs predominantly after fluctuation of energy. These processes occurring within $\Delta t_0 \simeq 10^{-12}$ sec may be termed "field induced statistical dissociations". In these cases statistical dissociation occurs preferentially at a specific bond where the positive charge is concentrated under the influence of the electric field.

A more detailed agreement between the experimental and the theoretical curves cannot be expected because of the numerous

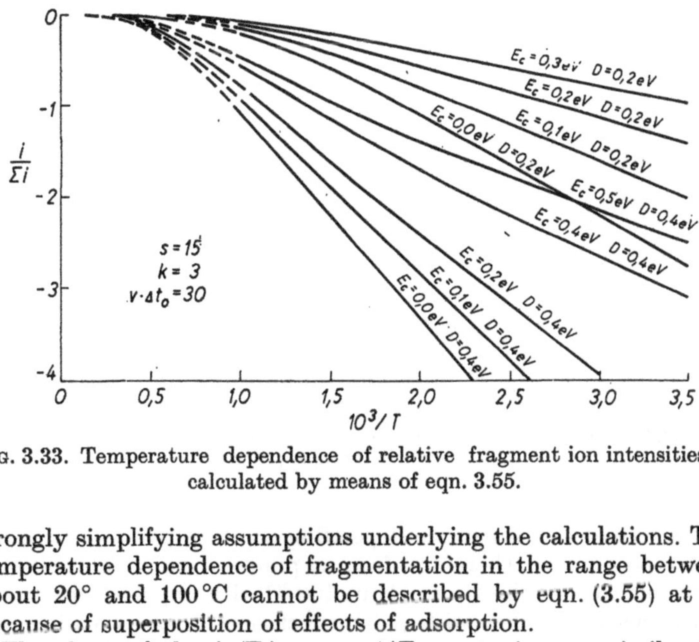

FIG. 3.33. Temperature dependence of relative fragment ion intensities as calculated by means of eqn. 3.55.

strongly simplifying assumptions underlying the calculations. The temperature dependence of fragmentation in the range between about 20° and 100 °C cannot be described by eqn. (3.55) at all because of superposition of effects of adsorption.

The slope of the $i_{fr}/\Sigma i$ versus $1/T$ curves is very similar for different fragment ions of n-hexane and also of other substances. This means that the activation energies for the different fragmentation processes are rather similar. Then the great differences of the relative fragment ion intensities can be explained only by the assumption that they correspond to the different ionization probabilities of different molecular orbitals of the molecule. This effect has been discussed in greater detail elsewhere [131].

The effect of an adsorbed layer on the temperature dependence of the FI mass spectrum is very pronounced with strongly polar substances such as diethyl ketone (see fig. 3.34.). The large permanent dipole moment of these substances causes a strong increase of particle density in front of the FI tip so that a condensed layer on the tip governs the fragment formation up to temperatures of 230 °C. Because of removal of this layer with increasing temperature, the relative fragment intensities decrease from 20°—230 °C. Neutral fragments formed during field dissociation of field ionized molecules are present in the adsorbed layer and may be field onized after a certain residence time at the tip surface.

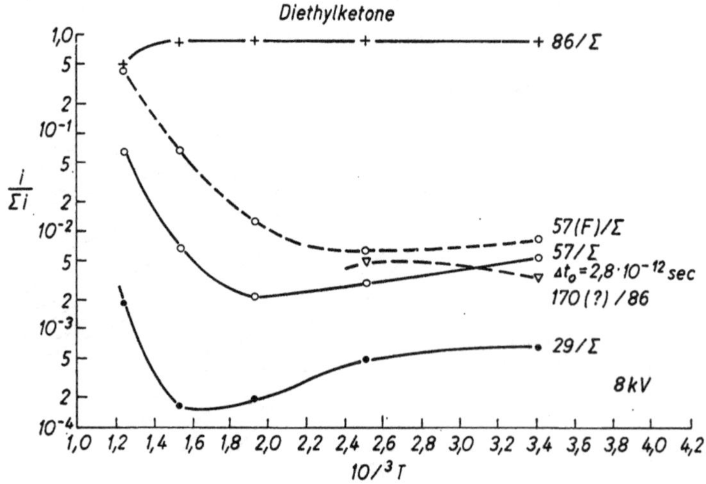

FIG. 3.34. Temperature dependence of the relative ion current in the FI mass spectrum of diethyl ketone.

The extent of adsorption, depending on many parameters, is not well reproducible. Therefore, from a practical point of view, it is advisable to heat the tip to about 100 °C with unpolar substances, and somewhat higher with polar substances in order to increase the reproducibility of the FI mass spectra and to decrease the short time fluctuations of the ion current.

The orientation of some molecules like n-pentanol influences the temperature dependence of the relative fragment ion intensity in a characteristic manner, as is discussed in detail by Knöppel [90, 131].

3.6. PROCESSES IN ADSORBED LAYERS ON FI EMITTERS

A variety of processes occur in the adsorbed mono- or multilayers on FI emitters, such as proton transfer reactions, reactions of radicals, association of molecules and charge transfer reactions.

3.6.1. Formation of Protonated Molecules

Strong evidence for reactions of field ionized molecules with layers adsorbed on FI emitters is provided by observation of protonated molecules (parent — plus — one peaks) in the FI mass spectra of many substances. One of the first processes of this type reported was the formation of ArH^+ which was observed in the FI mass spectrum of a mixture of water and argon [4]. Presently, the mechanism $Ar + H_3O^+ \rightarrow ArH^+ + H_2O$ (3.56) is assumed. A similar reaction is also observed in the EI mass spectrum of a mixture of argon and hydrogen [4]. However, this is a gas phase reaction, and the ratio ArH^+/Ar^+ is of the order of 10^{-3} whereas ArH^+/Ar^+ ratios greater than 10^3 can be observed in the FI mass spectra of water — argon mixtures on tungsten tips at moderate field strengths. The increase of the ArH^+/Ar^+ ratio by a factor of about 10^6 clearly indicates that the ArH^+ formation does not occur in the gas phase but on an adsorbed water layer on the FI tip. This is supported by the fact that a rise of the tip temperature from 300° to 400 °K reduces the ArH^+ intensity by a factor of about 100, and that the angular distribution of the ions changes appreciably [25].

The adsorption of mono- or multilayers is favoured by the increased density of particles near the FI tip due to the high electric field. In the case of complete equilibrium between the neutral particles approaching and leaving the tip (i.e. if the flux of neutral particles is very large as compared with the flux of ions produced at the tip), the density n_t near the tip is given by a Boltzmann factor

$$n_t = n_g \cdot e^{E_p/kT} \tag{3.57}$$

where

$$E_p = \frac{\alpha}{2} F^2 + \bar{\mu} F \tag{3.58}$$

(α = polarizability, $\bar{\mu}$ = average permanent dipole moment in the field direction, n_g = gas density at zero field far from the tip). For molecules having a large dipole moment like water, $\mu = 1.84 \times 10^{-18}$ e.s.u., and a polarizability $\alpha = 1.57 \times 10^{-24}\ \mathrm{cm}^3$ one obtains very large Boltzmann factors. At $F = 1 \times 10^8$ V/cm one obtains $e^{E_p/kT} = 10^{7.35}$ for water. Of course, eqn. (3.57) is no longer valid if the saturation vapour pressure is exceeded, as will be discussed in the next section. The proton attachment is strongly dependent on the field strength. At moderate fields the system is near to equilibrium conditions, and thus the increase of particle density according to eqn. (3.57) leads to strong adsorption and hence proton transfer by mechanisms similar to (3.56). At higher fields, however, the system is far from equilibrium conditions because of the very high rate of field ionization. The adsorbed layer is removed with increasing fields, and hence the proton attachment is strongly reduced. This is shown by fig. 3.35. which shows the strong variation of the $(P + 1)^+/P^+$ ratio ($m = 33/m = 32$) in the FI mass spectrum of methanol with the applied voltage [134]. Strong proton attachment occurs with many other

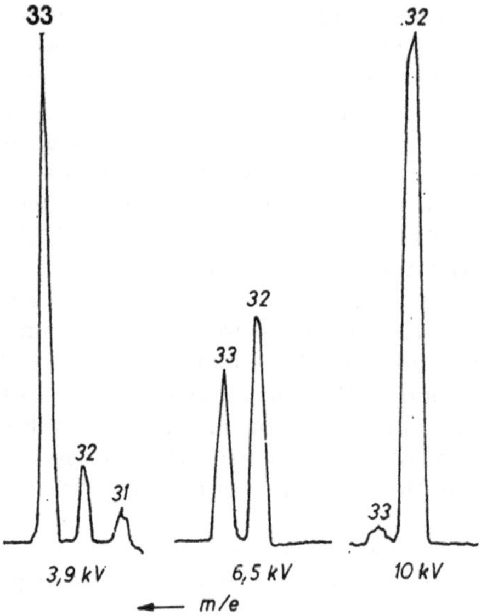

FIG. 3.35. FI mass spectrum of methanol on a tungsten tip at different tip cathode voltages.

substances in the presence of water. Thus N_2H^+ ions are formed from nitrogen (fig. 3.36.). Water ions appear predominantly as H_3O^+ ions ($m = 19$) and only to a small extent as H_2O^+ ($m = 18$) on tungsten FI tips [4]. The absolute water ion intensity is smaller on platinum FI tips, and the H_3O^+/H_2O^+ ratio is much smaller than on tungsten tips. This is the consequence of the different adsorption conditions for water on tungsten and platinum.

$(P + 1)^+$ ions are also found in the FI mass spectrum of methane, the intensity of which exceeds the ^{13}C contribution to the methane parent peak [105]. These ions are formed by the reaction

$$CH_4^+ + CH_4 \rightarrow CH_5^+ + CH_3 \qquad (3.59)$$

Addition of water increases the CH_5^+ intensity condiderably. This process was studied in greater detail by Block [123]. Strongly protonated peaks are formed by ethane in the presence of water:

$$C_2H_6 + H_3O^+ \rightarrow C_2H_7^+ + H_2O \qquad (3.60)$$

Proton transfer is also pronounced in the FI mass spectra of molecules having a strong dipole moment, like alcohols, amines, ethers, ketones, acids, esters, etc. These strong $(P + 1)^+$ peaks are

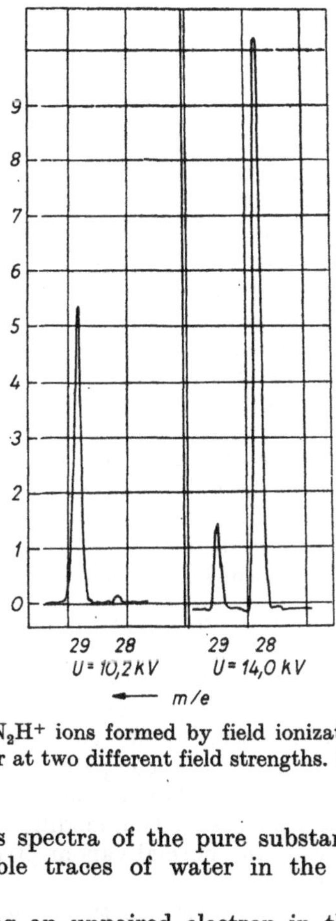

FIG. 3.36. N_2^+ and N_2H^+ ions formed by field ionization of nitrogen with traces of water at two different field strengths. (Tungsten tip.)

found in the mass spectra of the pure substances, even if there are no measureable traces of water in the background mass spectrum.

Molecules having an unpaired electron in the neutral ground state, like NO, ClO_2, NO_2 have a much smaller tendency of $(P + 1)^+$ formation because the unpaired electron is removed by field ionization. In the ionized molecules one unpaired electron is missing, and this would be required for formation of an electron pair valence bond with a hydrogen atom.

3.6.2. *Field Condensation and Ion Cluster Formation*

Gomer [16] has pointed out that multilayer adsorption may occur under certain conditions if the gas pressure — enhanced by polarization effects, see eqn. (3.61) — approaches the vapour pressure, which in turn is modified by the electric field. "Field condensation" may occur if the gas pressure tends to exceed the modified vapour pressure of the pure liquid.

Condensation is a cooperative phenomenon which occurs over a very narrow pressure range. Therefore, Gomer points out that field condensation can be taken as a step function of F. Field ionization is virtually also a step function of F. Let us assume first that the critical field strength for field condensation, F_c, is smaller for a certain substance than the threshold field strength F_i, for field ionization. (Qualitatively speaking, this condition would be applicable to a substance of low volatility and relatively high ionization potential, like water.) Gomer [16] discusses the change of saturation vapour pressure with field strength.

A consideration of that point based on investigations of Metzinger [135] and more detailed recent theoretical studies of the author may be presented here. An analysis will be made of whether true field condensation can occur at all, i.e. whether the actual pressure in the presence of a high field can exceed the modified vapour pressure of the liquid. Because of the condition $F \leqq F_i$ imposed at the first stage of this analysis, complete equilibrium between the gas at large distances from the emission tip and at the tip surface can be assumed. Therefore

$$P_t = P_g \, exp \, (E_{pg}/kT) \tag{3.61}$$

for pressures P_t at the tip surface which are below the vapour pressure of the liquid film. P_g is the gas pressure far from the tip. E_{pg} is the polarization energy of the gas at the tip surface.

The gas pressure P_t must be compared with the theoretically derived vapour pressure of a liquid film formed in the presence of a high field at a tip surface. This modified vapour pressure P_s can be correlated with the normal vapour pressure P_{s0} by means of two correction terms:

$$P_s = P_{s0} \cdot exp \left\{ \frac{E_{pg} - E_{pl}}{kT} + \frac{2\,M\,\sigma}{r_t k T \rho_l} \right\} \tag{3.62}$$

The first term in the exponent presents the difference of polarization energies in the gas phase and the liquid, the second term the vapour pressure enhancement due to the curvature of the surface. (σ = surface tension, M = molecular weight, ρ_l = density of the liquid, r_t = tip radius.) The second term represents only a correction of about 1% for a tip radius of 1000 Å and substances like water or acetone. This term may be neglected. Thus,

$$P_s = P_{so} \cdot exp \left\{ \frac{E_{pg} - E_{pl}}{kT} \right\} \tag{3.63}$$

where

$$E_{pg} = \frac{\alpha}{2} F_0^2 + \int_0^{F_0} \mu L_{\left(\frac{\mu F}{kT}\right)} dF \tag{3.64}$$

$$E_{pl} = \int_0^{E_0} P dE = \int_0^{E_0} \frac{(\varepsilon - 1) v_l E dE}{4 \pi N_L} \tag{3.65}$$

(P is the "molar polarization".)

F_0 is the field strength in the gas phase at the surface of the iquid film. $L_{(x)}$ is the Langevin-function, $L_{(x)} = \coth x - \frac{1}{x}$, $x = \frac{\mu F}{kT}$, ε is the dielectric constant of liquid, v_L the molar volume of the liquid, N_L the Avogadro number and E the macroscopic field strength in the dielectric medium. ($D = \varepsilon E$, where D = "dielectric displacement".)

The actual permanent dipole moment of a molecule in the gas phase, μ, has been replaced by μL in (3.64), and the dielectric constant has to be regarded as field dependent in (3.65) because possibly dielectric saturation effects could be of importance at the extremely high field strength considered here. These effects will now be discussed.

Debye [136] has treated the phenomenon of dielectric saturation effects at extremely high field strengths. The average dipole moment in the field direction in the gas phase is smaller than μ because of thermal disorientation. The average moment in the field direction is $\mu L_{(x)}$. However, at 5×10^7 V/cm the dipoles of strongly polar substances like acetone ($\mu = 2.8 \times 10^{-18}$) are ordered nearly completely parallel with the electric field lines in the gas phase. (For acetone: $\frac{\mu F}{kT} \simeq 12$ and $L_{(12)} = 0.92$.) Thus the second term of 3.64 may be approximated by μF_0 at very high field strengths:

$$E_{pg} \simeq \frac{\alpha}{2} F_0^2 + \mu F_0 \tag{3.64a}$$

The dielectric susceptibility $\frac{\varepsilon - 1}{4\pi}$ must decrease if the permanent dipoles of a liquid substances are ordered more and more parallel with the electric field. The dielectric constant ε of the liquid decreases appreciably if the field strengths is increased to values where the Langevinfunction tends to approach the limiting value $L_{(x)} = 1$. The magnitude of the field strengths where ε drops noticeably will now be analysed.

The Debye equation

$$\left[\frac{\varepsilon - 1}{\varepsilon + 2}\right] \frac{3 v_l}{4\pi N_L} = \alpha + \frac{\mu}{E} L_{(\mu E/kT)} \qquad (3.66)$$

(E = field strength)
is applicable only to polar substances in the gas phase or in diluted solutions of nonpolar solvents, but fails in case of pure polar substances in the liquid state because the hypothesis of Clausius and Mosotti does not hold (see Debye [136]).

Onsager [137] has developed a theory of dielectric polarisation applicable to polar molecules like acetone. The fundamental equation of Onsager, modified for extremely high external field strengths is

$$\frac{(\varepsilon - 1)(2\varepsilon + 1)}{12\pi \varepsilon N_L} v_l = \frac{\alpha}{1 - f\alpha} + \frac{1}{(1 - f\alpha)^2} \frac{\mu L_{\mu(\mu E_k/kT)}}{E_k} \qquad (3.67)$$

where

$$E_k = \frac{3\varepsilon E}{(1 - f\alpha)(2\varepsilon + 1)}$$

and f is a factor depending on ε and the average volume per molecule, $f\alpha \simeq 0.2 - 0.3$.

Up to $\mu E_k/kT = 1$ the Langevin-function may be approximated within a few percent by $L_{\mu(\mu E_k/kT)} \simeq \mu E_k/3kT$, and then ε is independent of the field strength. ε decreases at higher values of $\mu E_k/kT$. Substituting $E_k = \frac{1}{(1 - f\alpha)} \left(\frac{3\varepsilon E}{2\varepsilon + 1}\right)$ and replacing εE by $D \equiv F_0$ and setting $F_0 = 5 \times 10^7$ V/cm, one obtains for acetone ($\varepsilon = 21.4$)

$$\frac{\mu E_k}{kT} \simeq 1$$

This means that the dielectric constant of liquid acetone is virtually independent of the field strength up to 5×10^7 V/cm. Even if some correction terms for the Onsager equation would be required at these extremely high field strengths, the results obtained here should be qualitatively preserved.

In the case of water, alcohols and other associating polar liquids the Onsager equation needs further corrections. These have been introduced by Kirkwood [138]. For water one has to use a modified dipole moment $\mu' \simeq 1.26\,\mu$, and the Langevin-function has to be multiplied by a factor $g \simeq 2.5$. One again arrives at the conclusion that dielectric saturation becomes noticeable only if the argument of the Langevin-function is larger than one. Inserting the values for water at 5×10^7 V/cm, one obtains

$$\frac{\mu E_k}{kT} \simeq 0.21$$

In other words, dielectric saturation of water starts at about five times larger field strengths than in the case of acetone. $\varepsilon = 80$ may be used for water up to about 3×10^8 V/cm.

These results, which have shown the constancy of ε for highly polar substances like water and acetone at $F_0 = 5 \times 10^7$ V/cm simplify the evaluation of eqn. (3.65):

$$E_{pl} = \frac{(\varepsilon - 1)\, v_l E^2}{8\pi N_L} = \frac{(\varepsilon - 1)\, v_l}{8\pi N_L} \frac{F_0^2}{\varepsilon^2} \tag{3.65a}$$

because of the relation $\varepsilon F = D \equiv F_0$.

We now return to the problem whether field condensation of a polar substance could occur at a field strength F_c slightly below the FI threshold field. ($F_c \simeq 5 \times 10^7$ V/cm, for example.) Field condensation can occur if the gas pressure P_t at the tip surface exceeds the modified vapour pressure of the liquid film on the tip (eqn. 3.63):

$$P_t = P_g\, exp\,(E_{pg}/kT) \geqq P_s = P_{s0}\, exp\,[E_{pg}/kT - E_{pl}/kT]$$

or

$$P_g \geqq P_{s0}\, exp\,(- E_{pl}/kT) \tag{3.68}$$

where E_{pl} is given by (3.65a).

Gomer [16] points out that the term $exp\,(E_{pg}/kT)$ in the equation for the vapour pressure (3.62) must be omitted if the ionization probability in the gas phase above the liquid is small, and if the field decays to a small value in a distance short compared to the mean free path of a molecule. (This is because the evaporating molecules arrive in a region where $E_{pg} \simeq 0$ without previous collisions.) This condition is met at $F \leqq 2 \times 10^7$ V/cm at $p_g < 10^{-3}$ Torr. Under these conditions, the vapour pressure P_s at distances comparable to the mean free path must be compared with the gas pressure at that distances. P_t is approximately

P_g since the factor exp $(E_{pg}/kT) \simeq 1$ at a distance λ from the tip. Thus eqn. (3.68) remains valid, independently of the mean free path.

The minimum gas pressure P_g (far from the tip) for field condensation as calculated by means of eqn. (3.68) at $F_0 = 5 \times 10^7$ V/cm is 17.3 Torr for water ($P_{so} = 17.5$ Torr), and 143 Torr for acetone ($P_{so} = 165$ Torr). These pressures are orders of magnitude larger than the typical ion source pressure of 10^{-3} Torr. Field condensation cannot occur under these conditions, neither in the case of water and acetone nor in the case of any other substance having a vapour pressure $> 10^{-2}$ Torr at room temperature. It should be remembered here that the term "field condensation" is used for a gas pressure which is higher than the modified vapour pressure and a liquid film thickness of many layers, so that the normal value of ε can be used.

A phenomenon distinct from normal "field condensation" is the "multilayer adsorption" which can occur at pressures much below the vapour pressure of a liquid, according to the theory of Brunauer, Emmett and Teller [140] (BET). Under conditions where only one or a few layers of a substance are adsorbed on the surface of an FI emitter, eqn. (3.68) has to be modified:

$$P_t = P_g \, exp \, (E_{pg}/kT) \geqq P_s = P_{so} \, exp \, [E_{pg}/kT - E_{pads}/kT],$$

or

$$P_g > P_{so} \, exp \, (-E_{pads}/kT) \tag{3.68a}$$

Now, the depolarization field of a macroscopic liquid (characterized by a dielectric constant ε) vanishes if the layer is decreased to a few Å. Therefore, the polarization energy of such a thin film in a high electric field, E_{pads}, can be approximated by E_{pg} as suggested by Röllgen [139]. Thus we finally arrive at the condition

$$P_g \geqq P_{so} \, exp \, (-E_{pg}/kT) \tag{3.68b}$$

for multilayer formation, or $P_t \geqq P_{so}$. A numerical example for the case of water ($\alpha = 1.5 \times 10^{-24}$ cm^3, $\mu = 1.8 \times 10^{-18}$ e.s.u.) shows that multilayer formation can occur at $F = 7 \times 10^7$ V/cm, for example. The value of $exp \; E_{pg}/kT$ is 4.8. With the relation $P_t = P_g \, exp \, E_{pg}/kT$ we obtain with $P_g = 10^{-3}$ Torr (ion source pressure) that $P_t \simeq 63$ Torr, whereas the vapour pressure P_{so} of water at room temperature is only 17 Torr.

The assumptions underlying eqn. (3.68b) are only of approximate nature. However, the experimental observation of ion clusters $H_3O^+(H_2O)_n$, $n = 1, 2, 3 \ldots$ show that the conditions for BET-multilayer formation derived here must be qualitatively right. The multilayers are assumed to be formed on relatively smooth tips under the conditions discussed here.

In the case of strongly activated emitters having a very rough and porous structure of organic micro needles, multilayer formation is favoured even at lower field strengths as indicated above. According to considerations of Röllgen [139], a phenomenon related to the wellknown capillary condensation may occur under these circumstances. Ion cluster formation is then much favoured if hydrogen bonds can be formed as in the case of water. In other cases proton transfer to the parent molecules is strongly favoured.

Gas density near the tip surface at high FI rates

For these calculations, the rates of arrival and departure of molecules to and from the tip must be equalized. (For arriving at the qualitative conclusions at the end of this section, it is not necessary to use the improved supply function of Southon and Brandon [48].) The rate of arrival of molecules is:

$$\dot{n}_{in} = \frac{P_g}{(2\pi m k T)^{1/2}} \left(1 + \frac{E_p}{kT}\right) exp \left(-\int\limits_{x_c}^{\infty} \frac{dr}{\dot{r}_{in}\tau}\right) \qquad (3.69)$$

where the factor $(1 + E_p/kT)$ represents the enhancement of particle supply from the gas phase due to polarization (see eqn. 1.11). The exponential term arises from molecules which do not arrive at the tip because they are ionized during approach. $\dot{r}_{in} = (2E_p/m + kT/m)^{1/2}$. τ is the lifetime of the molecules with respect to field ionization. τ is nearly a step function of the distance from the tip. It has appreciable values only over a space interval d, near the critical distance for field ionization, x_c. Therefore $\int\limits_{x_c}^{\infty} \frac{dr}{\dot{r}_{in}\tau}$ may be approximated by $\frac{d}{\bar{v}_{in}\tau}$, where $\bar{v}_{in}$ is the average velocity of the incoming molecules.

Then

$$\dot{n}_{in} = \frac{P_g}{(2\pi m k T)^{1/2}} \left(1 + \frac{E_p}{kT}\right) \cdot exp \left(-\frac{d}{\bar{v}_{in}\tau}\right) \qquad (3.70)$$

Similarily, the number of molecules leaving the tip, $\dot{n}_{out}$ is given by

$$\dot{n}_{out} = \frac{N_t}{4} \left\{ 2 \int\limits_{v_{xc}}^{v_{Ep}} v \left[1 - exp\left(-\frac{d}{v\tau}\right)\right] f(v)\, dv + \int\limits_{v_{Ep}}^{\infty} v f(v) dv \right\} \qquad (3.71)$$

where $f(v)$ is the three-dimensional Maxwell velocity distribution which has to be used here because of the curvature of the tip. v_{EP} and v_{xc} are the velocities corresponding to the polarization energies $\frac{\alpha}{2} F_0^2 + \mu F_0$ and $\frac{\alpha}{2} (F_0^2 - F_{xc}^2) + \mu(F_0 - F_{xc})$, respectively.

N_t is the gas density at the adsorbed film on the tip. The first term in the brackets accounts for all particles having an energy $< E_p$ which cannot leave the tip as neutral particles, but which return to the tip due to the polarization forces. The factor of two allows for the chance of ionization occuring either during departure or return of the molecules to the tip. Thus the first term accounts for all particles being ionized after arrival to the tip surface from the gas phase. The second term accounts for all particles which have energies greater than the polarization energy E_p, and which therefore can leave the tip (either as neutral or ionized) molecules.

For order of magnitude calculations some further simplifications will be introduced. τ is assumed again to decrease stepwise at x_c and to be constant over the depth d of the ionization zone $\frac{d}{\tau v}$ is replaced by the constant factor $\frac{d}{\tau \bar{v}_{out}}$ before the integral, where v is replaced by the mean velocity $\bar{v}_{out}$ of the molecules accommodated to the tip temperature. After solving the integrals it can be shown that replacing the lower limit of the integral, $v_{(xc)}$ by $v = 0$ makes little difference.

Therefore, (3.70) may be approximated by

$$\dot{n}_{out} = \frac{N_t}{4} \left\{ 2 \left[1 - exp \left(-\frac{d}{\tau \bar{v}_{out}} \right) \right] \int_0^{E_p} v f(v) dv + \int_{E_p}^{\infty} v f(v) dv \right\}$$

Solving the integrals, one obtains

$$\dot{n}_{out} = \frac{P_t}{(2\pi m k T)^{1/2}} \left\{ 2 \left[1 - exp \left(-\frac{d}{\tau \bar{v}_{out}} \right) \right] \left[1 - \left(1 + \frac{E_p}{kT} \right) \right. \right.$$
$$\left. \left. \times\, exp \left(-\frac{E_p}{kT} \right) \right] + \left(1 + \frac{E_p}{kT} \right) exp \left(-\frac{E_p}{kT} \right) \right\} \quad (3.72)$$

Equating (3.70) and (3.72) and assuming that no field ionization would occur $\left(\frac{d}{\tau \bar{v}} = 0 \right)$ leads to the correct thermodynamic relation for complete equilibrium

$$P_t^{equ} = P_g\, exp \left(\frac{E_p}{kT} \right).$$

The pressure P_t near the tip drops if the equilibrium is disturbed by field ionization. If P_t drops below a certain value, P_c, multilayer formation will not be possible.

An appreciable drop of the pressure P_t occurs at high FI probabilities W_i, i.e.

$$W_i = \left[1 - exp\left(-\frac{d}{\tau\bar{v}}\right)\right] \simeq 1, \quad exp\left(-\frac{d}{\tau\bar{v}}\right) \to 0$$

Under these conditions (3.72) can be greatly simplified. $\left(exp\,(-E_p/kT) \ll 1\right.$ is also assumed$\left.\right)$.

Equating (3.70) and (3.72), one obtains

$$P_t \simeq \frac{P_g}{2}\left(1 + \frac{E_p}{kT}\right) exp\left(-\frac{d}{\tau\bar{v}_{out}}\right) \tag{3.73}$$

One can derive the value of τ from (3.73) where P_t drops below P_c. τ is a very steep function of the field strength. Therefore, the range of field strengths where multilayer formation could perhaps be observed by means of FI currents at the condition $F_m < F_i < F$, would be narrow. (F_m and F_i are the threshold fields for multilayer formation and field ionization, respectively.)

One of the most serious simplifications made here was neglect of the flow of physically adsorbed molecules from the shank to the centre of the tip. This effect does not change the conditions at complete equilibrium (i.e. without ionization) where (3.68) remains applicable. However, if field ionization becomes considerable, a correction for the flow of molecules by surface diffusion from the shank to the centre of the tip may be of great importance.

Dependence of the mean free path on the field strength.

The mean free path of a molecule in a high electric field dependents on many parameters, such as: tip radius, temperature, gas density, field strength, polarizability and dipole moment of the molecules. An equation for the mean free path was derived by Metzinger [135].

$$1 = \int_{r_t}^{\lambda} \pi\sigma^2 n_0\, exp\left\{\left[\frac{\alpha}{2} F_0^2 \left(\frac{r_t}{r}\right)^4 + \mu F_0 \left(\frac{r_t}{r}\right)^2\right] \cdot \frac{1}{kT}\right\} dr \tag{3.74}$$

Table 3.6 represents the mean free path of the molecules at different field strengths as calculated with a computer by means of eqn. (3.74), by Metzinger.

TABLE 3.6. MEAN FREE PATH FOR ACETONE AT $T = 293°K \cdot r_t = 5 \times 10^{-5}$ cm, $P_g = 10^{-3}$ Torr

F_0 (V/cm)	λ (cm)
2×10^7	0.7
6×10^7	4.3×10^{-7}

Ion clusters

Formation of ion clusters may occur due to interaction between polar molecules adsorbed on an FI emitter. The probability of formation of larger clusters will be enhanced if one, or a few molecules impinge statistically on top of a nearly filled monolayer in such a way that small pyramids of adsorbed molecules are formed.

Early measurements of ion clusters were made with water on tungsten tips [1] and with water, methanol and methanol-water mixtures on tungsten tips [5]. At room temperature, clusters of a proton and up to four water or methanol molecules could be detected, and mixed species were also found as shown by table 3.7.

TABLE 3.7. RELATIVE INTENSITIES OF HYDRATED AND POLYMERIC IONS IN THE FI MASS SPECTRA OF WATER, METHANOL AND METHANOL-WATER MIXTURES ON TUNGSTEN TIPS AT ROOM TEMPERATURE [5]

Water		Methanol		Methanol-Water Mixtures	
H_3O^+	100	CH_3OH	40	$H^+ \cdot CH_3OH$	100
$H_3O^+ \cdot H_2O$	100	$H^+ \cdot CH_3OH$	100	$H^+ \cdot CH_3OH \cdot H_2O$	200
$H_3O^+ \cdot (H_2O)_2$	3	$H^+ \cdot (CH_3OH)_2$	100	$H^+ \cdot CH_3OH \cdot (H_2O)_2$	10
$H_3O^+ \cdot (H_2O)_3$	0.1	$H^+(CH_3OH)_3$	0.1	$H^+ \cdot (CH_3OH)_2 \cdot H_2O$	10
		$H^+(CH_3OH)_4$	0.1	$H^+ \cdot (CH_3OH)_3 \cdot H_2O$	0.05
				$H^+ \cdot (CH_3OH)_2(H_2O)_2$	0.01
				$H^+ \cdot (CH_3OH)(H_2O)_3$	0.01

It is interesting to note that at low temperatures, the maximum relative intensity of water ions shifts from the monomeric and dimeric species in the direction of the tetrameric species at -195 °C, which is the $H_9O_4^+$ — complex, often discussed in electrochemistry [141]. This is shown in table 3.8. The tip of the FI source was cooled by liquid nitrogen.

TABLE 3.8. ION INTENSITY (ARBITRARY UNITS) OF POLYMERIC IONS FROM WATER, $H^+(H_2O)_n$ AT $T_{tip} = -195$ °C

(Typical result of a single measurement; effect of fluctuations in ion intensity is shown by the irregularity at $n = 6$ and 7)

n	1	2	3	4	5	6	7	8	9	10
I	6	50	500	400	180	180	190	125	40	20

Two different mechanisms for field ionization of water in the adsorbed layer on the tip have been considered [5]. With the first mechanism it is assumed that the primary step would be field ionization of a water molecule, which would then be followed by a proton transfer reaction:

$$H_2O^+ + H_2O \rightarrow H_3O^+ + OH. \tag{3.75}$$

The proton transfer does not occur by impact of a H_2O^+ ion on a H_2O molecule in the gas phase, but by transition of a proton from the one to the other of two associated water molecules in the adsorbed layer.

However, the ionization potential of water is too high (12.5 eV) for noticeable field ionization on a wire emitter. Therefore, this mechanism is not very probable.

The basis of the second mechanism is the interpretation of the FI of water from an electrochemical point of view. The dissociation of water into ions occurs to a small extent even if no external field is applied:

$$2H_2O \rightarrow H_3O^+ + OH^- \tag{3.76}$$

This dissociation process is influenced by the hydrogen bond between two water molecules which leads to a shift of charges. One electron can migrate more easily to one of the molecules, and the proton to the other molecule if the two molecules are connected by a hydrogen bond. If an external electrical field is applied, the dissociation rate will be increased, according to the theory of Onsager [142]. From this theory it can be shown that the dissociation rate is increased by several powers of ten if the field strength is of the order of 10^7 V/cm. The electron transferred to the OH radical should be bonded only for a short time and then be drawn to the emission tip due to the tunneling effect. Thus, reversing all fields in the FI mass spectrometer will not necessarily lead to detection of OH^- ions. Schmidt could not observe OH^- ions in the course of such experiments. With a positive tip, recombination of the OH radicals should occur

$$4OH \rightarrow 2H_2O + O_2 \tag{3.77}$$

Schmidt [28] reports on observation of O_2^+ ions during analysis of the FI mass spectrum of water on platinum and iridium tips. These O_2^+ ions have doubtless been formed by reaction (3.77). In the case of water on tungsten tips, one observes different tungsten oxide ions instead of free oxygen ions. The evaporation of these ions as observed by Schmidt will be discussed in the section on field evaporation. Another characteristic difference between the water FI mass spectra on tungsten and platinum (or iridium) tips consists in the finding that the abundance of H_2O^+ and H_3O^+ ions is about equal at $F \simeq 10^8$ V/cm with platinum tips, whereas the relative intensity of the H_2O^+ ions is negligibly small as compared with the H_3O^+ ion intensity with tungsten tips. Thus it is seen that the different chemical nature of the tip materials leads to different chemical reactions and different adsorption phenomena, and therefore the FI mass spectra are different. It should be noticed, however, that these differences obtained with clean tips are masked if micro needles are grown on emitter wires of different materials by means of the "wire activation" method using organic substances. The semi-conducting organic material of the micro needles grown on different metals is the same, and thus the differences of the FI mass spectra on different wires disappear.

Fig. 3.37. shows the intensity of different ions in the FI mass spectrum of water on platinum as a function of the field strength as measured by Schmidt [28]. Note the extremely steep rise of the ion intensity at the threshold field of about 6×10^7 V/cm. The low value of this threshold as compared to the value for FI of water in the gas phase indicates field ionization at the surface of an adsorbed layer, where the residence time of the molecules is high. The steep rise of the ion current curve can be explained in the following way. At low fields, the minimum distance d_{min} for FI is larger than the thickness d_f of the adsorbed layer. d_{min} decreases and d_f increases with increasing field strength, until both distances coincide. At this critical field strength, the FI probability will increase strongly.

Metastable decomposition of the polymeric water ions lead to broad peaks in the FI mass spectrum of water. The reactions are of the type

$$H_3O^+(H_2O)_n \rightarrow H_3O^+(H_2O)_m + \textit{neutral fragment}.$$

The reactions with

$$n = 3 \qquad m = 0$$
$$n = 2 \qquad m = 0$$
$$n = 2 \qquad m = 1$$
$$n = 1 \qquad m = 0$$

were observed by the author. Additional reactions were reported by Schissel [143]:

$$n = 3 \qquad m = 1$$
$$n = 3 \qquad m = 2$$

Later, Goldenfeld and associates [250] have also studied the FI mass spectra of water on tungsten, molybdenum, platinum and gold and also on wires of nichrome. They confirmed the observation of $H^+(H_2O)_n$ ions and of metastable ions. (For isotopic analysis by Goldenfeld see section 3.8.)

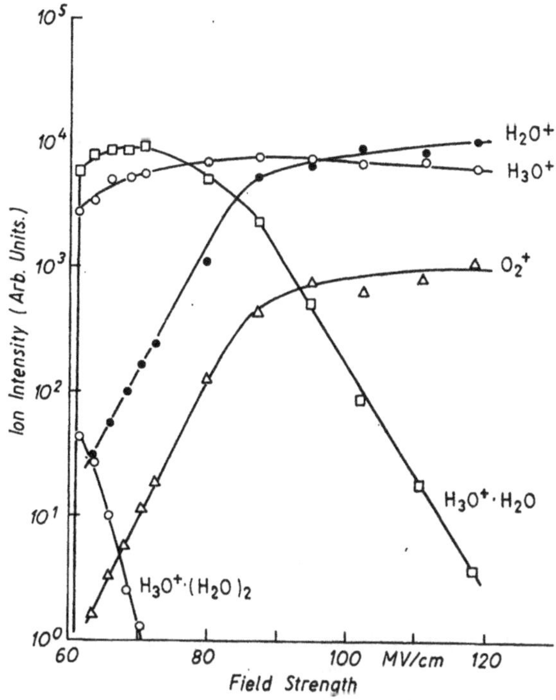

FIG. 3.37. Dependence of some ions observed in the FI mass spectrum of water on the field strength. FI anode: platinum tip (Schmidt [28]).

In a paper on field condensation of formic acid on platinum, gold, tungsten, molybdenum and nickel tips, Block [144] showed that the behavior of platinum and gold on the one hand and the other three metals on the other hand, was quite different. Field corrosion of Wo, Mo and Ni was observed, but not with Au and Pt. The $(P + 1)^+/P^+$ ratio was much larger with Wo and Mo than with Au and Pt. The parent ion $(m = 46)$ does not appear at all with tungsten tips, whereas it appears at high field strengths with platinum tips and decreases with decreasing field strength.

Fig. 3.38. shows the FI mass spectrum of the parent ions of formic acid on a platinum tip as measured by Block [144]. It has to be assumed that the broad peak shape of the $P + 1$ peak at low field strengths is due to the formation of ions on top of a multilayer where the potential is lower than at the tip surface. The variation of ion energy may be correlated with fluctuations of the multilayer thickness. Very probably — as mentioned before — there are not formed complete multilayers, but adsorbed molecular clusters of pyramidal structure.

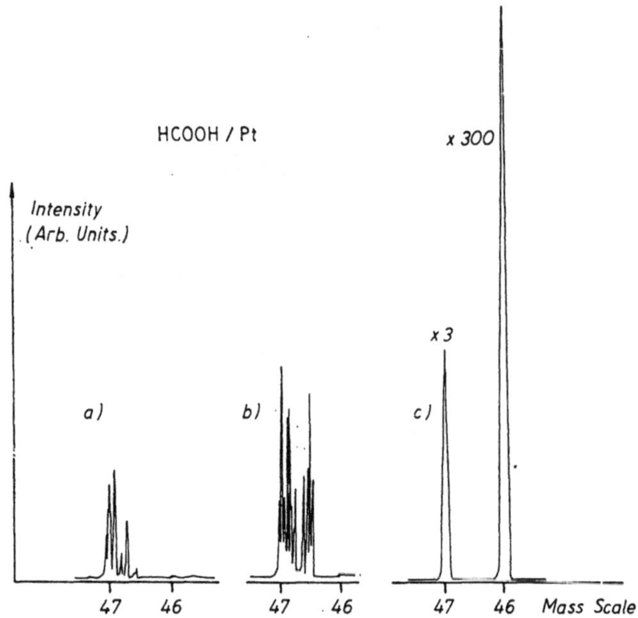

FIG. 3.38. Peak shapes of formic acid field ionized on a platinum tip ($m/e = 46, 47$).

a) near the threshold field; *b*) at a slightly higher field; *c*) at double the value of the threshold field. (Block [144].)

Large relative intensities of dimeric species of formic acid, $(HCOOH)_2 \cdot H^+$ are found on tips of different materials. However, the field strength dependence is different for gold and tungsten tips. There is a pronounced maximum of the dimeric ion intensity on gold tips at medium field strengths, whereas the dimeric ion intensity increases continuously with increasing field strengths on tungsten tips. This indicates the dependence of FI mass spectra on the FI emitter metals which will be discussed in a later section.

Beynon, Fontaine and Job [88] also reported on dimeric ions of polar substances field ionized on sharp stainless steel blades.

Job and Patterson [146] have observed dimeric and trimeric ions in the FI mass spectra of nitrobenzene and pyridine which appear at the mass numbers $[M_2^+]$ and $[M_3^+]$, in contrast to the alcohols and water which form polymeric ions at mass numbers $H^+[M]_n$, $n = 2, 3, \ldots$. Whereas the latter polymers are considered as linked through the proton by hydrogen bridges, Job and Patterson believe that the nitrobenzene and pyridine polymeric ions can be interpreted as charge transfer complexes. They point out that it is evident from experiments by Brown and Grayson [147] that the acceptor character of nitrobenzene is enhanced by co-ordination with aluminium chloride. Thus it has to be assumed that the positive nitrobenzene ion has acceptor character, whereas the neutral nitrobenzene behaves as a donor, thus permitting formation of a charge transfer complex of an ion and a neutral molecule which is observed in the FI mass spectrometer as M_2^+. The species M_3^+ can be explained by the assumption that the nitrobenzene positive ion can act as an acceptor to two neutral molecules simultaneously.

In pyridine the species M_3^{++} is observed instead of M_3^+ (besides M_2^+). It seems that the positive aniline ion does not act as an acceptor for a neutral aniline molecule. No species M_3^+ are found in the aniline FI spectrum.

However, the neutral aniline molecule has donor character. Therefore, in mixtures of aniline and either nitrobenzene or pyridine the positive ions of the latter will act as good acceptors for the aniline donor. Therefore, "hetero-dimer" ions are found in the FI mass spectra in which aniline is incorporated. The "hetero-dimers" are formed at larger intensities than the "homo-dimers" due to the greater donor character of aniline as compared to nitrobenzene and pyridine.

3.6.3. Shift of Chemical Equilibria by Electric Fields

The influence of electric fields on chemical equilibria was noticed relative early. Wien [145], for example, studied the shift of the degree of dissociation of electrolytes with the strength of electrical fields applied to the electrolytes. Linearity between the degree of dissociation of weak electrolytes and the field strength was observed (First Wien effect.) Onsager [142] studied these effects up to higher field strengths ($\simeq 5 \times 10^5$ V/cm) and found more complex theoretical relations under these conditions. The shift of the reaction $2H_2O \rightarrow H_3O^+ + OH^-$ with field strengths of the order several 10^7 V/cm was discussed by the author [134] on the basis of Onsager's theory. Further, he observed a shift of the equilibrium between monomeric and polymeric water ions in favour of the monomeric ions with increasing field strength [4, 134] (see fig. 3.39). Similar observations were reported for water ions and formic acid ions by Schmidt [28] and by Block [144].

An experimentally and theoretically more extended investigation on the shift of chemical equilibria with field strength has been carried out by Block [148]. This author pointed out that shifts

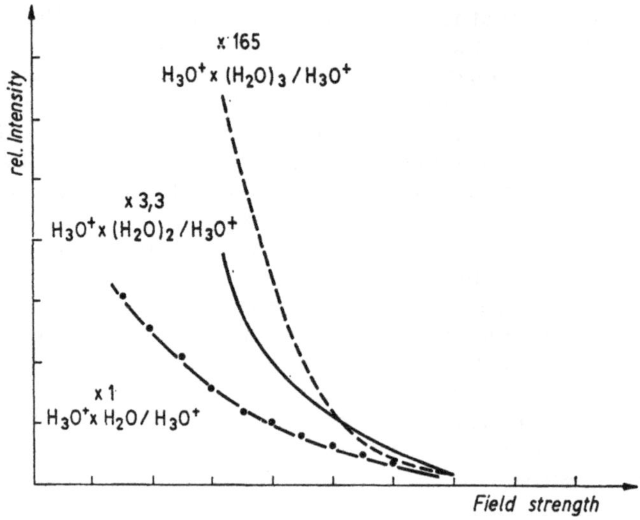

FIG. 3.39. Ratio of the $H_3O^+ (H_2O)_n/H_3O^+$ intensities of water as a function of the field strength. Field anode: tungsten tip.

of chemical equilibria by high electric fields do not only occur for ionic processes as indicated above, but also for equilibria between neutral molecules. Block calculated the shifts which are to be expected theoretically, and he compared this with observations on the equilibrium between monomeric and dimeric species of formic acid by means of an FI mass spectrometer. The theoretical and experimental results are as follows.

Chemical equilibria are shifted by external electrical fields if the reactions are connected with a change of the electric moments. By analogy with the van't Hoff equation one derives

$$\frac{\partial \ln K}{\partial F} = \frac{\Delta M}{kT} \tag{3.80}$$

where K is the equilibrium constant, and ΔM is the difference of the polarization energies of the reaction products and the reactants.

For the reaction

$$2\,HCOOH \rightarrow (HCOOH)_2 \qquad (3.81)$$

$$\Delta M = \frac{1}{2}(\alpha_{dim} - 2\,\alpha_{mon})F^2 + (\bar{\mu}_{dim} - 2\,\bar{\mu}_{mon}) \qquad (3.82)$$

where $\bar{\mu} = \mu \cdot L\left(\frac{\mu F}{k T}\right)$, and L is the Langevin-function as discussed earlier. At field strengths of about 10^7 V/cm one can assume α_{dim} approximately equal to $2\,\alpha_{mon}$, because of the additi-

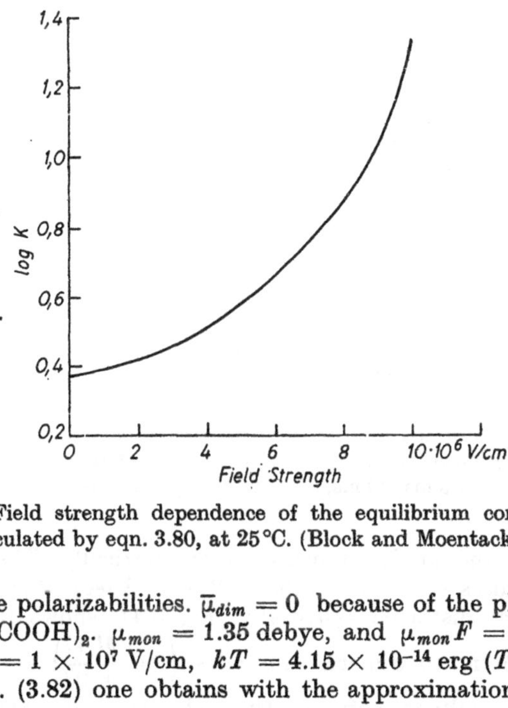

FIG. 3.40. Field strength dependence of the equilibrium constants K as calculated by eqn. 3.80, at 25 °C. (Block and Moentack [148].)

vity of the polarizabilities. $\bar{\mu}_{dim} = 0$ because of the planar structure of $(HCOOH)_2$. $\mu_{mon} = 1.35$ debye, and $\mu_{mon} F = 4.5 \times 10^{-14}$ erg at $F = 1 \times 10^7$ V/cm, $kT = 4.15 \times 10^{-14}$ erg ($T = 298$ °K). From eqn. (3.82) one obtains with the approximations used here

$$\Delta M \simeq -2\bar{\mu}_{mon} \qquad (3.83)$$

Fig. 3.40. shows log K as a function of F as calculated from the integrated form of eqn. (3.80). The ratio of monomers to dimers increases by about a factor of ten if F is increased from zero to 1×10^7 V/cm.

This result can be checked experimentally by applying the impulse desorption technique [148]. The emission tip of an FI mass spectrometer is set at a constant potential of $+4.5$ kV. The electrode opposite to the tip is provided with a d.c. voltage which can be varied from $+4.5$ kV down to negative values. Thus the total voltage is varied from zero to such values that the system remains below the threshold field for FI. Then, after preset time intervals t_r, short high negative impulses are applied to the cathode during a time interval $t_d \simeq 10^{-6}$ sec. The adsorbed monomeric and dimeric formic acid molecules are field ionized during

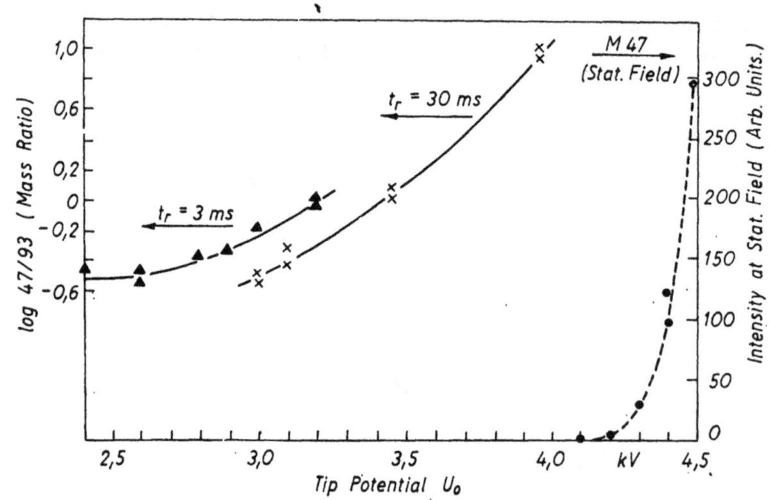

FIG. 3.41. Ratio of the intensities of monomeric and dimeric formic acid ions as a function of the field strength at two different times of reaction, $t_r = 3$ ms, 30 ms, $T = 25\,°C$, $t_d = 10^6$ sec. [148].

this period and analysed mass spectrometrically. The rise time of the impulses is so short (about 2×10^{-8} sec) that the equilibrium is not distorted essentially. Fig. 3.41. shows the ratio of the monomeric to dimeric molecules $\left(\log \frac{M\,47}{M\,93}\right)$ as a function of the d.c. voltage between tip and cathode for two different t_r — values. The monomeric and dimeric formic acid molecules are measured in a form where a proton is attached to the molecules ($M = 47$ and $M = 93$). The proton attachment does not influence the result to any noticeable extent.

Several side effects influence the measurements and cause some deviations of the experimental curves, fig. 3.41., from the theoretical curve, fig. 3.40., as pointed out by Block. However, the bulk effect of change of the equilibrium constants with field strengths has been proved by these experiments.

One of the points is that adsorption occurs to a greater extent with increasing residence time t_r of the molecules on the tip surface. Thus the intensity of the dimeric species is markedly increased at $t_r = 30$ msec as compared with $t_r = 3$ msec, as can be seen from fig. 3.41.

3.6.4. Field Desorption and Field Evaporation

First, the theory of field desorption will be briefly outlined and then some mass spectrometric observations will be discussed.

3.6.4.1. Theory

Desorption of electropositive metals from the surface of positively charged tungsten tips was discovered by Müller [152]. Later Müller showed that field desorption is not restricted to electropositive layers [34, 150]. He observed field desorption of substances like oxygen and nitrogen from positive tips. He also detected field desorption of tungsten atoms from their own lattice. This process may be termed field evaporation, although being basically the same as field desorption. Müller postulated a mechanism based on direct ionic evaporation over a Schottky potential saddle with a rate constant k given by

$$k = \nu \cdot e^{-Q/kT} \tag{3.84}$$

Here ν is a vibrational frequency of the order of 10^{13} sec^{-1}, and Q is a heat of adsorption, reduced from the thermodynamic value $Q°$.

$$Q° = H_a + I - \Phi \tag{3.85}$$

by a Schottky term $3.8\, n^{3/2} F^{1/2}$.

$$Q = Q° - 3.8\, n^{3/2} F^{1/2}. \tag{3.86}$$

H_a is the heat of adsorption with respect to a neutral adsorbed particle, I and Φ have the same meaning as before, F is the field in volts/Å and n is the charge of the evaporating ion. Gomer and Swanson [149] have considered field evaporation as a special case of field desorption in which the metal atom is evaporated as a multiply charged ion with an activation energy, Q, given by

$$Q = \Lambda + \sum_n I_n - n\,\Phi - (n^3 e^3 F)^{1/2} + \frac{1}{2}(\alpha_n - \alpha_i) F^2 \tag{3.87}$$

Λ is the heat of evaporation, $\sum_n I_n$ is the sum of the ionization potentials, n the charge of the ion, $1/2(\alpha_n - \alpha_i)F^2$ is a term introduced by Müller [37] accounting for the difference of polarization energies of the atom and the ion.

Inghram and Gomer [1] first demonstrated field desorption of organic substances from positive tips by mass spectrometric methods. They detected the CH_3O^+ ion in the FI mass spectrum of methanol which they proved to be a product of dissociative chemisorption of methanol on tungsten.

The theory of field desorption has been developed in greater detail for various cases by Gomer [151]. The problem will be discussed in terms of potential curves, following the treatment of Gomer.

Case A) $I \gg \Phi$.

Fig. 3.42.a. shows the potential curves for relatively strong covalent chemisorption, when $I - \Phi$ is large so that the ionic curve does not intersect the ground state. Ionic evaporation can occur only if energy is transferred to the adsorbed atoms by collision processes.

Fig. 3.42.b. shows the same system in the presence of a moderate, positive electric field. Desorption from the ground state may occur by vibrational excitation by an energy amount Q, which is approximately equal to the desorption energy H_a (modified by polarization terms). Desorption can be followed by field ionization at x_c. The rate constant for ionic desorption is given by

$$k_{ion} = \nu\, exp\left(\frac{-H_a + PF - {}^1/_2\alpha F^2}{kT}\right) \times \left\{1 - exp\left[-\left(\frac{m}{kT}\right)^{1/2}\int_{x_c}^{\infty}\frac{dr}{\tau_{(r)}}\right]\right. \qquad (3.88)$$

where ν is the vibrational frequency of the adsorbed particle along the desorption coordinate and PF the bond-dipole field interaction.

The rate constant for atomic desorption is

$$k_{at} = \nu\, exp\left(\frac{-H_a + PF}{kT}\right) exp\left[-\left(\frac{m}{kT}\right)^{1/2}\int_{x_c}^{\infty}\frac{d_r}{\tau_{(r)}}\right] \qquad (3.89)$$

The polarization term has been omitted here since the field decays over distances less than the mean free path of the desorbing atom, as shown in section 3.6.2.

It is seen from the figure that crossing of the neutral and ionic curves does not occur because generally no spin or symmetry differences of the states involved occur. Therefore, the two curves separate into the new states shown in the figure. The separation of the curves will usually be small, since there will be little overlap of the atomic and ionic wave functions at x_c.

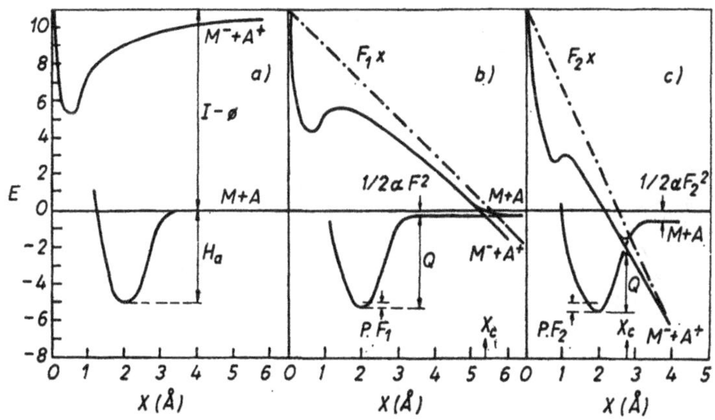

FIG. 3.42. Schematic energy level diagram for neutral and ionic adsorption at a metal surface, with large separation between states. I adsorbate (A) ionization potential; φ metal (M) work function; x adsorbate-surface distance; F applied field; $^1/_2\alpha F^2$ polarization energy of atom in field F; PF field bond interaction energy; H_a heat of adsorption of atom; Q activation energy of desorption; x_c intersection point of ionic and atomic potential curves.

a) zero field; *b*) moderate field, F_1, leading to field ionization after essentially normal desorption; (*c*) high field, F_2, leading to field ionization after partial desorption (Gomer [151]).

Fig. 3.42.c. shows the conditions for very high fields. The curves separate at a crossover point which lies on the attractive branch of the neutral adsorption curve. The activation energy for thermal desorption Q is reduced below the zero-field value. Q is given by

$$Q = V_{(x_c)} + H_a + PF - \frac{1}{2}\,\alpha F^2 - \frac{1}{2}\,\Delta E_{res} \approx V_{(x_c)} + H_a \quad (3.90)$$

ΔE_{res} is the resonance separation of the two states. This may be greater than in case *b*), but this term will not be further considered. $V_{(x)}$ is the (negative) potential of the neutral atom with the zero taken at infinity.

The rate for field desorption is, similar to (3.84), given by

$$k_{des} = \nu \cdot s \cdot e^{-Q/kT} \tag{3.91}$$

where s is the probability that a particle passing over the saddle point will be field ionized. s may be approximated by $s = (1 - e^{-t/\tau})$, where t is the time spent during one vibration in the region $x \geqq x_c$, and τ is the mean lifetime with respect to field ionization. (For a more exact expression see Gomer [151].) It was shown by Gomer introducing several approximations that $s \geqq 0.3$ in most cases.

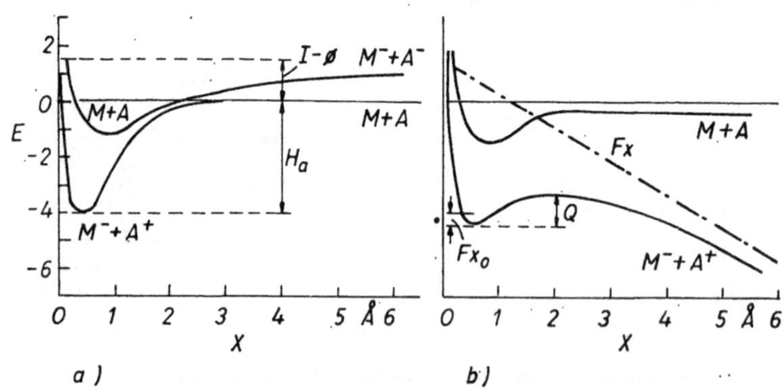

FIG. 3.43. Energy level scheme for neutral and ionic adsorption at a metal surface with $I - \varphi$ small. Notation as in fig. 3.42.; Fx_0 increase in binding due to field.

a) zero field; curves cross as indicated, leading to ionic ground-state; *b*) high field, this removes the level splitting, so that the ground-state remains purely ionic; desorption proceeds by straightforward vibrational excitation with activation energy Q (Gomer [151]).

It may be seen from fig. 3.42.b. that the applied field raises the potential energy of the tunneling electron by an amount

$$F \cdot x_c = I - \Phi - 3.6 \cdot n^2/x_c + |V_{(x_c)}| \tag{3.92}$$

This may be compared with the expression for a non-adsorbed particle, eqn. (1.13a).

Case B) $I - \Phi$ small

Fig. 3.43.a. shows the neutral and ionic curves when $I - \Phi$ is small. Crossing of the potential curves occurs even in the absence of an external field. As before, the curves will separate in most cases because of non-degeneracy of the states. The particles are adsorbed in ionic form and desorbed in neutral form if no vibrational excess energy is transferred to the particles.

Fig. 3.43.b. shows the conditions for such high fields that the two curves do not intersect at any distance. The ground state is purely ionic, and evaporation occurs by vibrational excitation over the Schottky saddle without need for field ionization.

The rate is given by the equation derived by Müller (3.84). Gomer [151] has also treated some intermediate cases ($I - \Phi$ is moderate).

Müller has shown that field strengths of about 100, 200 and 500 MV/cm are required for field desorption of barium, thorium and oxygen, respectively, from tungsten at room temperature. The evaporation field strength for tungsten atoms from their own lattices are 450 and 570 MV/cm, at $T = 273°$ and 77 °K, respectively [257, 48, 153]. These field strengths are usually not reached with wire emitters.

3.6.4.2. *Mass spectrometric observations under d.c. conditions*

The first mass spectrometric investigations of field desorption by Inghram and Gomer [1] and also those by Block [9] will be treated separately in the next section on the field impulse desorption method.

Very interesting observations on d.c. field desorption of a great variety of surface complexes of tungsten in the presence of water were made by Schmidt [28].

Table 3.9. shows the different species which appeared in the FI mass spectrum of water on a tungsten tip at a pressure of 3×10^{-4} Torr at room temperature.

The table lists a great variety of tungsten oxide ions at different degree of oxidation and ionization, some of them without, and others with added water molecules.

Quite obviously, the evaporation field strength is much smaller for tungsten oxides than for tungsten itself on its own lattice. The ions of the suboxides WO_2 and WO are detected only at higher field strengths than WO_3. One has to assume that the former two species are connected by stronger bonds to the tungsten lattice than the latter. Thus it is shown that quite generally one may obtain results on the bond energy for different surface complexes. Of course, partial evaporation of tungsten oxides leads to gradual roughing of the tips and thus to an increase of local field strengths.

The experiments of Schmidt may be considered in connection with observations of Mulson and Müller [154] on the corrosion of tungsten and iridium tips by field desorption of nitrogen and carbon monoxide. These observations were made by means of an FI microscope. Field desorption of a nitrogen atom from a single site

TABLE 3.9. TUNGSTEN OXIDE IONS PRODUCED BY FIELD IONIZATION OF H_2O ON A TUNGSTEN TIP [28]
Spectrum 1: F 2.2 × 10^7 V/cm; spectrum 2: F 3.0 × 10^7 V/cm. (Field strengths taken for smooth tips.)
$P_{H_2O} = 3 \times 10^{-4}$ Torr.
Intensity in arbitrary units. (For comparison: $I_{H_3O^+} = 2 \times 10^3$ units at $F = 2.2 \times 10^7$ V/cm, and $I_{H_3O^+} = 4 \times 10^3$ units at $F = 3.0 \times 10^7$ V/cm.)

Singly charged Ions	Abundance (arb. units) Spectrum 1	2	Doubly charged Ions	Abundance (arb. units) Spectrum 1	2	Triply charged Ions	Abundance (arb. units) Spectrum 1	2
WO^+		0.7	WO_2^{++}		0.7	$(4WO\ \cdot H_2O)^{+++}$		1.5
WO_2^+		0.7	WO_3^{++}		3	$4WO_3^{+++}$	1.2	15.0
WO_3^+	1.2	29	$(WO_3 \cdot H_2O)^{++}$		3	$(4WO_3 \cdot H_2O)^{+++}$	0.5	2.5
$(WO_3 \cdot H_2O)^+$	1.5	8.5	$(WO_3 \cdot 4H_2O)^{++}$		0.7	$(4WO_3 \cdot 2H_2O)^{+++}$	0.3	
$(WO_3 \cdot 2H_2O)^+$	0.6	1.8				$(4WO_3 \cdot 5H_2O)^{+++}$	0.5	
$(WO_3 \cdot 3H_2O)^+$	0.3	2.5	$(WO_3 \cdot WO_2)^{++}$	0.5	4.0			
$(WO_3 \cdot 4H_2O)^+$	0.6	7.5	$(2WO_3 \cdot H_2O)^{++}$	1.4	15.0	$5WO_3^{+++}$	0.5	
						$(5WO_3 \cdot H_2O)^{+++}$	0.5	
$2WO_3^+$	1.1		$3WO_3^{++}$	0.95		$(5WO_3 \cdot 2H_2O)^{+++}$	0.4	
$(2WO_3 \cdot H_2O)^+$	0.3		$(3WO_3 \cdot H_2O)^{++}$	0.8		$(5WO_3 \cdot 3H_2O)^{+++}$	0.5	
$(2WO_3 \cdot 2H_2O)^+$	0.3		$(3WO_3 \cdot 2H_2O)^{++}$	0.4		$(5WO_3 \cdot 4H_2O)^{+++}$	0.5	
$(2WO_3 \cdot 3H_2O)^+$	0.3		$(3WO_3 \cdot 3H_2O)^{++}$	0.4				
$(2WO_3 \cdot 4H_2O)^+$	0.3		$(3WO_3 \cdot 4H_2O)^{++}$	0.5				
$3WO_3^+$	0.4		$(4WO_2 \cdot 3H_2O)^{++}$	0.8				
$(3WO_3 \cdot H_2O)^+$	0.3		$(4WO_3 \cdot H_2O)^{++}$	0.65				
			$(4WO_3 \cdot 3H_2O)^{++}$	1.0				
			$(5WO_2 \cdot H_2O)^{++}$	0.8				
			$5WO_3^{++}$	0.3				
			$(5WO_3 \cdot H_2O)^{++}$	0.6				
			$(5WO_3 \cdot 2H_2O)^{++}$	0.2				
			$(5WO_3 \cdot 4H_2O)^{++}$	0.3				

removes also the underlying tungsten or iridium atom. However, field desorption of an adsorbed nitrogen molecule does not corrode the iridium substrate. The corrosion of a metal surface by certain gases and vapours in the presence of high electric fields may be termed "field corrosion". This process is assumed to be the result of field induced chemical reactions [18]. Field corrosion occurs during field desorption of hydrogen from Cu, Fe, Ni, Be, Si and other substrates. Water vapour was shown to remove surface atoms of tungsten as well as of platinum when the metal is kept at liquid nitrogen temperature.

Vanselow and Schmidt [156] have investigated the high temperature field evaporation of platinum both in ultra high vacuum (UHV) and in the presence of oxygen. At temperatures up to 1200 °K Pt^{+}, Pt^{2+} and Pt_3^{++} were observed during field evaporation of spectroscopically pure platinum under UHV conditions. The abundance of the Pt^{+} ions was much larger than that of the other ions. Impurities led to the formation of $PtFeO^{2+}$, $PtCuO^{2+}$, Pt_2FeO^{2+}, Pt_2CuO^{2+}, PtC^{+}, PtO^{+}, Pt_2O^{+}, Pt_2O^{2+}, Pt_3O^{2+} and Pt_4O^{2+}.

Oxygen can diffuse in large amounts into the platinum lattice if oxygen is added at high temperatures. A PtO-lattice can be formed at the surface under these conditions. An oxygen atom is surrounded tetrahedrally by four platinum atoms in this lattice, and a platinum atom is surrounded by four oxygen atoms in a planar structure. These coordination numbers are arrived but not exceeded, as can be seen from the observed ionic species: Pt_4O^{2+} is the ion with minimum oxygen content, PtO_3^{2+} the ion with minimum platinum content.

The relative intensities of the measured PtFeO and PtCuO complexes exceed the average abundances of the spurious Fe and Cu impurities by very large factors. Therefore, the method seems suitable for analysis of spurious impurities especially if the impurities are enriched at specific areas of the surface.

In a recent investigation Barofsky and Müller [155] have studied the low temperature field evaporation of Be, Fe, Cu and Zn by means of a mass spectrometer. Both the experimental techniques and the results showed interesting new features. The main difficulty of such experiments consists of the fact that the number of field evaporating ions per second is very small as compared to the number of ions corresponding to a continuous gas supply at 10^{-3} Torr to the tip. Field evaporation can be sustained at a maximum rate of about 10 atom layers per second for a period of only one minute. On the other hand, the field ionization current of a gas at a pressure of 10^{-3} Torr is equivalent to field evaporation of about 10^4 monolayers. Assuming a tip of 500 Å radius and an area of $(3.0\ Å)^2$ per atom, one has about 1.7×10^5 atoms on the tip surface. The transmission of a small mass spectrometer with large apertures (small resolving power) is of the order of 10^{-2}, that of a large machine of the order 10^{-4}. Taking an evaporation rate of 3 atom layers per second and a transmission of 10^{-3}, one

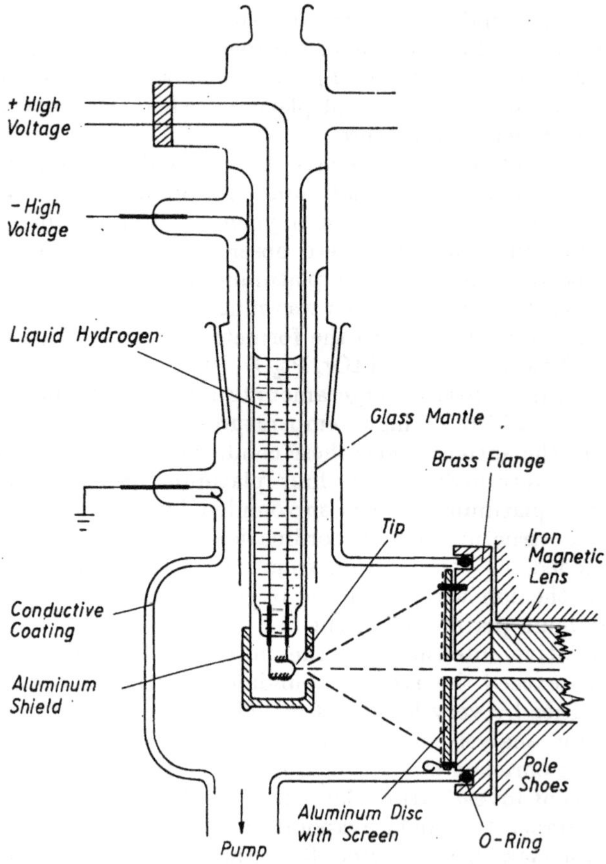

FIG. 3.44. Schematic presentation of the FI mass spectrometer used by Barofsky and Müller [155] for the study of low temperature field evaporation. *a*) Source construction.

will detect only 500 ions per second at the mass spectrometer detector. Extremely large fluctuations of the ion current have then to be expected, both because of the particle number statistics and because of the rapidly changing micro structure of the field evaporating surface and the corresponding fluctuations of the local field strengths.

In order to overcome these difficulties, Barofsky and Müller introduced two modifications as compared with the usual FI mass spectrometers: First, they constructed a rather small and compact FI mass spectrometer of very high transmission (about 1%) as shown in fig. 3.44., and second, they constructed a special ion detection system. The mass spectrometer shows the following features. The FI tip can be cooled by liquid hydrogen, because

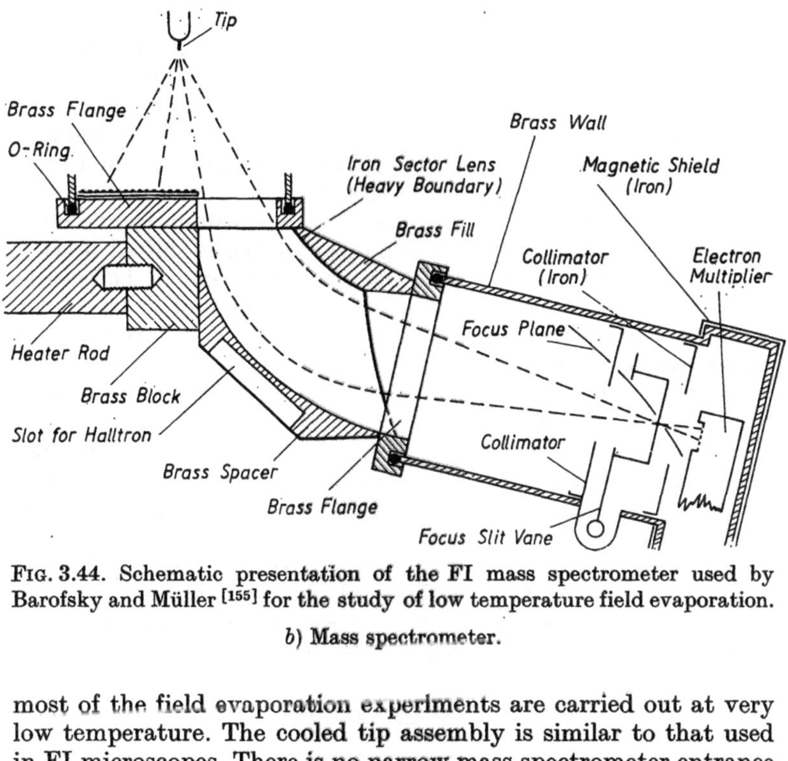

FIG. 3.44. Schematic presentation of the FI mass spectrometer used by Barofsky and Müller [155] for the study of low temperature field evaporation.

b) Mass spectrometer.

most of the field evaporation experiments are carried out at very low temperature. The cooled tip assembly is similar to that used in FI microscopes. There is no narrow mass spectrometer entrance slit because the tip itself acts as a sharp image point of the ion optical system. The FI image of the tip can be observed on a luminescent screen which is interrupted only by the mass spectrometer entrance channel. The magnet deflects the ion beam by 60°. The tip is set at a constant potential of +2.5 kV, the aluminium shield surrounding the tip is at a variable negative high voltage, the rest of the mass spectrometer at ground potential.

A saw — tooth signal of the Tektronix double beam oscilloscope type 555 is used in connection with a Hall probe for control of the magnetic field scan.

The ion beam detector consists of a Bendix electron multiplier, model 306, followed by a preamplifier and a Tektronix amplifier type 1A7.

Usually the horizontal beam scan serves as a time (and thus as a mass) basis during oscilloscopic recording; the vertical beam deflection is a measure of the intensity of the signals. In the present case, however, the ion detector signal is used for modulation of the intensity of the cathode beam of the oscilloscope. Normally, the intensity of the cathode beam is kept at zero. The intensity is increased to a definite value if a signal appears at the ion detector exit, and then a bright spot appears on the luminescent screen. A mass spectrum appears as a horizontal line of bright spots on the luminescent screen of the oscilloscope. Of course, the brightness of the spots serves only as an approximate measure of the ion intensities. The pattern is photographed by a camera. One horizontal line of bright spots appears at a constant total voltage between tip and cathode. The horizontal line is shifted upwards at a higher total voltage. Thus the mass spectrum can be recorded as a function of the total voltage in a maximum of 12 traces. The voltage is increased stepwise by 100 to 500 Volt intervals. The trace can be shifted upwards also manually by a potentiometer. Thus the mass spectrum may be recorded in different traces at different tip temperatures.

The true mass spots form a vertical column, whereas the spots representing spurious events scatter randomly about the photograph. It was possible to resolve two ions differing by one mass unit at mass 80.

The experimental results of these investigations of Barofsky and Müller are as follows. The velocity of field evaporation is given by eqn. (3.84) and (3.87). These equations without polarization term can be combined and resolved for F_n

$$F_n = (ne)^{-3} \left[H_a + \sum_n I_n - n\,\Phi + kT \ln \frac{k}{\nu} \right]^2 \qquad (3.93)$$

The smallest value of F_n is the predicted evaporation field for a given metal, and the value of n indicates the number of unit charges of the predicted ion species.

The values of F_n and the expected ions are given by table 3.10. for the investigated metals, according to calculations of Brandon [153].

Polarization effects have been included into the predictions of the ion species, which cause a shift for Ni and Zn from singly to doubly charged ions. The F_n values in the table are not corrected with respect to polarization effects. The required corrections increase the F_n — values by 20—50%. The observed ions are in good agreement with the predictions.

TABLE 3.10.
CALCULATED EVAPORATION FIELDS AND EXPECTED AND OBSERVED ION SPECIES

Element	F_n V/Å	Expected Ion	Observed Ion	Temperature °K
Be	4.6	Be^{++}	Be^{++}	21
Fe	3.6	Fe^{++}	Fe^{++}	21
Cu	3.1	Cu^{+}	Cu^{+}, Cu^{++}	21
Zn	3.5	Zn^{++}	Zn^{+}, Zn^{++}	21
Ni	3.5	Ni^{++}	Ni^{++}	78

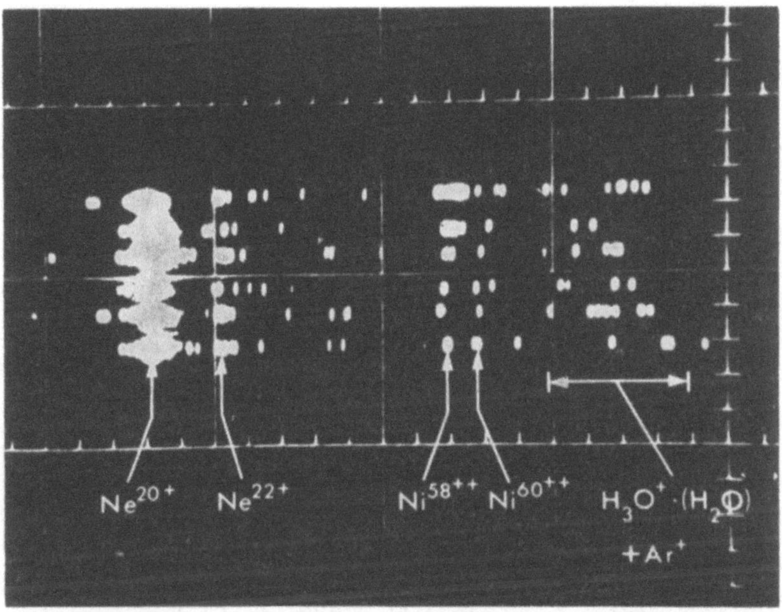

FIG. 3.45. Field evaporation mass spectrum of nickel; Ni + Ar — 2×10^{-6} Torr, tip temperature 78 °K [155]. Different traces for different tip voltages.

Fig. 3.45. shows the field evaporation mass spectrum of nickel in the presence of neon and argon at 2×10^{-6} Torr. $T = 78$ °K.

A completely unexpected effect is the change of the evaporation species of beryllium with temperature. At 21 °K beryllium field evaporates only as Be^{++}. At 78 °K, however, both Be^{++} and Be^{+} are observed, and at 240 °K beryllium evaporates only as Be^{+}. Fig. 3.46. shows the Be^{+}/Be^{++} ratio as a function of temperature as observed experimentally and predicted on the basis of thermal activation over a field reduced energy barrier. The ratio of the probabilities P^+/P^{++} is given by

$$P^+/P^{++} = exp\,[-(Q^+ - Q^{++})/kT] \tag{3.94}$$

Obviously, there are larger discrepancies between this theoretical ratio and the experimental one. Thus, the theory requires revision. It is possible that tunneling effects can be important in this case, as discussed by Gomer and Swanson [149] and by Brandon [157]. Tunneling from some thermally excited states will introduce a pre-exponential term into eqn. (3.94).

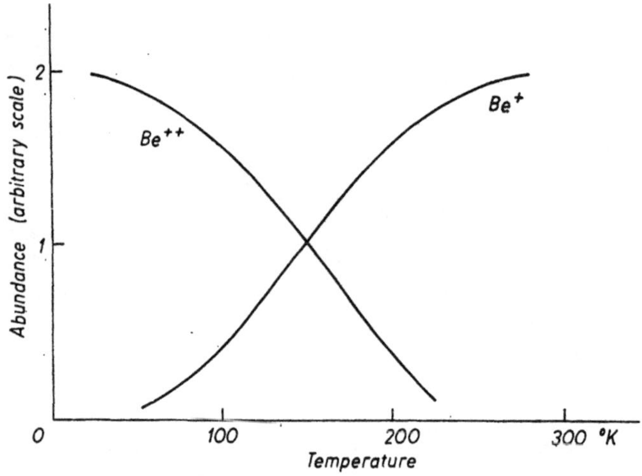

FIG. 3.46. Estimated abundances of the field evaporation species beryllium as a function of temperature with a constant evaporation rate [155].

Some time ago, Müller et al. [18, 158] discovered the "hydrogen promotion effect". If hydrogen is introduced into an FI microscope operated under nearly stable conditions of helium image formation, the field evaporation rate is increased to one to ten atom layers per second. This effect could be reconfirmed mass spectrometrically. The voltage has to be reduced up to 50% to obtain a stable image again. Metal hydride ions could be detected mass spectrometrically. Thus it can be seen that the promotion effect is due to the easier field evaporation of the hydrogen-metal complexes as compared with the pure metal atoms.

These and other observations prove the usefulness of field ion mass spectrometric studies of field evaporation.

Field desorption of organic radicals has been observed by a number of authors. Inghram and Gomer [1] first described field desorption of CH_3O radicals produced by dissociative chemisorption of methanol on tungsten. The author has extended this research together with Schulze [159] to a number of aliphatic alcohols and they have reconfirmed the mechanism for formation of the CH_3O^+ ion suggested by Inghram and Gomer. Fig. 3.47. shows

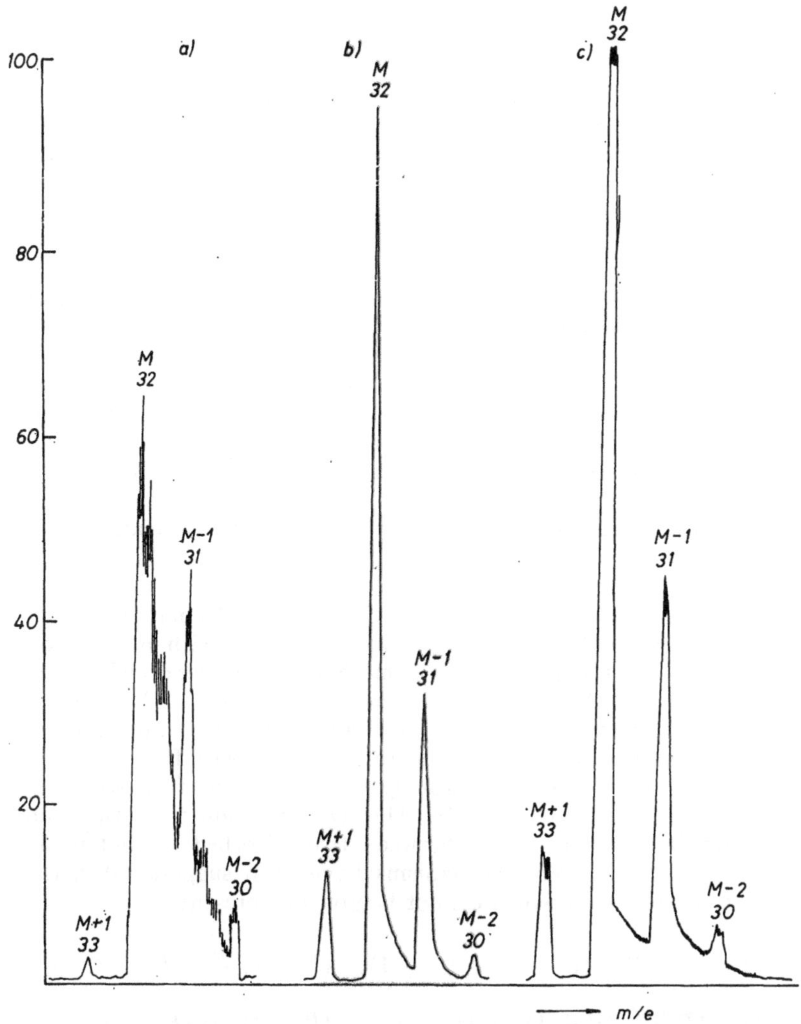

FIG. 3.47. FI mass spectrum of methanol on platinum tips [159].
a) extremely high field strength; *b*) somewhat lower field strength, same tip as with *a*); *c*) field strength comparable with measurement, *b*), but other platinum tip.

the FI mass spectrum of methanol on platinum tips at different field strengths. The parent peak ($m = 32$) is strongly broadened at high field strengths, indicating field ionization of the parent molecules in free space in front of the tip. The CH_3O^+ – peak ($m = 31$) is much sharper in case a), indicating a more localized place of origin (field desorption from the surface). At lower fields, the $P + 1$ ($m = 33$) exceeds the isotopic abundance of the ^{13}C atom. This peak is due to addition of a hydrogen atom to a methanol parent ion at the tip surface. This peak is sharp as compared with the parent ion peak. This again indicates field desorption of an ion species ($m = 33$) from the surface. The CH_3O^+ – peak ($m = 31$) is not completely sharp at the same field strength (case c)). This indicates that a fraction of the CH_3O^+ ions is produced by dissociation of the methanol ion in free space in front of the tip.

Ethanol and propanol FI mass spectra show strong $P - 1$ and $P - 2$ peaks at high field strengths. The intensity of these peaks increases strongly with field strength. The $P - 2$ ions are formed without any doubt by field desorption of surface complexes which have donated two hydrogen atoms to the tip surface. The alternative possibility, of a successive split of two hydrogen atoms in the gas phase, or a split of molecular hydrogen from the parent ions due to a rearrangement process in the gas phase, is extremely unlikely, as will be discussed in the section on structure analysis of organic molecules. Moreover, such delayed dissociation processes in the gas phase would lead to broadened mass lines which, however, are not found in these cases.

It seems that the mechanism for the $P - 1$ ion formation is different for the higher aliphatic alcohols, starting with n-pentanol. In contrast to the lower alcohols (methanol to propanol) where the relative $(P + 1)^+$ intensities decrease and the $(P - 1)^+$ increase with field strength, the $(P + 1)^+$ and $(P - 1)^+$ intensities of the higher alcohols change in a parallel manner with field strength. Both ion species decrease strongly with increasing field strength, as can be seen from fig. 3.48. The $(P - 1)^+$ ion formation can be attributed to reactions between two adsorbed alcohol molecules. Three different mechanisms have been suggested [159], the relative weights of which cannot be given at present:

$$\text{I)}\quad P^+ + P \rightarrow (P + 1)^+ + (P - 1); \quad (P - 1) \xrightarrow{field} (P - 1)^+ + e$$

$$\text{II)}\quad 2P \xrightarrow{field} (P + 1)^+ + (P - 1)^-; \quad (P - 1)^- \rightarrow (P - 1)^+ + 2e$$

$$\text{III)}\quad (2P + 1)^+ \xrightarrow{field} H_3O^+ + (P - 1) + (P - 17);$$

$$(P - 1) \xrightarrow{field} (P - 1)^+ + e.$$

The intensity of the dimeric alcohol ions decreases with field strength in a manner similar to that shown in fig. 3.48.

The much higher concentration of association products of the higher alcohols as compared with the lower ones is the reason for the predominance of mechanisms I—III for the $(P-1)^+$ ion formation as compared with the dissociative chemisorption mechanism.

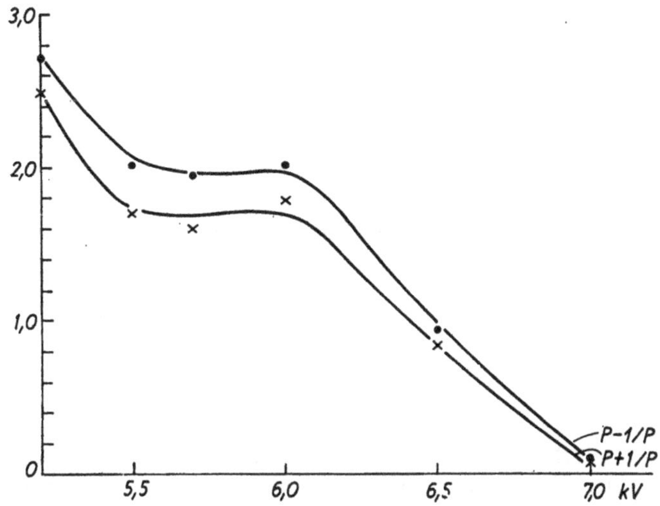

FIG. 3.48. Dependence of the $(P+1)/P$ and $(P-1)/P$ ratios of n-pentanol ions on the field strength. Field anode: platinum tip.

It is interesting to note that FI mass spectrometry is suitable for detection of phase transitions in the adsorbed layers [159]. Table 3.11. shows the relative intensity of the dimeric ions of propanol to n-hexanol on a platinum wire.

TABLE 3.11. RELATIVE INTENSITY OF THE DIMERIC IONS $(2M+1)^+$ OF ALCOHOLS ON PLATINUM WIRES. INTENSITY OF THE MOLECULAR IONS $M^+ = 100$

Propanol	*n*-Butanol	*n*-Pentanol	*n*-Hexanol
3.1	5.1	26.0	42.0

The intensity of the dimeric ions increases drastically from n-butanol to n-pentanol. This can be explained in the following manner. The alkyl chains of the lower alcohols from methanol to n-butanol are adsorbed parallel to the metal surface at low pressures (several 10^{-4} Torr). The alkyl chains of ethanol and propanol are ordered perpendicular to the adsorbent at higher pressure, according to Kemball [160], if mercury is used as adsorbent, for example. There is a pressure dependent two-dimensional phase transition. The critical pressure for this transition decreases with increasing chain length of the alcohol molecules. Whereas n-butanol is still adsorbed parallel to the metal surface at 10^{-4} Torr, n-pentanol is the first molecule of the alcohol series which is adsorbed with the hydrocarbon chain perpendicular to the surface at the same pressure. The formation of dimeric ions is favoured by the two-dimensional condensation of n-pentanol and the higher alcohols. The same phenomenon is observed with the protonated alcohol ions $(M + 1)^+$, formation of which is favoured by two-dimensional condensation at the metal surface. The relative intensities of the $(M + 1)^+$ ions of propanol, n-butanol and n-pentanol are 23, 27 and 200, respectively.

3.6.4.3. Field impulse desorption

The experiments using the field impulse desorption method as carried out by Inghram and Gomer [1], Block [9, 148] and by the author and Röllgen are discussed separately in this section. The method itself has been described already in section 2.8. The method is well suited for studying the kinetics of adsorption.

Inghram and Gomer [1] have introduced the impulse method and measured the field desorption of CH_3O^+ from tungsten. The conditions at 10^{-3} Torr are

$$(CH_3O)_{ads} \gg (CH_3OH)_{gas}, \quad \Theta \geqq 0.1.$$

$(CH_3OH)_{gas}$ corresponds to the number of methanol molecules in that part of the gas phase where the field is high enough for field ionization. Θ is the degree of coverage of the tip. If the field is switched on only for a very short period (about 1 μsec) as compared with the field free time (intervals of the order 0.1 to 100 nsec), the number of molecules approaching the tip is very small as compared with the very large number of adsorbed particles $(CH_3O)_{ads}$. The CH_3O^+ to CH_3OH^+ ratio at very high frequencies should be similar to that observed under d.c. conditions. However, the CH_3O^+ intensity should be much larger than the CH_3OH^+ intensity at low frequencies because the tip surface is a large reservoir for adsorbed particles which is filled in the interval between the impulses. This is confirmed experimentally, as can be seen from fig. 3.49. It is important that the rise time of the impulses is short as compared with the mean lifetime of field ionization. Otherwise the kinetic energy of the ions would not correspond to the tip potential.

A more detailed investigation using the field impulse method for the system cyclohexane — platinum was carried out by Block [9]. In this case one can differentiate between adsorption and chemisorption by observing the corresponding, different ionic species.

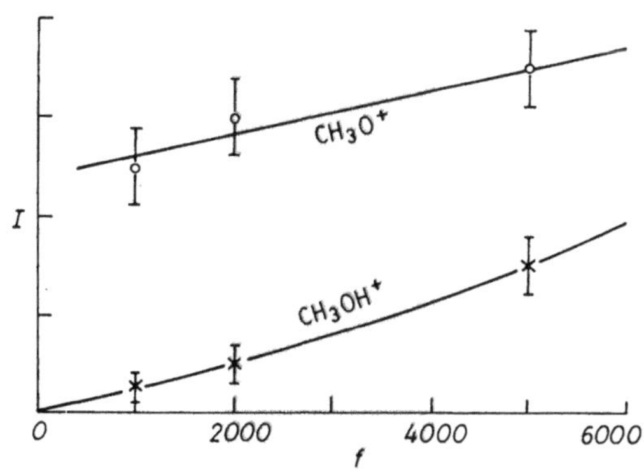

Fig. 3.49. CH_3O and CH_3OH ion current as a function of the pulse frequency f in sec^{-1} for pulses of 2 μsec plateau time [1].

Dissociative chemisorption of cyclohexane occurs with rupture of one or several C—H bonds. Thus $C_6H_{11}^+$ is found as the main product of chemisorption, whereas physically adsorbed cyclohexane is measured as $C_6H_{12}^+$. First, Block showed that the $C_6H_{11}^+$ ion (mass 83) is not formed by gas phase field dissociation of cyclohexane. In the latter case one would expect a strong increase of the $C_6H_{11}^+/C_6H_{12}^+$ ratio with the field strength. In contrast to this, it is observed experimentally that this ratio decreases strongly with field strength. This can be explained only by field ionization of chemisorbed complexes at a limited number of surface sites, on the one hand and by field ionization of the parent molecule in space on the other hand, which increases with field strength. Space ionization of the $C_6H_{12}^+$ ion was proved by broadening of the parent peak with field strength. Some further d.c. field experiments were made before application of the field impulse method. The ion current of cyclohexane ($m = 84$) is strictly proportional to the gas pressure in the ion source over the whole pressure range of 1 to 62×10^{-6} Torr. This indicates pure gas phase ionization (fig. 3.50.a.).

The pressure dependence of the ion current changes drastically if the field is pulsed in the same range of pressures. Fig. 3.50.b. shows the pressure dependence of $C_6H_{12}^+$ and $C_6H_{11}^+$ at field impulses of 1 μ sec duration and 1 m sec intervals. Both curves show saturation with increasing pressure. The upper curve represents a physical adsorption isotherm, the lower a chemisorption isotherm. Thus, it is seen that the bulk of the parent ions is formed from an adsorbed layer in the case of impulse operation of the FI source.

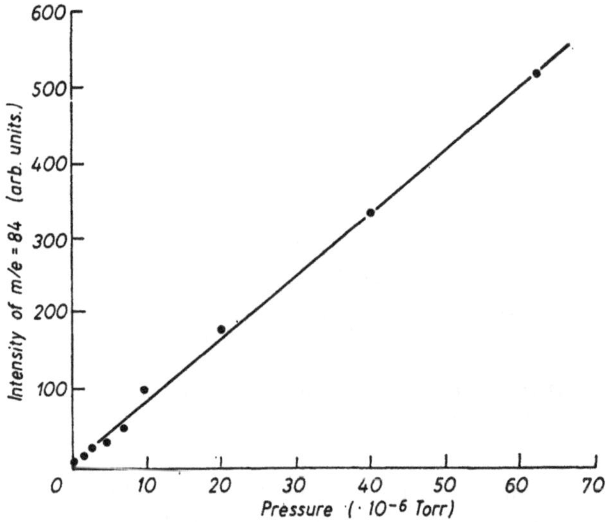

FIG. 3.50. Dependence of the parent peak of cyclohexane on the pressure. *a*) at stationary field according to Block [9] ($U_0 = 8.9$ kV; $T = 25$ °C).

The duration of the impulses does not change the mass spectrum of cyclohexane very much. However, change of repetition time leads to characteristic modifications of the intensities. This is shown by fig. 3.51. The intensities are related to the same number of impulses. A maximum intensity is found for the adsorbed molecules after 8 m sec, whereas the intensity of the chemisorbed radicals tends continually to saturation which is reached after 40 m sec. Order-of-magnitude gas kinetic calculations show that complete occupation of the originally free tip surface has to be expected after about 8 m sec. Thus the first part of the upper curve of fig. 3.51. represents a true rate of adsorption. Obviously, chemisorption is a consecutive reaction after adsorption of cyclohexane. The maximum of the upper curve can be explained in the usual way by the superposition of the rate curves for two consecutive reactions with different velocity constants. The mean life times for adsorption, τ_A, and for chemisorption, τ_c, can be derived from the curves. $\tau_A = 2 \times 10^{-3}$ sec and $\tau_c = 7.5 \times 10^{-3}$ sec.

The M_{83}/M_{84} ratios decrease with increasing pressure and decreasing temperature. The pressure dependence of chemisorption indicates a surface heterogeneity.

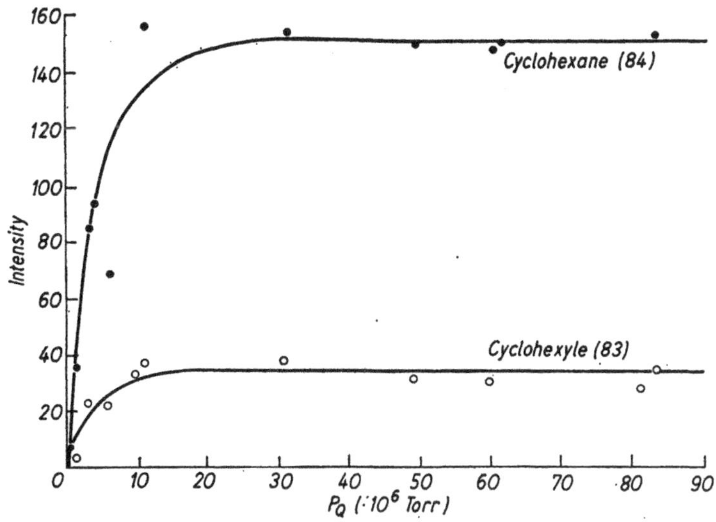

FIG. 3.50. Dependence of the parent peak of cyclohexane on the pressure b) at pulsed field (10^{-3} sec impulse interval; $T = 20\,°C$).

The chemisorption process studied by Block is an entirely irreversible process, the kinetics of which could not be investigated before by other methods. The field impulse desorption method is the first one which permits more detailed kinetic measurements of such processes. A cyclohexane molecule adsorbed in any position of the surface needs a time of about 10^{-3} sec for transfer of energy or for a shift into a specific position which is suitable for chemisorption.

Röllgen [139] has carried out studies with the field impulse desorption technique using 5 μm platinum wires or platinum foils as FI emitters. One of the reasons for using thin wires instead of tips was the need for larger ion current intensities. The second reason was that in the case of a wire position perpendicular to the slit in the cathode, the ion optical focusing conditions are much less sensitively dependent on variations of the cathode potential than with FI tips.

The FI wire has the shape of a loop because besides increase of its tensile strength — as mentioned before — the resonance vibrations of the emitter during impulse operation are reduced.

In contrast to the case of tips, one does not arrive at 100% field desorption after each field impulse if wires or sharp edges

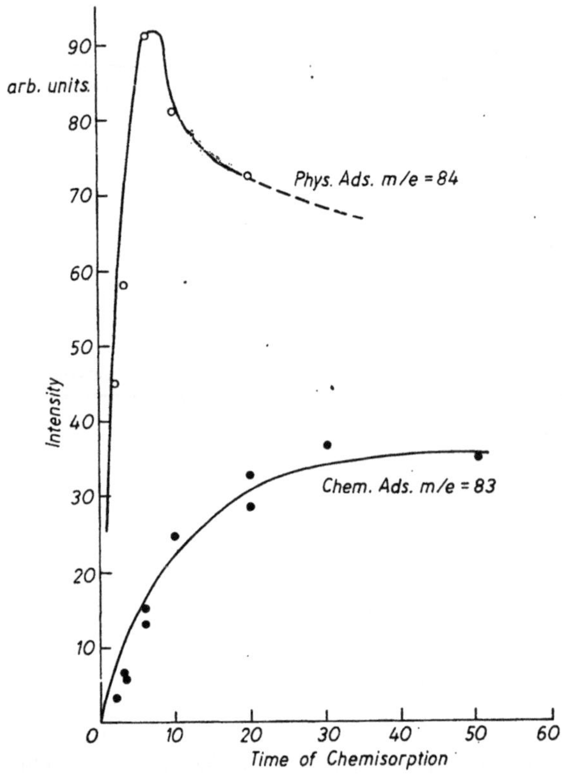

FIG. 3.51. Dependence of the physically and chemically adsorbed molecules on the adsorption time ($P = 5 \times 10^{-5}$ Torr, $T = 25\,°C$) [9].

are used as field ion emitters. However, the equilibrium of adsorption is strongly disturbed by the field impulses. Further, neutral products of surface reactions, which can undergo consecutive reactions during the intervals between the impulses, are formed during the impulse periods. One obtains information on such reactions by variation of different parameters, such as frequency and temperature.

Röllgen has carried out measurements on the catalytic activities of different field ion emitters on surface reactions of unpolar substances. The catalytic hydrogenation or dehydrogenation of paraffins, for example, is a reaction which sensibly depends on differences of the surface properties of the catalysts. It was observed that the catalytic dehydrogenation of n-heptane is much more intense on "pure" platinum emitters (Pt-foils) than on "activated" platinum emitters (Pt-wires) coated with needles of semiconducting organic material.

Table 3.12. shows the intensities of the parent and parent minus one or two hydrogen atoms peaks in the n-heptane impulse FI mass spectra on different emitters.

Table 3.12. Relative intensities of the P^+, $(P-1)^+$ and $(P-2)^+$ ions in the field impulse mass spectrum of n-heptane with different emitters under comparable experimental conditions

Peak	Pt-foil	"activated" Pt-wire
P	100	100
$P-1$	30–70	1
$P-2$	1–2	2

The peak corresponding to abstraction of one hydrogen atom from n-heptane increases from a fraction of a percent to 30–70 percent at impulse frequencies ranging from 1 k Hz to 250 Hz if a platinum foil is used. The corresponding peak intensity is very much smaller if an "activated" platinum wire (covered with needles) is used. The $P-1$ to $P-2$ ratio is reversed if platinum foils and activated platinum wires are compared.

The measured differences of catalytic activity of emission centres consisting either of carbon (or semiconducting organic material), or of platinum are in accordance with the well-known fact that the catalytic dehydrogenation depends strongly on the "percentage d-character" of the catalyst. Platinum has a high percentage d-character and consequently a high catalytic activity. The needles, on the other hand, consist of carbon (or a semiconducting organic material) which will have a negligibly small percentage *d*-character.

Table 3.13. shows the dependence of the relative intensity of the $P-1$ ions ($P = 100$) of n-heptane as a function of the impulse frequency.

At extremely high frequencies ($\gg 10^4$ Hz), the conditions are expected to be similar to the dc — conditions where the $P-1$ ion intensities drop below 1% of the parent ion intensity.

TABLE 3.13. FREQUENCY DEPENDENCE OF THE RELATIVE $P-1$ ION INTENSITY OF N-HEPTANE ON A PLATINUM FOIL. ($P = 100$). IMPULSE PLATEAU: 100 μ sec; AMPLITUDE: — 9 kV

m/e	10 k Hz	5 k Hz	2.5 k Hz	1 k Hz	500 Hz	250 Hz	90 Hz
99	17	18	19.5	31	43	73	105

The very strong increase of the $P - 1$ ion intensity in the frequency interval between 10^4 Hz and 90 Hz may be explained in the following way. In the first case, the adsorption time between two impulses is only 10^{-4} sec, in the second case, however, 1.1×10^{-2} sec. If one assumes that — as mentioned before by Block — the organic molecule has to stay first for a while in the physically adsorbed state until it is converted to a chemisorbed state at specific surface sites where it may react, then this time has to be regarded as too short at 10 k Hz, but large enough at 90 Hz impulse frequency. In this way, the kinetics of dissociative chemisorption can be studied.

3.6.5. CHANGE OF FI CURRENT WITH WORK FUNCTION

Metzinger and the author [161] investigated a new phenomenon, the change of the FI current of FI emitters with the work function of the emitters, as occurs during adsorption of different organic substances. Of course, change of work function is not the only parameter which influences the FI current of a specific substance if different adsorbates are added, as will be discussed later. But it is a parameter which produces pronounced effects.*)

The FI probability P of a substance is closely related to the tunneling probability $D_{(d_{min})}$ of the valence electron of the substance. P is proportional to $D_{(d_{min})}$ at $D_{(d_{min})} \leq 10^{-4}$, and $P \simeq 1$ at $D_{(d_{min})} \geq 10^{-4}$, at typical measuring conditions. Fig. 3.52. shows the tunneling probability as a function of the field strength at different work functions Φ of the emitter and at different ionization potentials of the adsorbate molecules. The tunneling probability was calculated according to the formulae given by Müller and Bahadur [3] for the hydrogen atom (see eqn. 1.1 and 1.3). The equation may also be applied within a certain degree of approximation for atoms similar to hydrogen, but with different ionization potentials. The work function enters into the equation as a parameter via the distance l of eqn. 1.1 and 1.3a

*) An increase of the work function is formaliy equivalent to a promotion the ionization probability by chemisorbed ions on the emitter surface [263].

$l = d_{min} \simeq \frac{I - \Phi}{F}$. Qualitatively speaking, a higher work function leads to a smaller minimum distance for FI, and hence to a larger tunneling probability, as can be seen from fig. 3.52. The work function can be changed by physical or chemical adsorption. Physical adsorption leads generally to a lowering of the work function by an amount $4\pi\sigma M$. (σ = surface concentration of the adsorbate and M = average dipole moment per adsorbed molecule perpendicular to the surface.) Chemisorption can lead to a rise as well as to a lowering of work function.

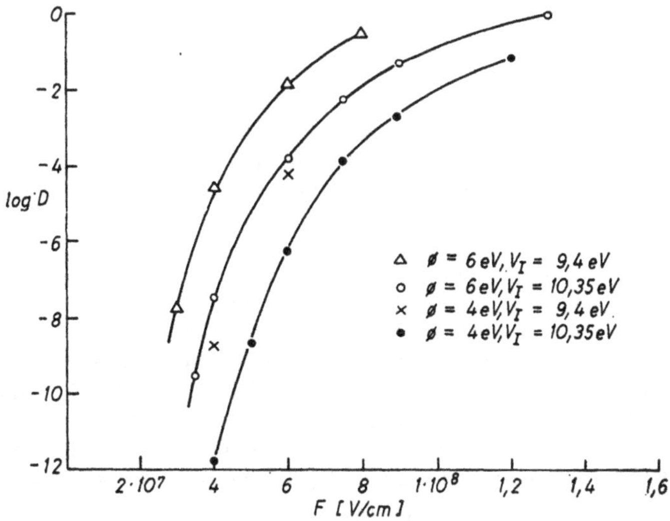

FIG. 3.52. Dependence of the tunneling probability D on the field strength F, the work function Φ and the ionization energy I [161].

Unpolar molecules are adsorbed as flat as possible to the metal surfaces at very low pressures, according to Kemball and Rideal [162]. Polar molecules are adsorbed in such a way that the polar groups — which contribute most to the adsorption energy — are in direct contact with the surface. The permanent dipole moment takes a preferential orientation under the influence of the surface field and the external field. The hydrocarbon chain of polar molecules is adsorbed onto the surface at very low pressures, as stated earlier. If the gas pressure is increased, two-dimensional clustering of the adsorbed molecules occurs, and the adsorbate density changes like a step function [163–166]. The pressure above the surface when phase transitions occur in the monolayer is between 1/100 and 1/1000 of the saturation vapour pressure P_s. Long hydrocarbon chains of polar molecules are arranged perpendicular to the surface and parallel with each other during two-dimensional clustering or "two-dimensional condensation". (Here, in contrast to three dimensional condensation, multilayer formation is excluded).

The adsorbate density is enhanced in the case of two-dimensional condensation of polar molecules. Therefore, the work function is changed appreciably, and hence the FI current-voltage curve is strongly changed.

In the case of mixed adsorption of several substances, a preferred adsorption of the component of highest adsorption energy occurs. Therefore, the work function is governed mostly by the substance having the highest adsorption energy. These effects may lead to a non-linearity of the mass-separated ion currents of an FI mass spectrometer with the partial pressures, as will be discussed in the section on quantitative analyses.

The shift of the FI current-voltage curves of a given substance with addition of another substance has been investigated by Metzinger and the author [161]. They used a simple FI diode system consisting of a platinum-Wollaston wire of 2.5 μm diameter and 10 mm length and a cylindrical cathode which served as an ion collector. The shift of an FI current-voltage curve with change of work function can be understood in the following way. Those protrusions of the wire are active as field ion emitters where a critical field strength F_0 is realized. The FI probability P_i is approximately equal to one at F_0. $P_i = 1$ is reached at higher F_0 values if the work function is lowered, as can be seen from fig. 3.52.

Therefore, the applied voltage $V_0(\beta)$ has to be increased because

$$\frac{V_0(\beta)}{r \cdot \ln R/r} \beta = F_0 .$$

Consequently, the whole current-voltage curve is shifted to the right, i.e. to higher voltages:

$$\Delta \log V = \log \{F_{0(\Phi-\Delta\Phi,I)}/F_{0(\Phi,I)}\} \tag{3.95}$$

The curve is shifted to lower voltages in the case of increase of work function due to a chemisorption process. (It should be borne in mind that only the shift of the curves with change of work function is analysed here, but that additional effects are effective in case of mixed adsorption.)

In an analogous manner, a change of the ionization energy of two substances (the other parameters being constant) leads to a shift of the current-voltage curves in the direction of the voltage axis:

$$\Delta \log V = \log \{F_{0(\Phi,I_2)}/F_{0(\Phi,I_1)}\} \tag{3.96}$$

On the other hand, the curves are shifted vertically if the particle supply to the FI emitter is changed.

The change of work function may occur as a relatively slow process (of the order of minutes) if two-dimensional condensed phases of polar substances are formed, for example. Therefore, the observed shifts of the FI curves extend over a time interval of several minutes.

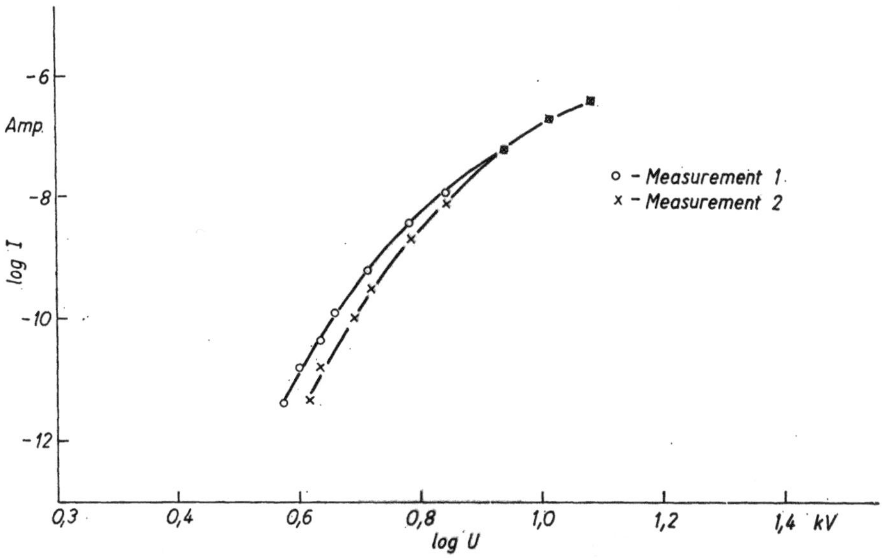

FIG. 3.53. Shift of the FI threshold voltage of n-heptane at two subsequent measurements. Field anode: 2.5 μm Pt-wire, $p = 10^{-4}$ Torr.

The shift of the FI curves of unpolar substances with time is much less pronounced than that of polar substances. Fig. 3.53. shows the current-voltage curve for heptane. The voltage was gradually increased with curve (1), and then after a couple of minutes, the curve was measured again, starting with a low voltage (curve 2). It can be assumed that two-dimensional condensation of the n-heptane occurs at high voltages under the influence of the polarization forces. If the voltage is decreased suddenly, this phase does not disappear instantaneously but only after about 15 min. This is due to the enhanced adsorption time, $\tau_A \sim exp. (E_A/RT)$, where the adsorption energy E_A is of the order 5—10 kcal for two-dimensional condensation of n-heptane, according to Jura, Basford, Loeser and Harkins [164].

Pentanol has the same ionization potential as n-heptane. But the FI current is smaller by a factor of 14 as compared with n-heptane, all other parameters being the same. The n-heptane intensity is reduced by the same factor if first a measurement with a clean surface is made, and afterwards a measurement of n-heptane is made using a surface pre-treated with n-pentanol. This phenomenon can be attributed to a strong lowering of the work function due to two-dimensional condensed n-pentanol layers.

The strong shift of the FI current-voltage curve of n-pentanol with time may be seen from fig. 3.54. The first curve ($M1$) was

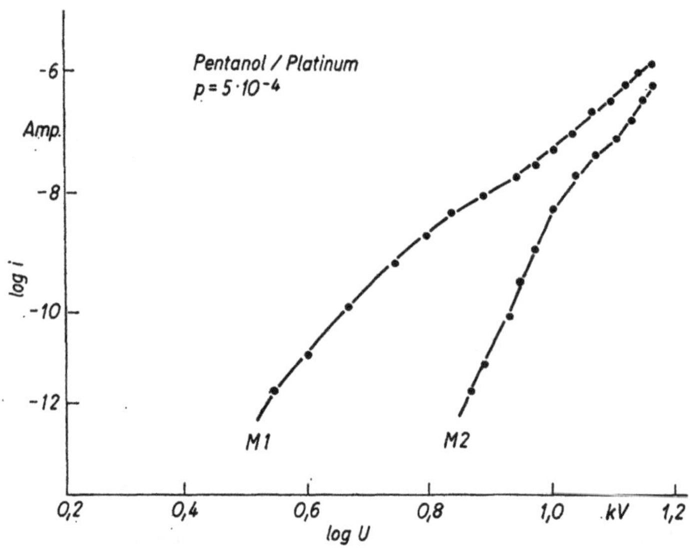

FIG. 3.54. Shift of the FI threshold voltage of n-pentanol at two subsequent measurements. Field anode: 2.5 μm Pt-wire, $P = 5 \times 10^{-4}$ Torr.

measured starting with low voltages. After a couple of minutes, curve ($M2$) was measured, also beginning with low voltages. This effect, which is explained by a lowering of the work function due to formation of a condensed phase, is confirmed by mass spectrometric observations which give strong evidence of such a condensed phase (see section 3.6.4.2., table 3.11.).

The shift of the current-voltage curve with time has the consequence that at constant voltage the ion current decreases with time. At the beginning, this decrease occurs quickly, later somewhat slower. After about 10 min half of the original intensity is lost.

The slowness of the process is explained by the fact that some time is required for statistical formation of clusters which serve as nuclei for two-dimensional condensation.

The current-voltage curves of ethanol, propanol and butanol do not shift with time. This is in accordance with the fact discussed earlier that these substances do not form condensed phases on the surface. It is also in accordance with the mass spectrometric observations which show many fewer dimeric ions (3%) than pentanol (57%) (see section 3.6.4.2.). So far, the shift of the current-voltage curves with adsorption of paraffins and alcohols to higher voltages has been discussed. The opposite effect, namely the shift to lower voltages, has been observed with unsaturated hydrocarbons like benzene, ethylene and acetylene. This effect can be attributed to an increase of the work function during chemisorption of these substances.

The ion intensity increases with time by a factor of about two at constant high voltage in the case of benzene. If the ion current of pure n-heptane is measured first and then small amounts of benzene, ethylene or acetylene are added at constant voltage, the ion intensity of n-heptane increases strongly. The whole current-voltage curve of n-heptane is shifted to lower voltages in this case.

The change of work function during adsorption of benzene, ethylene and acetylene on metal surfaces has been investigated by several authors, (Suhrmann, Krüger and Wedler [167], Mignolet [168], Culver and Tompkins [169], for example). These authors have carried out their experiments generally at low temperatures ($T = 90\,°K$). At these low temperatures, a lowering of work function with adsorption of the unsaturated hydrocarbons was generally observed.

However, at higher temperatures (about 300 °K) hydrogen is produced in the system benzene-platinum, according to Suhrmann and Wedler [170]. The adsorbed hydrogen atoms enhance the work function since they are polarized to H^-. This is in accordance with the observed shift of the FI current-voltage curves.

(For a new interpretation see [260]).

3.6.6. Dependence of FI Currents on Surface Diffusion

The particle supply by surface diffusion will be discussed for the case of a wire emitter in this section [171], although the derivations are applicable also for other emitter geometries.

The emitter wire consists, as discussed earlier, of a smooth part with about 10^6 protrusions of about 100 Å diameter. Ionization occurs only at the top of those protrusions with an FI probability $P_i \simeq 1$. The smooth part of the wire provides a very large reservoir for incoming particles as compared to the protrusions. The field strength at the smooth parts is relatively low, and the number of the incoming particles from the gas phase is approximately equal to the gas kinetic value

$$\dot{n}_0 = \frac{{}^1n c}{4} \qquad \text{per cm}^2$$

The particles are adsorbed and thermally accommodated to the surface. On the average, the residence time of the adsorbed particles at the surface is

$$\tau_A = \tau_0 \, exp. \, (E_A/RT) \tag{3.97}$$

where τ_0 is of the order of 10^{-13} sec, and E_A is the adsorption energy. The particles can move relatively freely over some distance on the surface during the residence time τ_A.

The surface concentration of the adsorbed molecules at the protrusions is enhanced as compared to that at the smooth part of the wire by a polarization term $exp \cdot (\Delta E_p/RT)$ at complete equilibrium. $\Delta E_p = 1/2\,\alpha\,(F_0^2 - F^2)$ where F_0 is the field at the protrusions and F that at the smooth parts $(F_0 \gg F)$. Therefore, $\Delta E_p \simeq 1/2\,\alpha\,F_0^2 = E_p$. Taking $\alpha = 13.6 \times 10^{-24}\,\text{cm}^3$ for n-heptane, $F_0 = 6 \times 10^7$ V/cm and $T = 300\,°\text{K}$, one finds an enhancement of surface concentration at the protrusions by a factor of 650. If in the case of strong field ionization $(P_i \simeq 1)$ the equilibrium concentration of the adsorbed molecules is to be maintained, the particle supply must also be increased by a factor 650 over the normal value $\dot{n}_0$. The enhancement of particle supply from the gas phase is only by a factor $\sqrt{1 + E_p/RT}$ for the wire geometry, a factor of 3 for this example. The particle supply from the surface must be much higher than the gas phase supply if the equilibrium concentration is to be nearly maintained when field ionization occurs.

The average transition time τ' for the motion of an adsorbed particle from one adsorption site to another is given by the approximate expression

$$\tau' = \tau'_0 \, exp. \, (E_D/RT) \tag{3.98}$$

Where $\tau'_0 \simeq \tau_0 \simeq 10^{-13}$ sec, and E_D is the activation energy for surface diffusion. $E_D \simeq E_{A/4}$ for a number of substances. $E_A \simeq 12$ kcal for n-heptane. Thus the following values result:

$$\tau_A = \tau_0 \cdot exp. \, (E_A/RT) = 5 \times 10^{-5}\,\text{sec} \tag{3.99}$$

$$\tau' = \tau_0 \cdot exp. \, (E_A/4RT) = 1.5 \times 10^{-11}\,\text{sec}. \tag{3.100}$$

The number of transitions between the adsorption sites is given by

$$n_s = \tau_A/\tau' = 3.3 \times 10^6 \tag{3.101}$$

Assuming with de Boer [172] that the adsorption sites are distant by a $\simeq 3$ Å, one obtains for the distance by which an adsorbed particle is displaced during the average adsorption time

$$S = a \cdot \dot{n}_s = 10^{-1}\,\mathrm{cm}.$$

S is not the displacement in a straight direction, because the adsorbed particles change direction randomly. The average displacement in one direction is given by

$$\overline{x^2} = 2\,Dt \tag{3.102}$$

The diffusion coefficient D is given by de Boer [173] by

$$D = a^2/4\tau' \tag{3.103}$$

For the given example, $D = 1.5 \times 10^{-5}$,

$$S_A = \sqrt{\overline{x^2}} = \sqrt{2D\tau_A} = 3.9 \times 10^{-5}\,\mathrm{cm}. \tag{3.104}$$

It is seen that the particles diffuse on the average by a distance over the surface which is about 40 times larger than the diameter of the emission centres.

Consider a micro tip on a plane. The FI probability at the micro tip may be assumed equal to one, that at the surroundings equal to zero. The micro tip is surrounded by a circular area of radius S_A which is characterized by the condition that all particles impinging within the area $\pi\sigma_A^2$ from the gas phase may diffuse to the micro tip in the centre of this area without desorption. According to eqn. (3.104) $\pi\sigma_A^2 = \pi \cdot 2Dt_A$ (3.105). Inserting (3.103) (3.99) and (3.100) in (3.105), one obtains

$$\pi\sigma_A^2 \simeq exp.\left(\frac{3E_A}{4RT}\right) \tag{3.106}$$

Qualitatively speaking, the particle supply to the micro tips by surface diffusion strongly increases with the adsorption energy of the diffusing particles. This result still holds if the approximate relation $E_D \simeq E_A/4$ is replaced by a more flexible relation. Next, the problem considered will be to what extent the differences of particle supply of various substances are reflected by differences of FI currents.

Some definitions will be given first.

$\dot{n} = dn/dt = \dot{n}_g + \dot{n}_s =$ total particle supply to an emission centre. ($\dot{n}_g =$ supply from the gas phase, $\dot{n}_s =$ supply by surface diffusion).

n_a° and n_a are the adsorbate concentrations at equilibrium and nonequilibrium conditions, respectively. Subscripts A, B refer to two different adsorbates of different adsorption energy.

$f \equiv \dfrac{n_{aA}^\circ}{n_{aB}^\circ} \Big/ \dfrac{\dot{n}_A}{\dot{n}_B}$ is the "enrichment factor".

k_i is the rate constant for field ionization of the adsorbed particles and k_d the rate of neutral desorption from the emission centre. It is assumed, for the moment, that field ionization occurs only within the adsorbed layer and not in free space. Generalizations will be discussed later.

The rates of arrival for particles A and B are $\dot{n}_A$ and $\dot{n}_B$. The rates of departure from the emission centre are $k_d n_{aA}^\circ$ and $k_d f n_{aB}^\circ$ in the case of equilibrium, and $(k_d + k_{iA})\, n_{aA}$ and $(f k_d + k_{iB}) n_{aB}$ in the case of noticeable field ionization. Equating the rates of arrival and of departure, and forming the quotient of the densities for A and B, one obtains

$$\frac{i_A}{i_B} = \frac{\dfrac{n_{aA}^\circ}{n_{aB}^\circ}\dfrac{k_{iA}}{k_{iB}} + \dfrac{k_{iA}}{k_d}\cdot\dfrac{\dot{n}_A}{\dot{n}_B}}{1 + k_{iA}/k_d} \tag{3.107}$$

Two limiting cases will be discussed.

1. $k_{iA,B} \gg k_d$, $\dfrac{n_{aA}^\circ}{n_{aB}^\circ}\dfrac{k_{iA}}{k_{iB}}$ is assumed to be only slightly different from $\dfrac{\dot{n}_A}{\dot{n}_B}$.

The result is:

$$i_A/i_B = \dot{n}_A/\dot{n}_B \tag{3.108}$$

2. $k_{iA} \ll k_d$. The result is:

$$i_A/i_B = n_{aA}^\circ/n_{aB}^\circ \cdot \frac{k_{iA}}{k_{iB}} \tag{3.109}$$

$n_{aA}^\circ/n_{aB}^\circ$ may be represented by an expression for mixed adsorption as given by Damköhler [174], extended by a polarization term $exp\cdot \dfrac{(E_{PA} - E_{PB})}{RT}$,

$$n_{aA}^\circ/n_{aB}^\circ = const.\ \frac{P_A W_B}{P_B W_A}\, exp.\left(\frac{E_{ads.A} + E_{PA} - E_{ads.B} - E_{PB}}{RT}\right) \tag{3.110}$$

where the constant comprises the quotient of the partition functions of the adsorbates, $P_{A,B}$ are the gas partial pressures and $W_{A,B}$ are surface areas per adsorbed particle.

It will now be shown that condition 2. is generally not realized. The field strength is rather low at the condition $k_i \ll k_d$. Then the minimum distance for FI, $d_{min} \simeq \frac{I - \Phi}{F}$ is large. For a number of substances, $F \leqq 0.4$ V/Å is a field strength where the FI current is much smaller than the particle supply current. Taking $I = 11$ eV and $\Phi = 5$ eV, one obtains

$$d_{min} \geqq 15 \text{ Å}$$

This is much more than the depth of the adsorbate layer (except for large chain molecules arranged perpendicularly to the surface). The assumption underlying eqn. (3.107) (field ionization occurring within the adsorbate layer, $i = k_i n_a$) is no longer valid. The rate of ionization is proportional to the particle concentrations n_t at the distance d_{min} ($\geqq 15$ Å in the present case), $k_i \ll k_d$:

$$i_A/i_B = \frac{n_{tA}^\circ \cdot k_{iA}}{n_{tB}^\circ \cdot k_{iB}} = \frac{n_A^g \cdot k_{iA}}{n_B^g \cdot k_{iB}} \cdot exp. \left(\frac{E_{PA} - E_{PB}}{RT}\right) \quad (3.111)$$

The ion currents are no longer dependent on the adsorption energies, in contrast to the conditions underlying eqn. (3.107), but they depend on the polarization energies E_{PA} and E_{PB}. The general case ($k_i \approx k_d$) is given by eqn. (3.107) under the present conditions, and n_a° in (3.107) has then to be replaced by n_t°.

If the field strength is increased, the minimum distance for FI approaches more and more closely to the boundary of the adsorbate. However, FI still can occur only in the gas phase, and the probability for FI remains relatively low ($k_i \ll k_d$). A stepwise increase of FI probability occurs if d_{min} coincides with the minimum distance for field ion desorption, after further increase of the field strength. The residence time t of the adsorbed particles in the neutral state is large as compared with the average lifetime τ_i with respect to field ionization. Thus, the FI probability $P_i = (1 - e^{t/\tau_i})$ suddenly becomes equal to one. Then the first limiting case (3.108) of eqn. (3.107) becomes effective, i.e.

$$i_A/i_B = \frac{k_{iA}}{k_{iB}} \cdot \frac{\dot{n}_A}{\dot{n}_B} = \frac{k_{iA}}{k_{iB}} \frac{(\dot{n}_A^g + \dot{n}_A^S)}{(\dot{n}_B^g + \dot{n}_B^S)} \quad (3.112)$$

Suppose that the densities of substances A and B in the gas phase are equal. Considerable differences in the FI currents may then be observed, because the particle fluxes by surface diffusion, $\dot{n}_A^S$ and $\dot{n}_B^S$ may be quite different if the adsorption energies are different (see eqn. (3.106)).

Summarizing the situation for field ion desorption: The differences of FI currents of different substances do not depend mainly on the equilibrium concentrations of the adsorbates (because $k_i \gg k_d$), but mainly on differences of particle supply by surface diffusion.

3.7. FIELD-INDUCED RESONANCE STATES AT A SURFACE

Observation of field-induced quantum states at a surface has been reported by Jason, Burns, Parr and Inghram [177,178]. A detailed description of the experimental results and a thorough theoretical treatment of this phenomenon has been given by Jason [179]. The discovery of field-induced resonance states by Jason opens a basic new insight into some properties of a surface.

The most essential experimental observation is represented in fig. 3.55. In the FI mass spectrum of hydrogen at relatively high resolution it is found that the H_2^+ ion peak shows some structure corresponding to groups of ions with different kinetic energies. H_3^+ ions are found only as a relatively sharp single peak. They are formed by a surface reaction. The H^+ peak also shows some structure, but the peaks are more diffuse. The H_2^+ peaks will be discussed first. Field ionization of the H_2 molecules cannot occur at distances smaller than a minimum distance from the surface, d_{min}, which is given by the relation $eFd_{min} = I - \Phi - \frac{e^2}{4d_{min}}$, as discussed earlier. The energy deficit of ions formed at d_{min} as compared with ions which are assumed to have started from the tip surface is $e\Delta E_c = eFd_{min} + \frac{e^2}{4d_{min}}$ where ΔE_c is the potential difference between the tip surface and d_{min}. Combining these two relations one obtains

$$e\Delta E_c = I - \Phi \qquad (3.113)$$

The energy deficit of the first H_2^+ peak represented in fig. 3.55. is thus adequately explained. The energy analysis is made here in a differential manner by using the magnetic analyser. (Energy resolution about 1/1000.) Energy analysis was also performed integrally by using a retarding potential technique. The ions leaving the tip through an aperture in a screen are focused by a quadrupole lens system onto a slit in a retarding electrode. (The slit is covered with a fine mesh.) The portion of the ion beam which is transmitted is refocused to the mass spectrometer entrance slit. (Energy resolution of the retarding system about 1/10000).

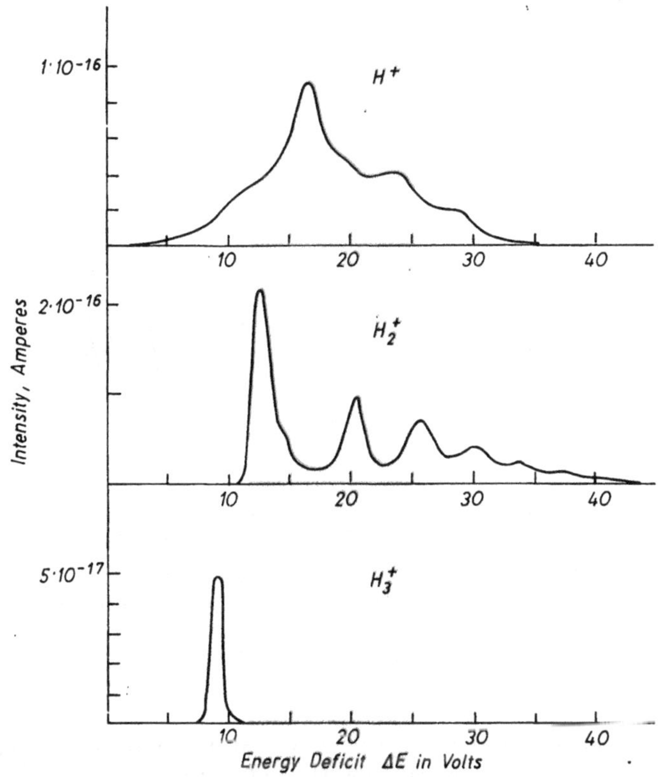

FIG. 3.55. Energy deficit ΔE of different ions from full tip energy for hydrogen on a tungsten tip in a field of 2.5 V/Å. Resolution of the magnetic analyser about 2 V. $T = 77\,°K$ [177].

The ion source is surrounded by a cryostat and can be cooled with liquid hydrogen or helium. When the system is well baked, pressures are in the range of 10^{-10} Torr if the source is cooled to 20 °K. The mass spectrometer is a 12-in.-radius, 60° magnetic sector field instrument.

The observed structure is not of instrumental origin, which can be concluded (among others) from the following facts. The same structure is observed with the two different types of analyses.

Structure is independent of focusing conditions but depends only on field strength. All peak intensities vary linearly with pressure from 10^{-3} to 10^{-6} Torr ruling out gas collisions. Structure is independent of tip treatment and residual vacuum. Peak separation is independent of tip radius.

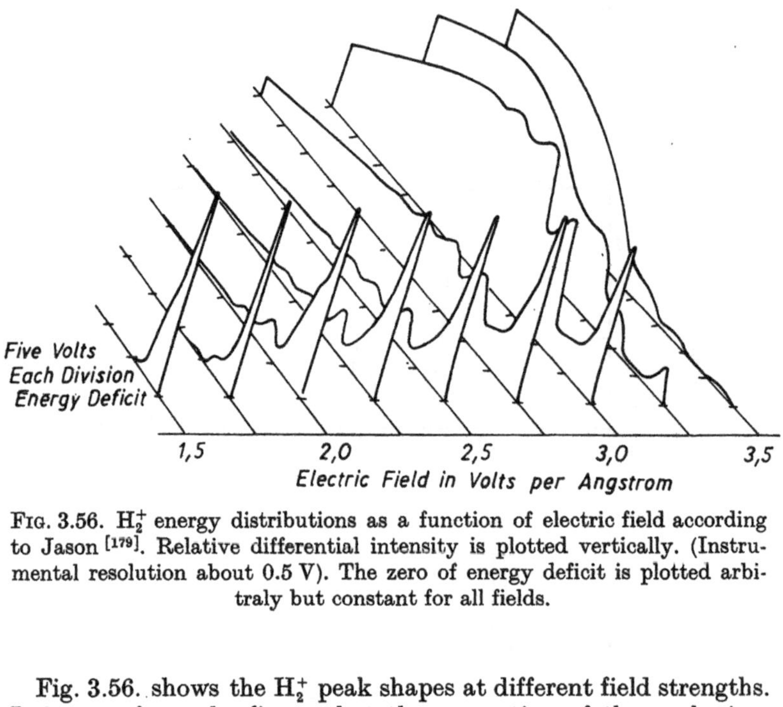

FIG. 3.56. H_2^+ energy distributions as a function of electric field according to Jason [179]. Relative differential intensity is plotted vertically. (Instrumental resolution about 0.5 V). The zero of energy deficit is plotted arbitraly but constant for all fields.

Fig. 3.56. shows the H_2^+ peak shapes at different field strengths. It is seen from the figure that the separation of the peaks is a (linear) function of the field strength. Peak width and energy separation decrease with increasing energy deficit. The ratios of the intensities of the energetically deficient peaks to the first peak decrease with field. It is also seen from the figure that at high fields the peaks of high energy deficit predominate to form a broad gas phase ionization structure. The theoretical interpretation of the observed phenomena as given by Jason is as follows. Fig. 3.57. shows the potential distribution at the surface with an atom near the surface in the presence of a high electric field. The surface and the electric field potential form a nearly triangular potential well. Consider an electron which tunnels into the potential well at an energy level above the Fermi level. There exists a certain probability that the electron will be reflected by the metal surface. Assume that the reflectivity is unity and that the probability of tunneling back to the atom is very low. Then well defined electron energy levels will exist within the well. If the reflectivity is smaller, the lifetime of the electrons in the stationary states

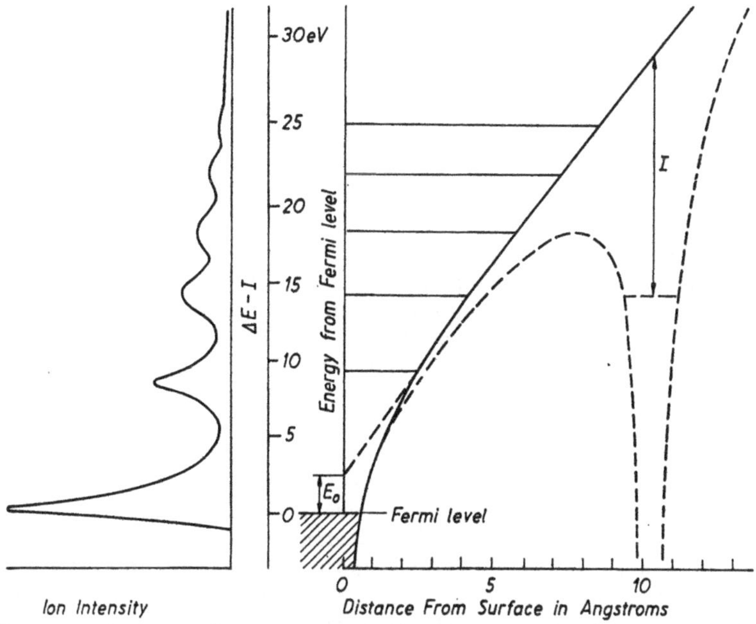

FIG. 3.57. Plot of surface potential configuration on the right and observed ion energy distribution on the left for H_2 molecules with an applied field of approximately 2.5 V/Å, according to Jason [179]. The ion distribution is shifted from ΔE (the energy deficit from surface potential) by I, the ionization potential, so that the features are aligned with the pertinent level. The potential distribution with an atom (of $I = 15.6$ eV) present is drawn in dashed lines. Extrapolation of the potential energy (not including the atomic field) to the surface defines E_0. Locations of the first five stationary states of the infinite triangular well are also shown. The scale in the center is in electron volts and applies to both sides.

will be shorter, and level broadening will occur. The highest probability of electron transfer from the atom to the continuum will be realized if the atomic levels are energetically aligned with the well levels. This alignment of the atomic levels occurs at certain distances between the atom and the surface because the atomic levels are raised by the field potential with increasing distances. In this way, the relatively sharp states of the atom "scan" the energy axis of the well.

The wave function at the boundary of the atom, the potential well and the surface must be matched for solution of the problem. The amplitudes of the wave functions in the potential well are given in terms of Airy-functions [240]. It is found that high tunneling rates are to be expected only if the atomic levels are aligned with the well levels. The presence of the atomic and image potentials introduces a small change in the well levels. The resulting potential may be replaced by a closed and an open square well for approximate calculations.

The eigenvalues E_n of an infinite triangular well are given by the relation

$$-(2m/\hbar^2 e^2 F^2)^{1/3} E_n = a_n \tag{3.114}$$

where a_n is the n th root of the Airy-function which is finite for all real values of its argument. The quantity E_0 must be added to E_n, which is the difference between the Fermi energy and the energy at which the extrapolated field curve intersects the surface. In fig. 3.57. both the results of these calculations and the experimental energy levels are shown. The agreement is reasonable. The energy scale is shifted so that the first peak coincides with the Fermi level. The calculated variation of the peak distances with the strength of the electric field agrees well with the experimental results. The dependence is virtually linear.

An only partially resolved fine structure of the first peak is found in addition to the main structure discussed so far. One can expect that one will obtain further interesting information on the surface — especially on the density of surface states — from further analysis of this fine structure.

3.8. ISOTOPIC EFFECTS IN FI MASS SPECTRA

Isotopic effects in FI mass spectra can be studied from two different aspects. First, one can investigate whether the high external electric field enhances the differences of dissociation rates between molecules containing light or heavy isotopes, respectively. Such an effect could perhaps be expected because the tunneling rate of an atom or a radical through a potential barrier narrowed by the high electric field depends strongly on the mass (see eqn. 1.3).

Second, one can use the isotope labelling technique for elucidation of the mechanisms of decomposition of organic ions, a widely applied technique in the electron impact mass spectrometry.

Isotope effects in FI mass spectra have been studied from both of these aspects by Dahmen and the author [229,230]. A very brief summary of some results will be given here. The ratio of the CH_2OH^+ to the CD_2OH^+ fragment ions from 1-propanol and 1-propanol-D_2 was found on the average to be about three. A similar observation was made for the analogous butanol molecules.

The ratio of the 1-butanol minus one hydrogen to 1-butanol-D_2 minus one deuterium ions is also about three, on the average.

Both of the isotope effects described here can be interpreted as the well-known regular kinetic isotope effects as described by Biegeleisen and Wolfsberg [231]. These authors were able to show, on the basis of the transition state theory of Eyring [232], that the ratio of the dissociation rates of two isotopic molecules is dependent on the ratio of the partition functions of the two molecular species in their normal and their activated states. The difference between the observed and calculated rates due to the regular kinetic isotope effect usually does not exceed a factor of ten but is lower in most cases. Therefore, the observed difference of rates of a factor of three in the two cases of field ionization of isotopic alcohols described above seems to be within the limit of the regular kinetic isotope effect. Tunneling of a proton era D^+ ion through the potential barriers in the presence of a very high electric field may increase the reaction rate. The tunneling rate is strongly mass dependent, as stated before.

A surprisingly prominent effect is the difference of rates for the elimination of CH_3 from normal propanol and 1-propanol-D_2, which is equal to a factor of about ten. The "secondary kinetic isotope effect" described in the literature is much smaller. It is observed with molecules where the isotopic atoms are not connected with the two atoms which are linked by the dissociating bond (the β bond of $CH_3^{\beta}-CH_2^{\alpha}-CD_2OH$ in our example). Biegeleisen and Wolfsberg [231] give a value of $k_H/k_D = 1.74$ for the ratio of the rate constants for the case of a H—D exchange on a C—H bond which does not involve the carbon atoms of the dissociating bond.

Therefore, there must be an additional reason for the relatively large isotope effect observed in the case of CH_3 elimination of alcohols. One possible reason is the "mass effect" of field dissociation which was discussed earlier (see section 3.2., eqn. 3.10a). It was shown that the probability of field dissociation is higher, the smaller the mass of the ionized fragment is relative to the mass of the neutral fragment, all other parameters being the same. The relative mass difference between CH_2CH_2OH and CH_2CD_2OH is about 4%. An equivalent difference of 4% of the effective field strength according to eqn. 3.10a can lead to a noticeable difference of the dissociation probability of a bond, as can be seen from table 2.2. (tip voltages at 4.4 and 4.6 kV). It is improbable, on the other hand, that the "mass effect" would be the only reason for the large observed isotope effect. Further processes may be effective which are presently unknown.

A remarkable value effect is found for the ratio of the 1-butanol minus H_2 to 1-butanol-D_2 minus D_2 peaks. This ratio can range up to about one hundred, depending on the surface conditions of the FI emitter. This phenomenon can be explained in the following way. Kemball [233] has shown that in the case of simultaneous adsorption of the hydroxyl oxygen and the butyl group of 1-butanol only one of the two α-hydrogen atoms (or α-deuterium atoms) can be in contact with the surface. Consequently, only one of the two α-hydrogen or α-deuterium atoms can be eliminated from the butanol molecule by dissociative chemisorption. In the case of 1-butanol it is very probable that one of the two hydrogen atoms eliminated by dissociative chemisorption is an α-hydrogen, and the other one a β-hydrogen atom. In the case of $CH_3CH_2CH_2CD_2OH$ it is extremely inprobable that first one of the two deuterium atoms is eliminated by dissociative chemisorption, and then the second deuterium atom is eliminated by a dissociation process in the gas phase. Thus, a simple explanation for the large differences in the $P{-}H_2$ and $P{-}D_2$ rates is given by the assumption that the mechanisms for these two processes are entirely different, the latter one being very improbable. Such an observation should not be termed "isotope effect" because eliminations of heavy and light hydrogen atoms from different parts of the molecules are involved.

Another pronounced isotope effect is observed for the ions of higher mass than the molecular ions of alcohols. These may be termed condensation complexes. Fig. 3.58.a. shows that in the case of 1-propanol very intense peaks are found in the FI mass spectrum between the parent peak ($m/e = 60$) and a peak of $m/e = 128$. The relative intensity of these peaks drops drastically if deuterium is introduced in the α-position of 1-propanol (fig. 3.58. b). Deuteration at the terminal C atom does not lead to such an effect. Very probably, the first step of these reactions is the 1.2 elimination of water from the adsorbed neutral alcohol molecule. In this reaction, an α-hydrogen atom is involved, the elimination of which strongly depends on substitution with deuterium. In analogy with the mechanism for the gasoline synthesis by H_2 and CO on a catalyst (Eidus [234]) it can be assumed that the olefin molecules formed by catalytic dehydration react with a CH_2O — adsorption complex at the metal surface. The major part of the isotope effect of the condensation complexes can then be assumed to be due to the 1.2 dehydration of the alcohol molecules via an α-hydrogen C—H bond rupture, as, for example, discussed by Rideal [235].

The technique of isotopic labelling has been used by Dahmen in order to elucidate the mechanism of water elimination from alcohol ions. This process has been carefully studied by a number of authors [236–239] using the electron impact technique for ioni-

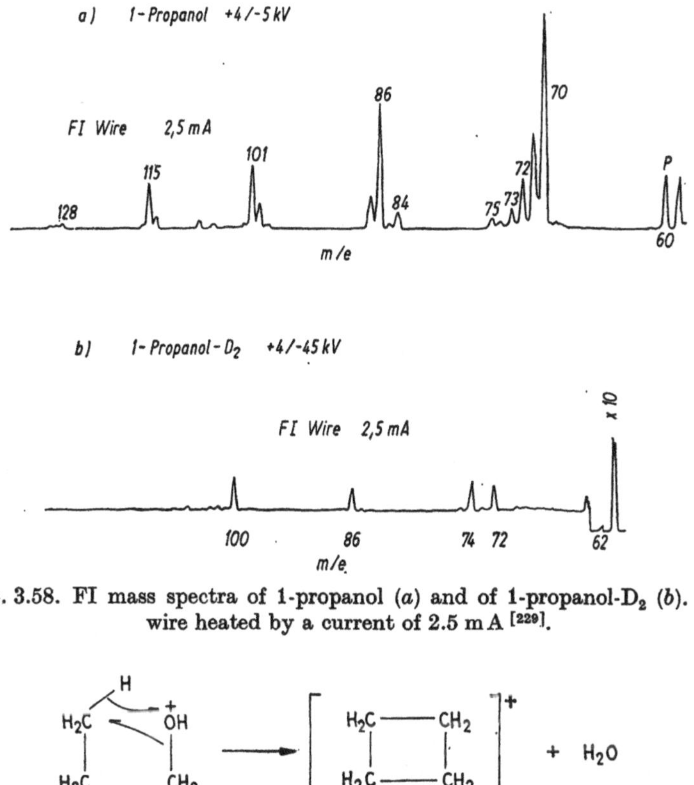

FIG. 3.58. FI mass spectra of 1-propanol (*a*) and of 1-propanol-D_2 (*b*). FI wire heated by a current of 2.5 mA [229].

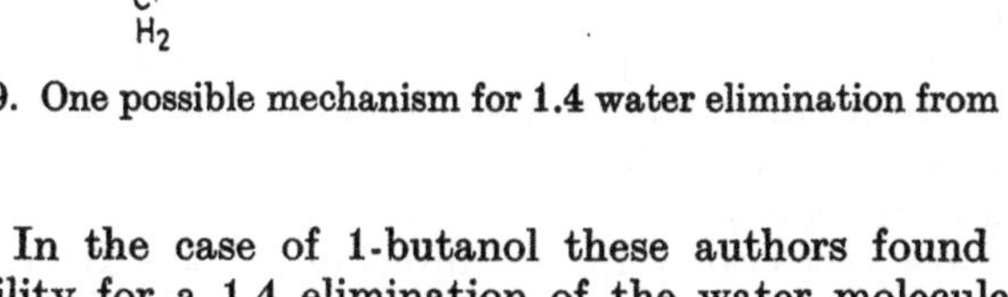

FIG. 3.59. One possible mechanism for 1.4 water elimination from 1-butanol.

zation. In the case of 1-butanol these authors found that the probability for a 1.4 elimination of the water molecule is 80%, that for a 1.3 elimination in the order of 10% (see fig. 3.59.). Dahmen using $CD_3C_3H_6OH$ could show that the $(P - 19)^+$ and $(P - 18)^+$ had approximately equal abundance. The 1.2 elimination from the M^+ ion is sterically very much disfavoured in contrast to the above-mentioned thermal 1.2 elimination of water from the neutral adsorbed alcohol molecules. Therefore, it can be assumed that the 1.4 and the 1.3 elimination of water from 1-butanol are about equally probable. A specific orientation of the butanol molecule in the high electric field may favour to some extent the 1.4 elimination as compared to the case of electron impact.

Korostyshevskii and Goldenfeld [251] have shown that quantitative determination of the isotopic ratios of organic substances can be carried out with fair accuracy by FI mass spectrometers. The protonation of the parent ions at the emitter surface constitutes a certain problem. However, it could be shown that the protonation can be greatly reduced under specific conditions for many organic substances. Protonation of water, alcohols and organic acids cannot be avoided. This does not, however, prevent analysis of the content of oxygen isotopes.

A formula for the theoretical distribution of the mass lines of H_2O-D_2O mixtures has been given by Goldenfeld and his associates [261]. The experimental FI peak distribution agrees with this formula.

Richter and Schmidt [252] have also shown that the H/D ratio of H_2O/D_2O and NH_3/ND_3 mixtures can be determined by an FI mass spectrometer with fair accuracy.

3.9. NEGATIVE ION EMISSION FROM FIELD EMITTING METAL SURFACES

The emission of negative ions from field emitters is an interesting problem from a basic point of view, although at present it seems to promise less analytical application than positive field ion emission. Little research has been done so far on this problem, except for an extended investigation of Robertson and Williams [175, 176].

The conditions during field emission of negative ions are more complicated than those during positive ion emission for several reasons. A large electron current is emitted simultaneously with negative ions from a negative field emitter. The electrons collide with molecules in the gas phase, thus forming positive ions. The positive ions are accelerated in the direction of the field emitter. Negative ions may be released from the emitter by impingement of positive ions on its surface. This process is generally known as "sputtering". Negative ions are formed as a secondary process of a field emission phenomenon, in this case. Two primary processes may lead to formation of negative ions as a result of a high electric field: i. Field desorption of negative ions from the surface. ii. Attachment of electrons to neutral molecules in the gas phase in the presence of a high electric field. The first of these processes is analogous to field desorption of positive ions from a surface, as discussed in section 3.6.4. The second process is a resonance phenomenon with strong energy restrictions. This is a much less probable process. In addition to these processes for negative ion formation, thermal ionization must be considered due to resistive heating of whiskers grown in the field emitter.

Robertson and Williams used a 180° magnetic deflection mass spectrometer similar to that described earlier [10], but equipped with 20 μm platinum or tungsten wires for ion production. Field electron currents of several μA were emitted at potentials of 2—3 kV and wire — cathode distances of 0.5—1.0 mm. The experimental conditions are much more difficult as compared to positive field ion work because the measured ion currents are much lower, and the statistical fluctuations of the currents are much higher (as much as several hundred percent). A number of projections grew up on the wire in the presence of substances like CCl_4 or C_2H_2 at high fields.

Ions were observed from most of the electronegative gases studied, some of which are listed in table 3.14.

TABLE 3.14. NEGATIVE IONS PRODUCED IN THE PRESENCE OF A HIGH ELECTRIC FIELD ON NEGATIVE TUNGSTEN OR PLATINUM WIRES. (IONS LISTED IN ORDER OF ABUNDANCE)

Gas	Ions observed
CCl_4	Cl^-, Cl_2^-, CCl^-, C_2Cl^-, CCl_2^-
SF_6	F^-, F_2^-, S^-, SF_2^-, SF_5
Cl_2, Br_2, I_2	Cl^-, Br^-, I^-,
C_2H_2	C_2H^-, C_2^-
O_2	O^-, OH^-
$C_2(CN)_4$	CN^-, $C_2(CN)_4^-$

The ratio, I_i/I_e, of the resolved ion current to the electron current was generally about 1×10^{-7} for the most abundant ions.

Fig. 3.60. shows the mass spectrum of ions produced from CCl_4 on a cold, field emitting platinum wire. The mass spectrum was scanned by using a permanent magnet and varying the ion accelerating voltage. It is seen from the figure that the mass peaks have tails corresponding to ions with kinetic energies of tens to hundreds of electron volts. This effect is characteristic for formation of ions by sputtering. Moreover the Cl^- ion current could be increased by the admission of C_2H_2 or Ar to the ion source. This again shows that sputtering is an important process in this case. It is difficult to say whether the sputtering yields were large enough to account for the whole of the observed ion currents. From the experiments one can derive a sputtering yield of about 0.02 or 0.2 negative ions per incident positive ion for C_2H_2 or Ar additives, respectively. The lower limit seems to be in accordance with theoretical considerations, whereas the upper limit seems to be greater than the theoretical value. Therefore, other processes like field desorption may possibly be present in addition to sputtering.

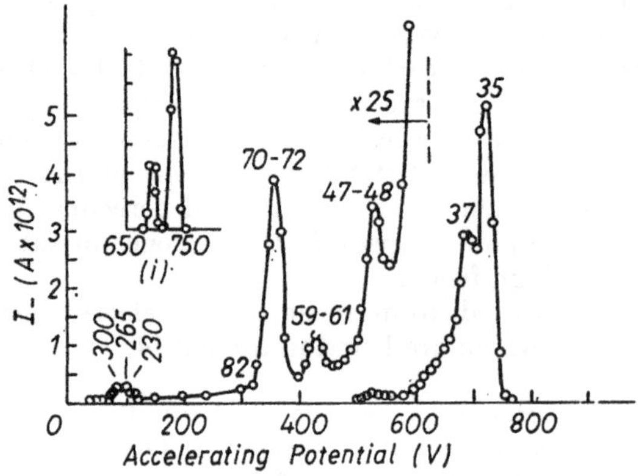

FIG. 3.60. Mass spectrum of ions produced from CCl_4 on a cold, field-emitting platinum wire, according to Robertson and Williams [175]. Insert (*i*): Cl^- peaks from heated, thermionically-emitting wire.

Field attachment could, in principle, give negative CCl_4 parent ions. However, very probably the CCl_4^- ion is not stable, and therefore some of the observed fragments could have been produced by dissociation of unstable CCl_4^- ions formed by field attachment. This could be checked by energy analysis of the ions. The parent ions could be formed only at a minimum distance

$$d_{min} \simeq \frac{\Phi - S}{F} \tag{3.115}$$

in analogy with the condition $d_{min} \simeq \frac{I - \Phi}{F}$ for positive ions. (S is the electron affinity of the molecule.) These would be deficient in energy by an amount $\Phi - S$ as compared with ions produced with zero kinetic energy at the surface. Energy analysis was carried out by means of a retarding potential technique for

Cl^- and Cl_2^- produced from CCl_4^-. No sign of ions deficient in energy could be found. Therefore, field attachment in the gas phase does not seem to be a probable process here.

Admission of C_2H_2 to CCl_4 generally enhances the sputtering process, as noticed by measurement of the pressure dependence of the Cl^- current. In certain cases, however, C_2H_2 seems to influence the Cl^- current by a mechanism other than sputtering. This effect occurs if platinum wires are heated to 500 °C in C_2H_2 at 10^{-4} Torr prior to projection growth in CCl_4. In this case the pressure dependence was $I = k \cdot p^n$ with $n = 0.2-0.7$, whereas $n = 1$ has to be expected for a pure sputtering process. Conclusions drawn from some further experiments are indicative for a field desorption process of Cl^-.

Negative parent ions could not be detected with normal wires in the case of NO_2, SF_6 and $C_2(CN)_4$. All of these molecules are known to be stable. However, $C_2(CN)_4^-$ ions could be detected with a "super-efficient" wire. This was a platinum wire heated in C_2H_2 prior to projection growth in CCl_4. The wire was heated to 700 °C in $C_2(CN)_4$ after pumping out of the CCl_4. After this treatment, small $C_2(CN)_4^-$ parent ion signals could be observed. Robertson and Williams carried out some further experiments which gave evidence that the $C_2(CN)_4^-$ ions are formed by sputtering. Theoretical consideration by those authors show that the probability of detection of negative ions formed by field attachment is extremly low. Energy restrictions make the probability of a field attachment process already low. Moreover, there exists a certain probability that the negative ion can lose the electron by field detachment into the vacuum away from the surface. This probability depends on the time which the ion spends in the high field region and on molecular parameters. Estimations of the field detachment probability by Robertson and Williams show that Cl^-, NO_2^- and especially $C_2(CN)_4^-$ would be observable at all fields likely to be encountered in the ion source.

It must be concluded from non-observation of these ions that the probability of the initial field attachment is very low. A crude estimation of this probability shows that this should be in fact the case.

4. QUALITATIVE ANALYSES WITH THE FI MASS SPECTROMETER

FI mass spectra can serve for identification of unknown substances and for determination of their structure, especially in combination with EI mass spectra. Combination of EI and FI data results in additional information which may in some cases facilitate evaluation of the molecular formula and structure. This is based on the fact that the EI and FI mass spectra differ greatly in several respects:

a) The relative parent ion intensity is much larger in the FI than in the EI mass spectra under normal working conditions. The molecular ion peaks are often found in the FI mass spectra even if they are undetectable in the EI mass spectra taken at low electron energy and temperature.

b) The relative fragment ion intensities are generally greatly different in EI and FI mass spectra.

c) The kinetics of decomposition of ions formed by FI or EI, respectively, are largely different. Therefore, combination of the two methods may lead to additional insight into the kinetics and mechanism of decomposition of organic ions.

4.1. FI MASS SPECTRA OF HOMOLOGOUS SERIES OF ORGANIC SUBSTANCES

The FI mass spectra of a number of homologous series of organic substances have been measured in the past mainly with a view to obtain information on the charge distribution of organic ions in high electric fields and to get insight into the mechanism of field dissociation. These aspects have been discussed in detail in sections 3.1. and 3.2. On the other hand, investigations on FI mass spectra of homologous series may serve also as a guide for the general fragmentation behavior of certain classes of substances and may thus facilitate identification of unknown compounds.

4.1.1. *Alkanes*

The homologous series of normal and branched paraffins up to C_{16} has been investigated by the author and Schulze [94,180]. Some general features of the FI mass spectra of n-paraffins concerning the charge distribution and field dissociation have already been discussed in sections 3.1. and 3.2. In this section the FI mass spectra of a number of n-paraffins will be compared.

Fig. 4.1. shows the FI mass spectra of some n-paraffins from

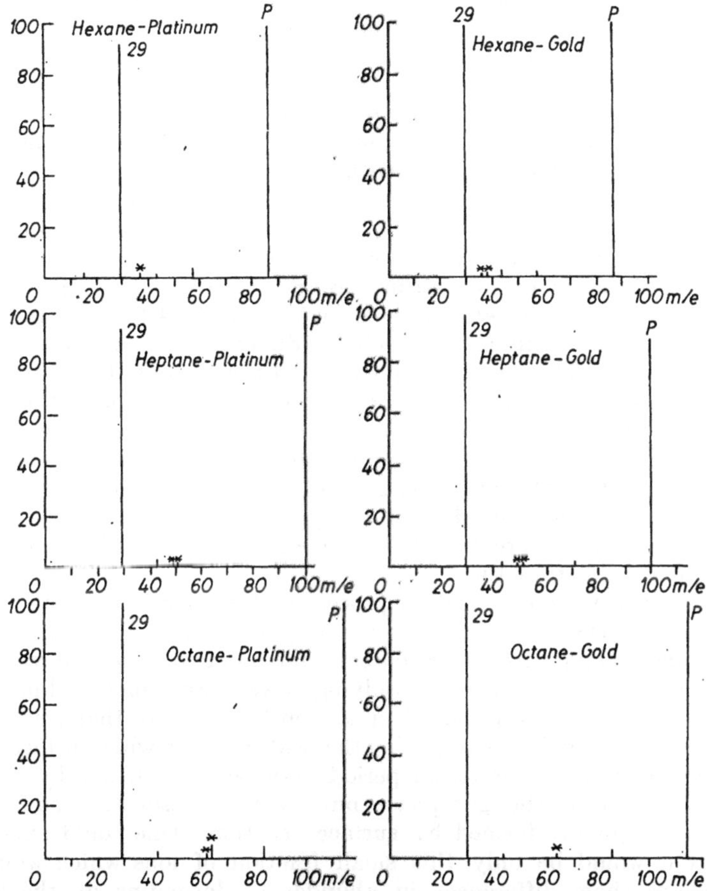

Fig. 4.1. FI mass spectra of n-paraffins on platinum tips (left) and gold tips (right). Only peaks of more than 1% rel. intensity are plotted, the ^{13}C peaks are omitted. Metastable peaks are characterized by asterisks.

hexane to octane on platinum and gold tips. Values for the field strength were chosen for every substance such that the intensity of the $C_2H_5^+$ fragment ions was about equal to the parent ion intensity. (This field strength decreases with the molecular weight, as discussed in section 3.1.). It is seen from the figure that the $C_2H_5^+$ — fragment is by far the most intense fragment ion in the spectra of all n-paraffins. Therefore, the fragment with $m/e = 29$ may be taken as characteristic for field ionization of molecules containing an unbranched paraffinic chain in the course of identification of unknown substances. (It will be shown later that this fragment is characteristic also for molecules containing a polar group at the end of a long paraffin chain.)

Most of the analytical work is presently done with FI emitters consisting of thin wires, sharp edges or multi-tips arrays at lower field strength than that giving the spectrum in fig. 4.1. The relative intensity of the $C_2H_5^+$ ions is typically of the order of one percent under the conditions with wires, etc.

The figure shows that the FI mass spectra of the n-paraffins on platinum and gold tips are very similar under the conditions underlying the figure. The reason for this is twofold. The actual fragment peaks of n-paraffins are not sharp and symmetrical (different from the schematic drawings of fig. 4.1.), but rather are unsymmetrically broadened in the direction of smaller apparent mass values. The ions corresponding to the tails of the fragment peaks are formed by field dissociation of the parent ion in the gas phase. This field dissociation process is only a function of the field strengths and not of the field emitter material (except for distances of the order of one angstrom from the tip). Therefore, the relative intensities of different fragments corresponding to the tails of the peaks are expected to be equal for different tip metals at equal field strengths.

The gas phase dissociation processes also contribute greatly to the peak maxima which comprise fragments with a spectrum of lifetimes ranging from one to about one hundred molecular vibrational periods. (A low mass resolving power corresponds to limited time resolution, as discussed in section 3.4.1.) Only that fraction of the fragment ions comprising the peak maxima which is formed after about one vibrational period may be formed, not by field dissociation in the gas phase but by field desorption as fragments originally formed by surface reactions. One could expect that it would be only this small fraction of ions which would perhaps show differences in abundances depending on the tip material.

An additional reason for the similarity of the n-paraffin spectra on platinum and gold tips may be the following one. It has to be assumed that tightly bound chemisorbed films are formed on the tip surfaces in the presence of organic material. Possibly these films are similar on platinum and gold tips, and the field strengths underlying fig. 4.1. may be insufficient for field desorption of these films. Then the n-paraffin molecules impinge upon similar surfaces in both cases. Therefore, that fraction of the ions contributing to the peak maxima which are not field dissociated in the gas phase, but by surface reactions, may be comparable with platinum and gold tips.

Table 4.1. shows the intensities of the most important peaks in the FI mass spectra of methane, ethane, propane and butane on platinum tips. The field strengths are rather high, and therefore the relative fragment intensities are much higher than those observed on wire emitters at low field strengths.

TABLE 4.1. FI MASS SPECTRA OF SOME PARAFFINS ON PLATINUM TIPS. THE FIELD STRENGTH IS LARGER THAN THE THRESHOLD FIELD BY A FACTOR OF 1.8 IN THE CASE OF METHANE, AND BY A FACTOR OF 2.4 IN THE CASE OF ETHANE, PROPANE, AND BUTANE (^{13}C ISOTOPES ARE NOT LISTED)

Substance		Methane	Ethane	Propane	Butane
Tip-cathode voltage		13.8 kV	1.33 kV	8.9 kV	7.2 kV
Formula	Mass				
CH_3	15	1.4	1.8	0.5	0.1
CH_4	16	100			
C_2H_3	27		2.7		
C_2H_4	28		3.9		
C_2H_5	29		2.0	2.3	16.0
C_2H_6	30		100		
C_3H_7	43				1.0
C_3H_8	44			100	
C_4H_{10}	58				100

Table 4.2. shows the FI mass spectra of n-paraffins from C_5 to C_9 at such field strengths that the intensities of the $C_2H_5^+$ ions are approximately equal to those of the parent ions.

Again it should be noted that the relative $C_2H_5^+$ fragment ion intensities are much smaller with FI wires (of the order of one percent) and also the intensities of the other fragments are lower.

TABLE 4.2. FI MASS SPECTRA OF *n*-PARAFFINS FROM C_5 TO C_9 ON PLATINUM TIPS AT FIELD STRENGTHS AS CHARACTERIZED IN THE TEXT (ONLY THE IMPORTANT PEAKS ARE LISTED, AND THE ^{13}C ISOTOPES HAVE BEEN OMITTED)

Substance		Pentane	Hexane	Heptane	Octane	Nonane
Tip-cathode voltage		8.7 kV	8.2 kV	7.4 kV	6.2 kV	5.9 kV
Formula	Mass					
CH_3	15	1.4	1.3	0.2	0.1	0.1
C_2H_5	29	90	100	94	100	100
C_3H_7	43	8.0	1.0	3.5	2.5	4
C_4H_9	57	0.15	4.6	0.3	0.3	1.4
C_5H_{10}	70			0.8	0..8	
C_5H_{11}	71			1.0	0.7	0.2
C_5H_{12}	72	100				
C_6H_{12}	84				50	2.0
C_6H_{13}	85				0.5	0.2
C_6H_{14}	86		92			
C_7H_{14}	98					2.5
C_7H_{15}	99					0.1
C_7H_{16}	100			100		
C_8H_{18}	114				100	
C_9H_{20}	128					100

Table 4.3. shows the relative intensities of the most important peaks in the FI mass spectrum of n-heptane taken with a 2.5 μ platinum wire at 10 kV, using a carefully cleaned mass spectrometer. Previous measurement of compounds containing double bonds increases the relative fragment intensities of n-heptane (see also table 4.14.).

TABLE 4.3. FI MASS SPECTRUM OF N-HEPTANE ON A 2.5 μm PLATINUM WIRE AT 10 kV (*i* = isotope peak). Intensities related to peak areas

m/e	Formula	rel. Intens.
29	C_2H_5	0.01
42	C_3H_6	0.02
43	C_3H_7	0.01
57	C_4H_9	0.01
100	C_7H_{14}	100.00
101	C_7H_{14} (i)	7.80

Metastable Peaks $m^* = m^2/M$

m^*	Process	rel. Intens.
91.4	$M^+ – C_3H_8$	0.08
32.5	$M^+ – C_3H_7$	0.08
49.0	$M^+ – C_2H_6$	0.43
50.4	$M^+ – C_2H_5$	0.34

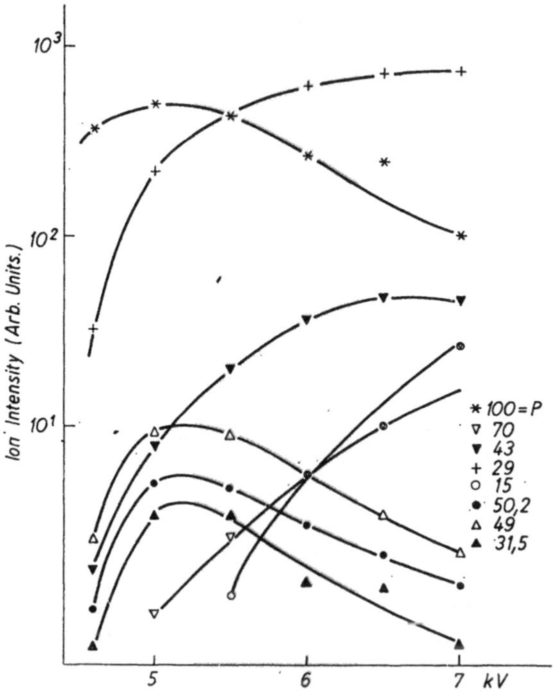

FIG. 4.2. Field strength dependence of the most important peaks in the FI mass spectrum of n-heptane on a platinum tip [180].

Fig. 4.2. shows the field strength dependence of the n-heptane FI peaks produced by a platinum tip. It may be seen from the figure that — in accordance with the consideration on the charge distribution, section 3.1. — the CH_3^+ ions ($m/e = 15$) increase much more strongly with the field strength than the other ions. The metastable (asterisked) peaks vary largely in parallel with the parent ion peak intensity.

Relatively intense metastable peaks are found in the FI mass spectra of paraffins. Table 4.4. shows the intensities of the metastable peaks formed by elimination of neutral C_2H_6 or C_2H_5, respectively, from n-paraffin parent ions.

TABLE 4.4. RELATIVE INTENSITIES OF METASTABLE IONS FORMED FROM N-PARAFFIN PARENT IONS ON Pt-TIPS AS A PERCENT OF THE PARENT ION INTENSITIES

Substance	Pentane	Hexane	Heptane	Octane	Nonane
Tip-cathode voltage	8.7 kV	8.2 kV	7.4 kV	6.2 kV	5.9 kV
Apparent mass m^*	24.5	36.6	49.0	61.9	75.03
Intensity of P-30	5.0	2.0	1.0	2.0	1.5
Apparent mass m^*	25.7	37.8	50.4	63.4	76.5
Intensity of P-29	1.0	0.5	1.7	5.0	6.0

It is seen from the table that at lower molecular weight of the n-paraffins the C_2H_6 — molecule is eliminated preferentially, whereas at higher molecular weights the C_2H_5 radical is eliminated with greater abundance. This effect is similar to that observed by Steiner, Giese and Inghram [181] in their photo ionization work on paraffins. They found a relative intensity of the P–30 peak from n-pentane of 4% and a much smaller intensity of the P–29 peak from n-pentane. In the case of n-octane they found about 8% P–29 and 3.2% P–30. (The photon energy was 10.7 and 10.9 eV, respectively.) The ratio of the $(P–29)/(P–30)$ peaks from n-octane is 2.5 both with field ionization and with photo ionization at 11.9 eV. In view of the large delay time between ionization and metastable decomposition (about 10^8 vibrational periods) it is not surprising that the relative abundances of competing metastable processes are rather independent of the mode of ionization.

The change of the ratio of eliminated molecules and radicals from n-pentane to n-nonane is continued with higher n-paraffins [180]. Characteristic changes occur also with elimination of neutral fragments of higher mass. These effects are summarized by fig. 4.3. [180]. The results from independent measurements of Wagner [182] are also indicated in fig. 4.3.

The intensity ratios of competing metastable processes are virtually independent of the field strength. It can also be seen from fig. 4.2. that the fraction of the molecular ions which decompose as metastable ions is rather independent of the field strength. However, the metastable ions decrease relative to

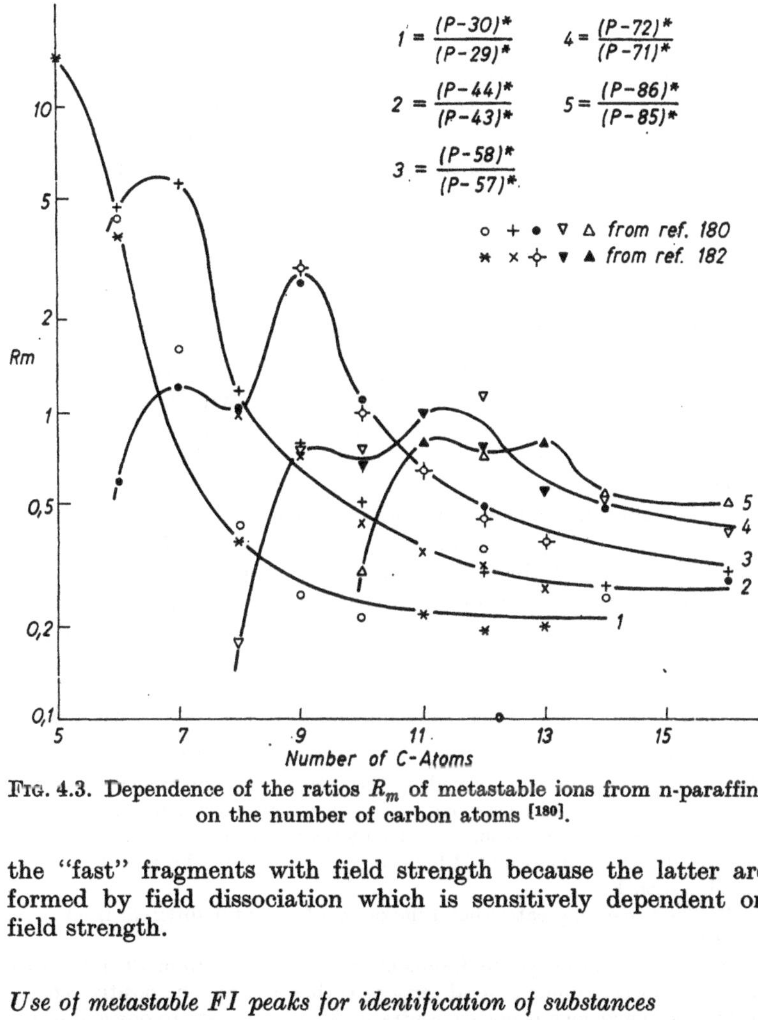

FIG. 4.3. Dependence of the ratios R_m of metastable ions from n-paraffins on the number of carbon atoms [180].

the "fast" fragments with field strength because the latter are formed by field dissociation which is sensitively dependent on field strength.

Use of metastable FI peaks for identification of substances

The largest fragment peaks above $m/e = 43$ in the FI mass spectra of molecules containing long paraffinic or olefinic chains are frequently some metastable peaks (see for example, fig. 4.5. of next section.) Wanless and Glock [183] have pointed out that this fact could be useful for identification of unknown pure compounds, or even of compounds in mixtures. The apparent, non-integral mass numbers of the metastable ions are characteristic for a compound, and the mass numbers are not redundant for different members of a homologous series. In contrast to electron impact mass spectra, the mass range between $m/e = 49$ and the parent ion mass is occupied only by weak "fast" fragment peaks, so that this region is left relatively clear for the metastable peaks. Isomers of the same molecular weight can be differentiated by the different abundances of comparable metastable decomposition processes. This is demonstrated by fig. 4.4. The relative abundances of these metastable ions will vary with the structure of the isomers. The peak intensities of this "metastable cracking pattern" can be rationalized by means of simple cleavage rules.

Comparison of FI and chemical ionization mass spectra

Field and Munson [184,185] have discovered a feature of chemical ionization mass spectra which is common with field ionization mass spectra. This is the relatively high parent ion intensity in the mass spectra of the majority of compounds. The chemical ionization (ChI) and electron impact mass spectra of eight isomeric decanes have been compared in another paper of Field et al. [186]. The author has made a complementary study on the same series of isomeric decanes with an FI mass spectrometer [187]. Some characteristic differences between the spectra originating from the two different modes of ionization became apparent. (It should be noted that the FI source was at room temperature, whereas the chemical ionization source was at 150 °C, see table 4.5.)

Some general results derived from the table are the following ones.

(1) The stability of the molecular ions formed by FI is much greater than that of the P–1 ions formed by ChI. Carbonium ions are formed with ChI (even number of electrons), and molecular ions are formed with FI (uneven number of electrons). An energetically favourable bond fission is possible if ChI occurs at a C–H bond in a β position of a branch. This is not possible with FI.

(2) The places of greatest bond fission in FI are quite different from those in ChI.

(3) A weak and unspecific fragment spectrum arises from FI in the case of molecules containing several tertiary carbon atoms. The positive charge of the molecular ions is not concentrated at a point near to one end of the molecular ions, as in normal paraffins, but the charge is distributed over the points of the branches. This reduces the fragmentation probability per branch as compared with a singly branched or a normal paraffin. On the other hand, the bonds of quaternary carbon atoms in a molecular ion are so weak that field dissociation is possible to a large extent, even if there are several points of branching.

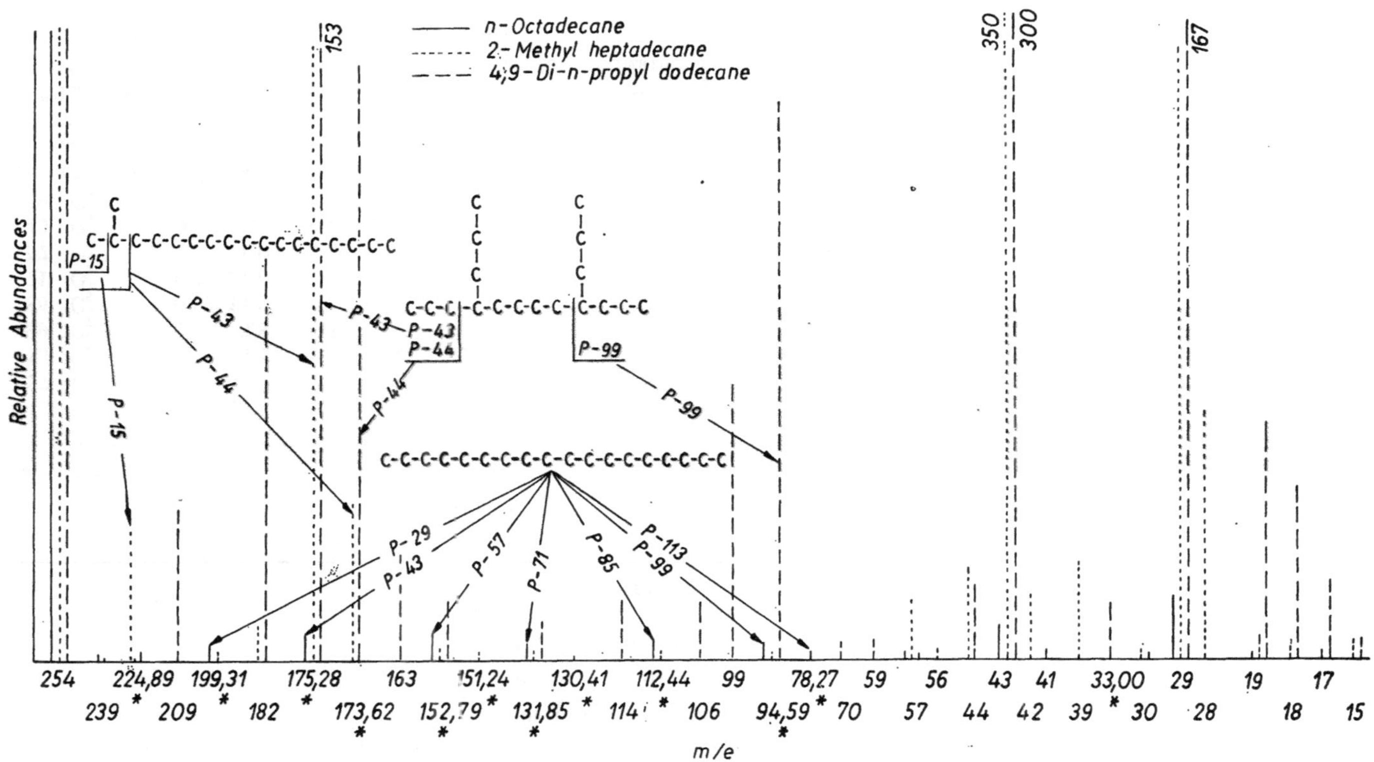

FIG. 4.4. FI mass spectra of octadecane isomers showing abundances of primary metastable ions, according to Wanless and Glock [183].

TABLE 4.5. RELATIVE ION INTENSITIES IN THE FIELD IONIZATION AND CHEMICAL IONIZATION MASS SPECTRA OF EIGHT ISOMERIC DECANES

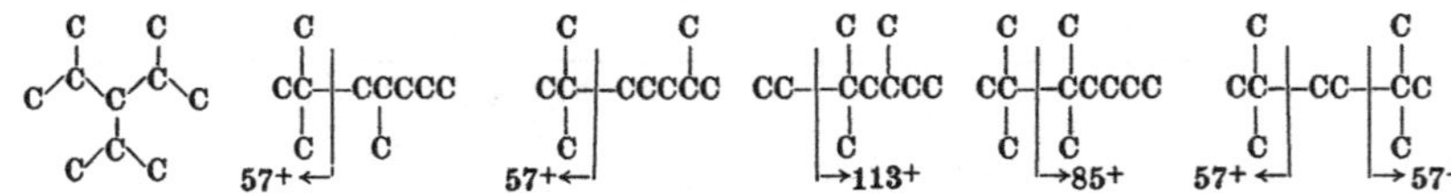

		6		7		8		10		11		12		13		14	
		2-Me-5-Et-C_7		2,3,6-Me-C_7		2,4-Me-3-*i*-Pr-C_5		2,2,4-Me-C_7		2,2,6-Me-C_7		3,3,5-Me-C_7		2,2,3,3-Me-C_6		2,2,5,5-Me-C_6	
m/e	Ion	FI	ChI	FI	ChI	FI	ChI	FI	ChI	FI	ChI	FI	ChI	FI	ChI	FI	ChI
142[b]	C_{10}^+	98.2	20.0	98.0	9.4	95.7	1.9	50.0	5.7	59.2	9.2	42.6	4.5	14.4	0.5	26.3	2.0
127	C_9^+	0.2	8.1	...	10.3	...	1.3	3.7	12.0	4.2	16.2	2.7	9.5	1.6	8.9	3.8	13.0
113	C_8^+	...	4.8	0.2	0.4	1.0	...	0.5	0.4	...	...	37.0	4.8	...	0.1	0.9	...
99	C_7^+	...	3.8	0.1	2.6	1.0	3.9	...	3.6	...	1.9	3.6	1.5	6.4	12.0	...	0.4
85	C_6^+	0.6R	14.0	0.3R	19.0	...	23.0	2.5	17.0	1.1	17.0	5.7	14.0	70.6	16.0	1.0	16.0
71	C_5^+	0.1	21.0	0.1	26.0	...	27.0	0.5	19.0	...	24.0	5.1	44.0	...	27.0	0.1	30.0
57	C_4^+	...	11.0	...	17.0	...	34.0	42.8	32.0	34.6	21.0	1.5	13.0	6.7	28.0	68.0	28.0
43	C_3^+	0.3	...	0.5	...	1.3	...	0.3	...	0.8	...	0.8	...	0.3	...	...	...
29	C_2^+	...	...	...	...	...	...	...	...	...	...	0.1	...	...	...	...	...
Σ minor ions		0.6	4.3	0.8	4.9	1.0	3.2	...	3.8	...	4.0	0.8	2.5	...	4.1		4.9
	ΣI	100.0	87.0	100.0	89.6	100.0	94.3	100.1	93.5	99.9	89.1	99.9	93.8	100.0	96.6	100.1	94.3

[a] The compounds are numbered as found in ref. [186]. R = rearrangement peak with *m/e* 86 instead of 85. Isotopic peaks are not included into the table. Great bond fission with field ionization is indicated by arrows in the schematic molecular formulas.

[b] *m/e* = 141 stands for 142 in the case of ChI.

4.1.2. Alkenes

Systematic studies of the FI mass spectra of olefins on platinum tips have been carried out by the author with Wagner [182] and with Schulze [96]. The olefins have been investigated up to C_{14}. The most important regularities found in the olefin spectra are the following ones.

1. Field dissociation of the C=C bonds of unbranched olefins does not occur even at very high field strengths.

2. In general, no rupture of a C–C bond next to a C=C bond is observed. The reason for this is the partial double bond character of this bond which may be explained in terms of resonance structures, as discussed in section 3.1.2.

3. The stability of the C_3H_5 radicals and the $C_3H_5^+$ ions has an influence on many fragmentation processes (see section 3.1.2.).

The main features of the FI mass spectra of branched olefins can be interpreted by the simple cleavage rules discussed under 3.1.2. and the rules (1)—(3) noted above.

The $C_3H_5^+$ — fragment ion ($m/e = 41$) may be regarded as the most specific fragment in the FI mass spectra of olefins larger than butene. Its relative intensity starts with about 15% for 1-pentene, runs over a maximum value for 1-hexene with about 45% and drops to about 3% for 1-heptene, then staying approximately at the same value up to 1-tetradecene. (The data are related to platinum tips at such field strengths that the $C_2H_5^+$ fragment from the paraffinic chain is approximately equal to the parent ion intensity). The high intensity of the $C_3H_5^+$ peak in the case of 1-hexene is due to the high stability of the $C_3H_5^+$ ions and C_3H_5 radicals which are formed simultaneously by bond rupture.

The fragment spectrum becomes more similar to the spectrum of the paraffin chain with increasing molecular weight of the olefins. For example, in the 1-decene spectrum only the $C_2H_5^+$ and $C_3H_7^+$ ions have abundances above 1%, whereas all other ions have lower abundances. Only the metastable ions are noticeable at $m/e > 43$, as can be seen from fig. 4.5. The metastable fragments of olefins differ from those of the paraffins in that elimination of a neutral CH_3 radical is a relatively intense process up to 1-heptene. Moreover, small neutral olefin molecules are eliminated. Ethylene is eliminated preferentially from 1-pentene to 1-octene. The relative intensity of this metastable process can go up to 15% with 1-hexene and 1-heptene. Elimination of propene from 1-heptene is also an abundant process. The metastable ion mass spectrum of the olefins also becomes more similar to that of the paraffins with increasing chain length, i.e. elimination of

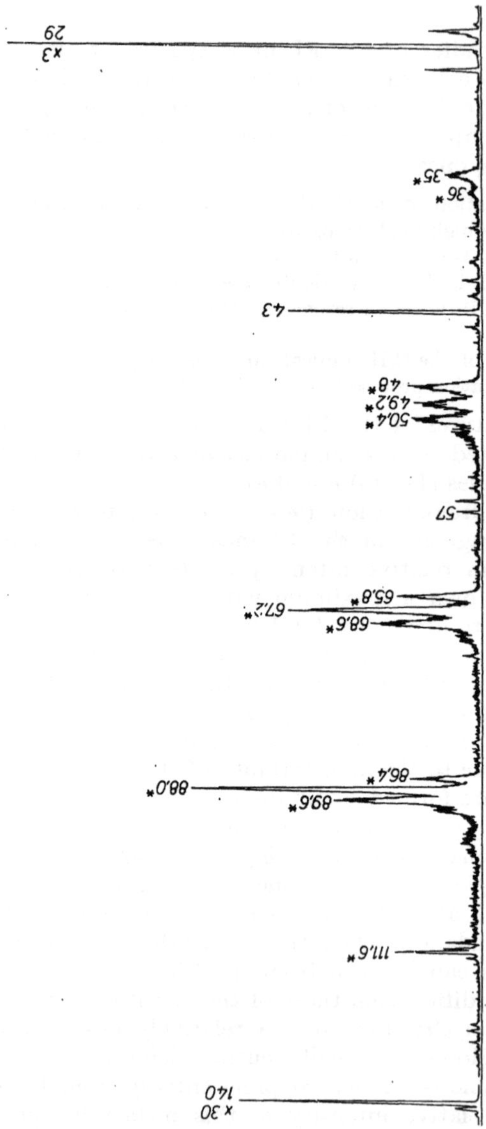

FIG. 4.5. FI mass spectrum of 1-decene at low field strength [96].

larger alkyl radicals becomes predominant. In addition to these processes, the metastable formation of small olefin molecules is continued also with large olefins.

4.1.3. Alkynes

The FI mass spectra of alkynes have been investigated by Patterson and Seakins [188]. A platinum wollaston wire of 2.5 μm diameter was used by these authors. In general, the alkynes obey well the rules developed for field dissociation of paraffins and olefins. In addition to these, Patterson and Seakins noticed some features which are specific for the alkynes. The FI spectrum depends strongly on the position of the triple bond in the molecule. In the isomeric series of 1-, 2-, 3- and 4-octyne considerable field dissociation occurs with 1-octyne and it decreases markedly through the series. The ratio of $[P^+ + (P+1)^+]$ to the total ion intensity is 0.044, 0.43, 0.62 and 0.80 for 1-, 2-, 3- and 4-octyne, respectively. This effect is a consequence of the charge distribution of the molecules in the presence of high fields, as discussed in section 3.1.2.

The rule that no C—C bonds adjacent to the multiple bonds are field dissociated is even more strictly obeyed for alkynes than for olefins.

The great stability of the C_3H_3 radical leads to relatively strong fragment peaks if elimination of this radical from the alkyne molecules is possible. The stability of the C_6 hydrocarbons with respect to field dissociation decreases in the following order: 2-hexene > hexane > 3-hexyne > 1-hexene > 1-hexyne.

The metastable decomposition processes observed in the FI mass spectra of alkynes are similar to those occuring in the alkene spectra. In general, the charge remains with the larger mass. An extremely large metastable peak is observed in the FI mass spectrum of 1-hexyne: $C_6H_{11}^+ \rightarrow C_4H_7^+ + C_2H_4$, the relative intensity of the corresponding peak being 28.6%. Usually, only primary metastable processes are found in FI mass spectrometry, i.e. metastable decomposition of parent ions. In the case of alkynes, however, Patterson and Seakins report a few weak metastable decompositions of fragment ions. These processes are elimination of C_2H_2 or C_2H_3 from $C_5H_7^+$. However, it has not yet been decided experimentally whether the $C_5H_7^+$ ions are formed as fragments from the parent ions in the gas phase or by field desorption of $C_5H_7^+$ ions which have been formed previously as neutral C_5H_7 radicals by dissociative chemisorption. In the latter case, the $C_5H_7^+$ ions have to be regarded as "parent" ions, and the decomposition process had to be considered as a "primary metastable process". Similar arguments hold for any case where weak metastable decomposition processes from fragments are observed in FI mass spectra.

Peaks at about double the molecular weights are found in the FI mass spectra of some alkynes. The relative intensities of these $2P$–2, $2P$–1 or $2P$ peaks may be as high as 1.4%. Very strong $P + 1$ and $P - 1$ peaks are observed in many cases.

It is interesting to note that relatively strong peaks at $m/e = 31$, 45 and 59 appear in the FI mass spectra of most alkynes. These peaks can neither be explained by impurities nor by field dissociation of the parent ions. Patterson and Seakins were able to show by adding increasing amounts of water to the alkynes that these peaks are formed by association between a CH, C_2H_3 or C_3H_5 ion and a water molecule.

4.1.4. Amines

The FI mass spectra of a number of normal and branched amines up to decylamine have been measured by the author and Wagner [189] with platinum tips and wires. The most important features of these spectra will be summarized by a few rules.

1. "The $C_2H_5^+$ and the $CH_2NH_2^+$ ions are the most abundant fragment ions. The abundance of the $C_2H_5^+$ ions relative to the $CH_2NH_2^+$ ions increases from n-propylamine to n-heptylamine".

It has to be assumed that the strongly polar CH_2NH_2 group is adsorbed onto the field emitter surface whereas the hydrocarbon chain of the amines is relatively mobile. In the case of orientation of the hydrocarbon chain parallel with the field lines, a split of a $C_2H_5^+$ fragment from the parent ion is favoured as in the case of n-paraffins (see fig. 4.6., position *I*). If the hydrocarbon chain is ordered parallel with the surface, the positive charge will be located mainly at the NH_2 group because of the relatively low ionization potential of a lone pair electron of the N-atom. The field force acts on that ionized part of the molecule. The molecule is then turned from position *II* to position *III* where the NH_2^+ group is on the tip-distant side of the molecule. Split of a $CH_2NH_2^+$ group from the molecular ion is favoured in this position of the molecule. The high stability of the $CH_2NH_2^+$ ion and the corresponding weakness of the adjacent C—C bond enhances the probability of this process.

The increase of the $C_2H_5^+$ fragments over the $CH_2NH_2^+$ fragment with increasing size of the amine is in accordance with this model. Similarly to the alcohols, it is to be expected that the lower amines up to n-butyl amine are adsorbed preferentially with the hydrocarbon chain parallel with the emitter surface (position *II*). Thereafter, position *III*, and hence split of a $CH_2NH_2^+$ fragment, becomes predominant. The higher amines are expected to form

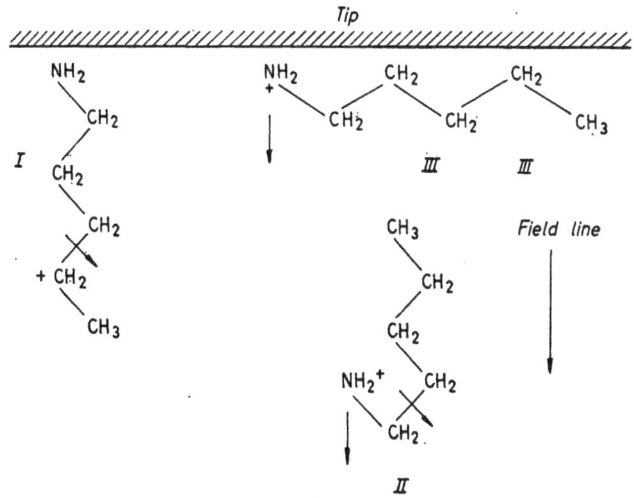

FIG. 4.6. Different orientations of n-pentylamine molecules immediatelly after field ionization near to the field anode. A change of the orientation of the molecules occurs in the direction of the dashed anodes.

condensed adsorbed phases, similar to the alcohols (see section 3.6.5.), with the alkyl chains ordered parallel with each other and perpendicular to the emitter surface (pos. *I*). This explains the enhanced $C_2H_5^+$ ion formation of the higher amines. This view is supported by the fact that very intense $P + 1$ peaks are observed in the FI mass spectra of higher amines which are characteristic for formation of two-dimensional condensed phases.

2. "The strong broadening of the $CH_2NH_2^+$ peak ($m/e = 30$) is characteristic for all amines".

Fig. 4.7. shows a characteristic fragment peak group of $m/e = 29$ and 30 of a n-amine molecule. The $m/e = 29$ peak is slightly broadened due to delayed dissociation of parent ions in the gas phase. It is seen from the figure that the peak at mass 30 is broadened much more strongly. This means that elimination of a $CH_2NH_2^+$ fragment still occurs at large distances from the field emitter, where the field strength is too low for pure field dissociation. The reason for this delayed dissociation is the weakness of the

C—C bond adjacent to the $CH_2NH_2^+$ group which makes statistical decomposition possible after a large number of vibrational periods.

3. "Protonation increases with increasing chain length of the amines".

The "ammonium" ion intensity can go up to become as high as three times the parent ion intensity in the case of n-octylamine,

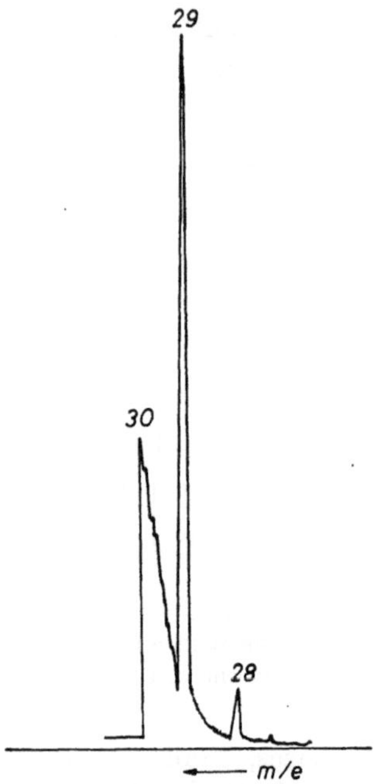

FIG. 4.7. Shape of the peaks at $m/e = 29$ and 30 in the FI mass spectra of amines.

for example. The effect has just been explained as a consequence of formation of two-dimensional condensed phases at the emitter surface. Formally, protonation can be considered as a proton transfer reaction in the condensed phase

$$RNH_2^+ + RNH_2 \rightarrow RNH_3^+ + RNH.$$

4. "The probability of formation of NH_2^+ by FI of amines is very low".

The split of a CH_3^+ ion is at least two orders of magnitude more probable. The stability of the C—N bond in the CNH_2^+ ions is due to its double bond character. The peak observed at mass 16 is mainly due to the ^{13}C isotope of CH_3. Only a small fraction of this peak can be attributed to NH_2^+ ions.

5. "The most abundant metastable decomposition process for n-amines above n-pentylamine consists of elimination of a $C_2H_4NH_2^+$ fragment".

The maximum relative intensity of this process is about 6.2% (peak height) with n-heptylamine. The same fragment is the most abundant in the EI mass spectrum of n-heptylamine, although its abundance is only 0.06%. Elimination of a $CH_2NH_2^+$ fragment is preferred with the lower n-amines, however, with lower abundances.

6. "A small percentage of the amine parent ions and of some abundant fragment ions are doubly ionized".

The fraction of the doubly ionized amine molecules in the EI mass spectra is apparently lower. Collin [190] does not mention observation of doubly charged amines, and the API Mass Spectral Data [191] record only very low abundances of doubly charged amine ions. The EI mass spectrum of n-heptylamine [189] shows only 0.02% total intensity of doubly charged ions as compared with 1.3% intensity in the FI mass spectrum. This fraction increases with increasing length of the amines. Only the ions appearing on half integral mass numbers are considered here. In addition to these, doubly charged ions are expected to appear on integral mass numbers which, however, can be differentiated from other fragment ions only by high resolution mass spectrometers, in some cases. Thus the total intensity of doubly charged ions is expected to be higher by a factor of about two than that observed at non-integral mass numbers.

The relatively intense double ionization may be regarded as a consequence of the orientation of the negative pole of the NH_2 group to the tip and the resulting strong polarization of the lone electron pair of this group. Doubly charged $(M + 1)$ ions are not observed. This arises from the fact that the affinity of the hydrogen atom to molecules having one unpaired electron (singly charged parent ions) is larger than to those having no unpaired electron (doubly charged parent ions).

4.1.5. Alcohols

The most important features of the FI mass spectra of alcohols have been discussed already in sections 3.6.2. and 3.6.5. Therefore, only one additional feature will be described here.

"The $C_2H_5^+$ and the CH_2OH^+ peaks ($m/e = 29$ and 30) are the most intense ones in the FI mass spectra of unbranched alcohols at medium field strengths".

The formation of these two fragments depends on the orientation of the alcohol molecules in a manner quite analogous to that discussed under the amines. The CH_2OH^+ peak is more broadened than the $C_2H_5^+$ peak, but less than the $CH_2NH_2^+$ peak of n-amines.

4.1.6. Cyclic Compounds

Opening of the ring of cyclic compound requires rupture of two bonds. This is an unprobable process in the case of field ionization. Therefore, the FI mass spectra of unsubstituted cyclic compounds are very simple [182, 192]. Besides the singly charged parent ion group they show mainly doubly charged parent ions (see fig. 4.8). Long aliphatic substituents of aromatic compounds lead to fragments similar to those found in the spectra of the separate substituent molecules.

4.1.7. Other Compounds

The FI mass spectra of some long chain esters have been measured by Rohwedder [253].

The spectra were taken with a stainless steel razor blade. Molecular ions could be detected unambiguously for all esters which were studied. (The parent peak of methyl linolenate, for example, is the most intense peak in the spectrum.)

The absolute intensity of the FI spectra increases continuously for all long chain methyl esters studied in the range between C_6 and C_{22}. An increase of two carbon atoms on a chain 16 carbon atoms long almost doubles the intensity. This is a consequence of adsorption phenomena, which were discussed in section 3.6.6.

4.1.8. Sensitivity Factors for FI and EI Mass Spectra

Wanless and Glock [183] have determined the sensitivity "gain" factors for FI as compared to EI mass spectra. The gain is defined as follows:

$$\text{“Gain”} = \frac{\textit{Ratio parent ion / fragment peak (FI)}}{\textit{Ratio parent ion / fragment peak (reference EI spectrum)}}$$

In this definition "fragment peak" stands for the largest fragment ion above $m/e = 43$. Normally, the parent ion intensity is expressed as a percentage of the total ionization, which is the sum of all ions in the spectrum. The new definition is used because the

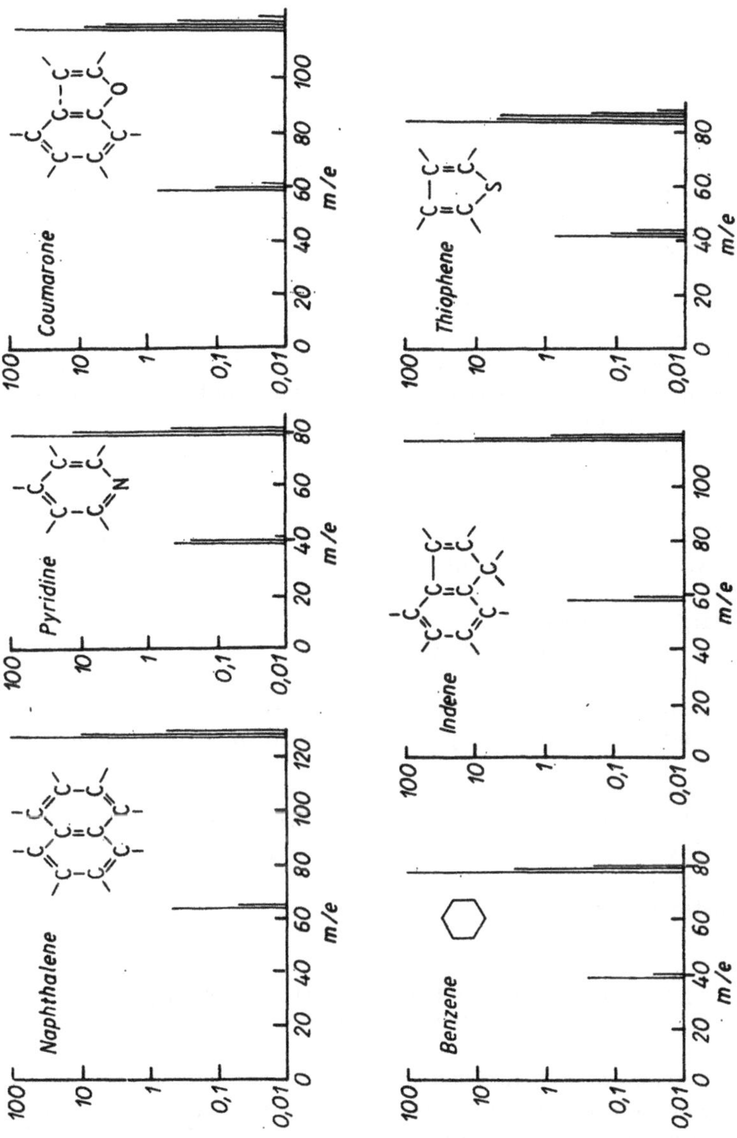

FIG. 4.8. FI mass spectra of some aromatic and heterocyclic compounds [182].

fragment peaks at $m/e = 43$ and lower cause little interference in interpretation of most spectra, and because the $m/e = 43$ and 29 peaks are prominent in many FI mass spectra.

The gain factors are a measure of the improvement of the relative parent ion intensity which may be important for identification of unknown substances, especially of mixtures of such substances.

The results obtained by Wanless and Glock for 92 substances belonging to 17 classes of compounds are represented in table 4.6.

Column (*A*) lists the ratios of the parent ion divided by the largest fragment above ion $m/e = 43$, as taken from the API–44 Mass Spectral Data Catalogue for electron impact work [191]. The same ratio is listed in column (*B*) for field ionization. The masses of the parent ions and of the most intense fragment ions above $m/e = 43$ are listed first in column (*B*), and then the intensity ratios are listed. Fractional mass numbers of fragment ions are listed in those cases where a metastable fragment peak is the most intense one. Column (*C*) lists the "gain" factors. Column (*D*) represents sensitivity data for field ionization. The experimental conditions underlying the sensitivity data are exactly defined by Wanless and Glock; however, the figures of column (*D*) will be used here only for comparison of the relative sensitivities for field ionization of different substances since the experimental conditions in the different FI mass spectrometers at present in use are largely different.

The "gain" varies widely from one class of compounds to another. For the normal paraffins C_5 to C_{16} the field ionization "gain" ranges from 57 to 1821. Molar sensitivities of linear paraffins and olefins generally increase with increasing molecular weight.

Note the extraordinarily high sensitivity of acetone with respect to field ionization which was observed already in the course of the first investigation by the author on FI mass spectrometry [193]. Acetone is recommended to be used for making the first test adjustments of new FI sources or fresh FI emitters.

4.2. DETECTION OF FREE RADICALS BY FI MASS SPECTROMETRY

Detection of free radicals produced in the course of chemical reactions can be carried out by means of an electron impact mass spectrometer. This technique was pioneered by Eltenton [256] who applied low electron energy techniques for differentiation of radicals which are produced in the EI source by electrons at the usual energy of 70 eV and of radicals which are produced in a

TABLE 4.6. SIMPLIFIED MASS SPECTRA OF ALIPHATIC COMPOUNDS. A COMPARISON WITH API-44 REFERENCE SPECTRA

	Compound	(*A*) API–44	(*B*) MS–1 (*b*)	"Gain" B/A (*d*)	(*C*) Dual Field Ion Source (*b*) Ratio (*a*)		"Gain" C/A (*d*)	Field Ion Sensitivity (*c*)
(*1*)	Propane	1.3(a)	—	—	44/0	—	—	55.
(*2*)	n-Butane	5.1	—	—	58/0	—	—	55.
(*3*)	n-Pentane	0.70	—	—	72/43	= 40.	57×	1100.
(*4*)	n-Hexane	0.14	0.90	6.4×	86/57	91.	645×	1700.
(*5*)	n-Heptane	0.27	1.05	3.9×	100/50.50	35.	129×	2800.
(*6*)	n-Octane	0.20	0.62	3.1×	114/63.40	8.9	44×	1950.
(*7*)	n-Nonane	0.095	—	—	128/76.50	12.	126×	2000.
(*8*)	n-Decane	0.075	1.16	15×	142/90.49	17.	227×	2050.
(*9*)	n-Undecane	0.045	—	—	156/103.39	26.	578×	2600.
(*10*)	n-Dodecane	0.082	—	—	170/94.88	29.	353×	2950.
(*11*)	n-Tridecane	0.053	—	—	184/108.05	33.	623×	3550.
(*12*)	n-Tetradecane	0.028	—	—	198/143.95	51.	1821×	4700.
(*13*)	n-Hexadecane	0.021	4.0	190×	226/126.38	25.	1190×	5500.
(*14*)	n-Octadecane	0.010			254/152.79	22.	2200×	3300.
(*15*)	Neopentane	0.0001	0.0064	64×	72/57	0.0063	63×	8.3
(*16*)	2,2 di me butane	0.0006	0.0073	12×	86/57	0.025	42×	82.
(*17*)	2,3 di me butane	0.23	0.26	1.1×	86/58.65	14.	60×	200.
(*18*)	3,3 di me pentane	0.0015	0.015	100×	100/71	0.014	93×	114.
(*19*)	2 me heptane	0.067	—	—	114/85.97	4.8	72×	600.
(*20*)	3 me heptane	0.045	—	—	114/57	2.3	51×	950.
(*21*)	4 me heptane	0.059	—	—	114/71	9.0	152×	1550.

	Compound	(*A*) API–44	(*B*) MS–1 (*b*)	"Gain" B/*A* (*d*)	(*C*) Dual Field Ion Source (*b*) Ratio (*a*)		"Gain" C/A (*d*)	Field Ion Sensitivity (*c*)
(*22*)	3 ethyl hexane	0.057	—	—	114/61.89	7.7	135×	900.
(*23*)	2,3 di me hexane	0.027	—	—	114/71	3.2	110×	1700.
(*24*)	2 me, 3 ethyl pentane	0.025	—	—	114/71	2.6	104×	1450.
(*25*)	2,4 di me hexane	0.023	—	—	114/57	2.9	128×	800.
(*26*)	2,5 di me hexane	0.048	—	—	114/110	2.42	50×	800.
(*27*)	3,4 di me hexane	0.022	—	—	114/57	1.9	86×	500.
(*28*)	2,3,4 tri me pentane	0.0054	—	—	114/71	0.46	85×	65.
(*29*)	2,2 di me hexane	0.0003	—	—	114/57	0.03	100×	85.
(*30*)	3,3 di me hexane	0.0002	—	—	114/71	0.03	143×	25.
(*31*)	2,2,3 tri me pentane	0.0003	—	—	114/57	0.01	33×	70.
(*32*)	3 me, 3 ethyl pentane	0.0002	—	—	114/85	0.02	100×	35.
(*33*)	2,2,4 tri me pentane	0.0002	—	—	114/57	0.02	100×	70.
(*34*)	2,3,3 tri me pentane	0.0002	—	—	114/71	0.01	50×	90.
(*35*)	2,2,3,3 tetra me butane	0.0002	—	—	114/57	0.0056	28×	2.5
(*36*)	2 me pentadecane	0.006	—	—	226/148.15	1.5	250×	1120.
(*37*)	7 n propyl tridecane	0.9	—	—	226/148.40	3.6	19×	1140.
(*38*)	5,8 di ethyl dodecane	0.0014	—	—	226/70	6.0	4286×	2000.
(*39*)	5 ethyl, 5 n propyl undecane	0.0	—	—	226/169	0.06	—	45.
(*40*)	2,2,4,4,6,8,8 heptamethyl-nonane	—	—	—	226/57	0	—	—
(*41*)	2 me heptadecane	0.0064	—	—	254/175.28	1.0	156×	950.
(*42*)	4,9 di n propyl dodecane	0.00007	—	—	254/175.28	1.7	24000×	1300.

	Compound	(A) API–44	(B) MS–1 (b)	"Gain" B/A (d)	(C) Dual Field Ion Source (b) Ratio (a)		"Gain" C/A (d)	Field Ion Sensitivity (c)
(43)	Propylene	0.69	—	—	42/41	21.3	31×	3600.
(44)	1-Butene	0.41	—	—	56/55	18.6	45×	6000.
(45)	1-Pentene	0.51	—	—	70/69	55.5	109×	6550.
(46)	1-Hexene	0.33	—	—	84/56	18.	55×	2250.
(47)	1-Heptene	0.19	—	—	98/57	6.7	35×	1500.
(48)	1-Octene	0.14	—	—	112/56	7.1	51×	1400.
(49)	1-Nonene	0.10	—	—	126/56	7.5	75×	1400.
(50)	1-Decene	0.083	3.10	37×	140/88.03	9.3	112×	1500.
(51)	1-Undecene	0.11	—	—	154/101.46	9.4	85×	1300.
(52)	1-Dodecene	0.028	—	—	168/56	8.3	296×	1400.
(53)	1-Tetradecene	0.053	4.2	79×	196/56	11.0	208×	1800.
(54)	1-Hexadecene	0.016	—	—	224/171.50	13.8	862×	2050.
(55)	3,3 di me-1-pentene	0.09	1.06	12×	98/69.00	3.6	40×	1050.
(56)	Cyclopentene	0.43	—	—	69/68	30.	70×	450.
(57)	Cyclohexene	0.39	—	—	96/70.50	111.	285×	1750.
(58)	Cyclooctene (e)	0.31	—	—	110/55	46.	148×	4300.
(59)	1,3 Cyclooctadiene (e)	0.40	—	—	108/80.17	71.	177×	2350.
(60)	1,5 Cyclooctadiene (e)	0.061	—	—	108/59.42	13.	213×	700.
(61)	1-Pentyne	0.17	0.34	2.0×	68/66	0.5	3.0×	150.
(62)	2-Pentyne	1.1	2.9	2.6×	68/67	21.	19.×	1600.

	Compound	(*A*) API–44	(*B*) MS–1 (*b*)	"Gain" B/A (*d*)	(*C*) Dual Field Ion Source (*b*) Ratio (*a*)		"Gain" C/A (*d*)	Field Ion Sensitivity (*c*)
(*63*)	Acetal	1.8	—	—	44/43	53.	·30×	600.
(*64*)	Propanal	3.5	22.1	6.3×	58/56	111.	32×	1100.
(*65*)	Acetone	14.	46.	3.3×	58/43	1333.	95×	39000.
(*66*)	Methyl ethyl ketone	2.9	6.3	2.2×	72/55	455.	157×	32300.
(*67*)	3-Heptanone	0.08	2.0	2.5×	114/63.43	50.	625×	6000.
(*68*)	Diethyl ether	0.65	2.1	3.3×	74/73	56.	86×	250.
(*69*)	1-Decanol	0.00035	0.40	1140×	158/124.37	0.11	310×	4.2
(*70*)	Silyl ether of 1 decanol (*f*) (*g*)	2.0	22.7	11.4×	215/161.07	18.	9×	2500.
(*71*)	Hexanoic acid	0.0004	—	—	117/99	2.7	6700×	19.
(*72*)	Silyl ether of 1-hexanoic acid (*f*) (g)	0.18	—	—	173/73	2.4	13×	74.
(*73*)	1-Decanoic acid (*e*)	0.094	0.512	5.4×	172/97	11.3	112×	280.
(*74*)	Silyl ether of 1-Decanoic acid (*f*) (*g*)	1.5	—	—	244/73	1.04		2350.
(*75*)	1-Hexane nitrile	0.011	0.43	39×	98/68	0.83	75×	3650.
(*76*)	Succinonitrile (*e*)	0.14	0.75	5.2×	81/75.5	200.	1389×	2550.

	Compound	(*A*) API–44	(*B*) MS–1 (*b*)	"Gain" B/A (*d*)	(*C*) Dual Field Ion Source (*b*) Ratio (*a*)		„Gain" C/A (*d*)	Field Ion Sensitivity (*c*)
(*77*)	1-Bromohexane (*e*)	0.046	0.20	4.4×	164/85	0.61	13×	167.
(*78*)	2-Bromohexane (*e*)	0.0012	0.0051	4.3×	164/85	0.0051	4.3×	29.
(*79*)	3-Bromohexane (*e*)	0.0026	0.012	4.6×	164/85	0.0033	1.3×	16.
(*80*)	Ethyl thiol (*f*)	1.2	—	—	62/61	18.	15×	4800.
(*81*)	n-Propyl thiol (*f*)	0.70	—	—	76/75	59.	84×	3900.
(*82*)	n-Butyl thiol (*f*)	0.54	—	—	90/88.6	11.	20×	4900.
(*83*)	n-Pentyl thiol (*f*)	0.69	—	—	104/102.5	10.	14×	4900.
(*84*)	n-Hexyl thiol (*f*)	0.25	—	—	118/59.80	11.	42×	5400.
(*85*)	n-Heptyl thiol (*f*)	0.35	—	—	132/72.76	4.4	13×	4600.
(*86*)	n-Octyl thiol (*f*)	0.14	—	—	146/85.92	3.8	27×	3600.
(*87*)	n-Nonyl thiol (e)	0.20	4.75	24×	160/9.923	4.6	23×	2900.
(*88*)	n-Decyl thiol (*f*)	0.03	—	—	174/112.64	2.6	87×	1600.
(*89*)	n-Dodecyl thiol (*f*)	0.022	—	—	202/139.72	1.6	73×	900.
(*90*)	n-Tetradecyl thiol (*f*)	0.016	—	—	230/167.03	1.4	88×	1100.
(*91*)	n-Hexadecyl thiol (*f*)	0.012	—	—	258/194.48	1.2	100×	1300.
(*92*)	n-Decylamine (*f*)	0.025	—	—	158/56	16.1	640×	1180.

(a) The parameters reported are the ratios of the parent ion divided by the largest fragment ion above *m*/*e* 43.
(b) Conditions defined in experimental section.
(c) Sensitivities given in chart divisions per micromole. See experimental section for details of inlet system.
(d) "Gain" defined in detail in results and discussion (I).
(e) Uncertified mass spectrum. Origin will be provided by authors.
(f) Esso research spectrum.
(g) See experimental section for preparation.

chemical reactor prior to ionization. This technique has been successfully applied to investigation of many radical reactions. However, the main difficulty connected with this method consists in the fact that the equilibrium radical concentrations in steady state experiments are usually extremely low, whereas the intensities of the radicals with the same mass numbers produced by electron impact are very large. (Larger than the parent ion intensities in many cases.) This difficulty can be partially overcome by using the field ionization method, although difficulties of different nature are connected with this method.

4.2.1. Photochemically produced free radicals

Photochemical investigations by means of a mass spectrometer have been carried out by the author and Groth [193] and by Okabe, Groth and the author [194].

The apparatus used for these studies of free radicals produced by a mercury quartz lamp is shown by fig. 4.9. The gas to be irradiated by photons passes through a Suprasil*) tube which is transparent at the resonance wavelengths 1849 Å and 2537 Å of mercury. The pressure within the tube is normally a few microns. A quartz lamp of the Lossing type [195] which produces a mercury discharge surrounds the Suprasil tube. (The inner tube of the lamp is also made of Suprasil because of its greater transparency for light of 1849 Å as compared to ordinary quartz.) The intensity of the lamp is fairly high at this wavelength, approximately 10^{17} quanta/sec in the 2 cc reaction volume [196]. The intensity of the 2537 Å line is about ten times larger.

After flowing through the irradiated part of the Suprasil reactor tube, the gas passes through a hole of about 2 mm diameter into the FI source. The hole is about 60 mm distant from the end of the mercury lamp. The gas beam leaving the hole is directed exactly onto an FI tip (or an FI wire in some cases). The residence time of the gas in the illuminated zone was about 10 msec.

The intention in these experiments was to detect radicals formed in unsensitized primary photochemical reactions. This is almost impossible by conventional mass spectrometry since the concentration of radicals under steady state conditions is extremely small with these unsensitized reactions. Kistiakowsky and Kydd [197] used the flash photolysis technique in conjunction with a time-of-flight mass spectrometer in order to produce a much larger concentration of radicals. Lossing and his cowor-

*) Quartz of high purity.

kers [198] took advantage of mercury ($^3P_1$) sensitization for enhancement of radical concentrations. Since the absorption coefficient of mercury vapour at 2537 Å is extremely large for the organic substances to be investigated, it was possible to produce large concentrations of radicals which could be detected by a conventional electron impact mass spectrometer.

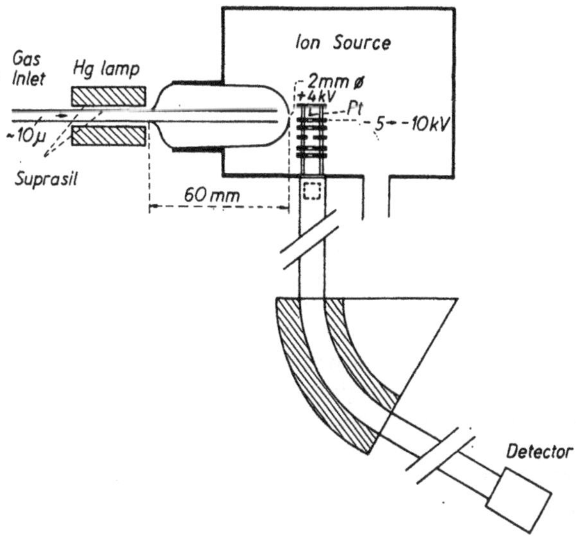

FIG. 4.9. FI mass spectrometer with reactor for production of photo radicals.

First consider the unsensitized primary photochemical decomposition of acetone at 2537 Å. The absorption cross section of acetone is large at this wavelength. The reaction is:

$$CH_3COCH_3 + h\nu \rightarrow CH_3CO + CH_3$$

The relative intensity of the CH_3CO^+ ions, $m/e = 43$, produced in an EI source

$$CH_3COCH_3 + e \rightarrow CH_3CO^+ + CH_3 + 2e$$

is four times larger than the parent ion intensity. In the case of field ionization, however,

$$CH_3COCH_3 \xrightarrow{field} CH_3CO^+ + CH_3 + e$$

the relative intensity of the CH_3CO^+ amounts only 3.5×10^{-4} of the parent ions. Thus, the "background" of CH_3CO^+ ions in the FI mass spectrometer is reduced by a factor of 10^4, and the mass region at $m/e = 43$ is virtually free for detection of extremely small concentrations of neutral CH_3CO radicals from photochemical reactions. The intensity of the CH_3CO^+ peak increased by a factor of four after illumination of the acetone in the reactor, as compared with the unilluminated acetone.

The CH_3 radicals which are produced simultaneously with the CH_3CO radicals could not be detected in the unsensitized decomposition of acetone because of the high ionization potential of the CH_3 radical and its correspondingly low FI probability at medium field strengths.

Sensitization of the reaction with mercury increased the concentration of the radicals to such an extent that the CH_3 radicals could also be detected by field ionization.

It turned out that FI tips are more suitable than thin wires for detection of photo-radicals by means of an FI mass spectrometer. The reason for this is the larger "background" of fragment peaks of nonirradiated substances which is much larger with wires than with tips. These peaks originate from chemical reactions within adsorbate layers on wires which are reduced with the higher field strengths produced by tips. This is illustrated by fig. 4.10. At the same time it was found that the background peaks arising from wires can be reduced by resistive heating of the wires. Therefore, it may turn out in the future that wires could be used as well as tips for detection of free radicals. The advantage of wires over tips consists in their larger emission area and the corresponding higher FI currents, as discussed in detail earlier.

Fig. 4.11. shows the FI mass spectra of 1-butene with and without illumination by the 1849 Å line of the mercury lamp. The absorption cross section of 1-butene is large for this wavelength, but negligibly small for the 2537 Å line. The three most probable primary processes are

$$C_4H_8 + h\nu \rightarrow C_3H_4 + (CH_3 + H) \text{ or } CH_4$$

$$C_4H_8 + h\nu \rightarrow C_3H_5 + CH_3$$

$$C_4H_8 + h\nu \rightarrow C_3H_6 + CH_2$$

corresponding to the peaks at $m/e = 40, 41$ and 42 observed in the FI mass spectrum.

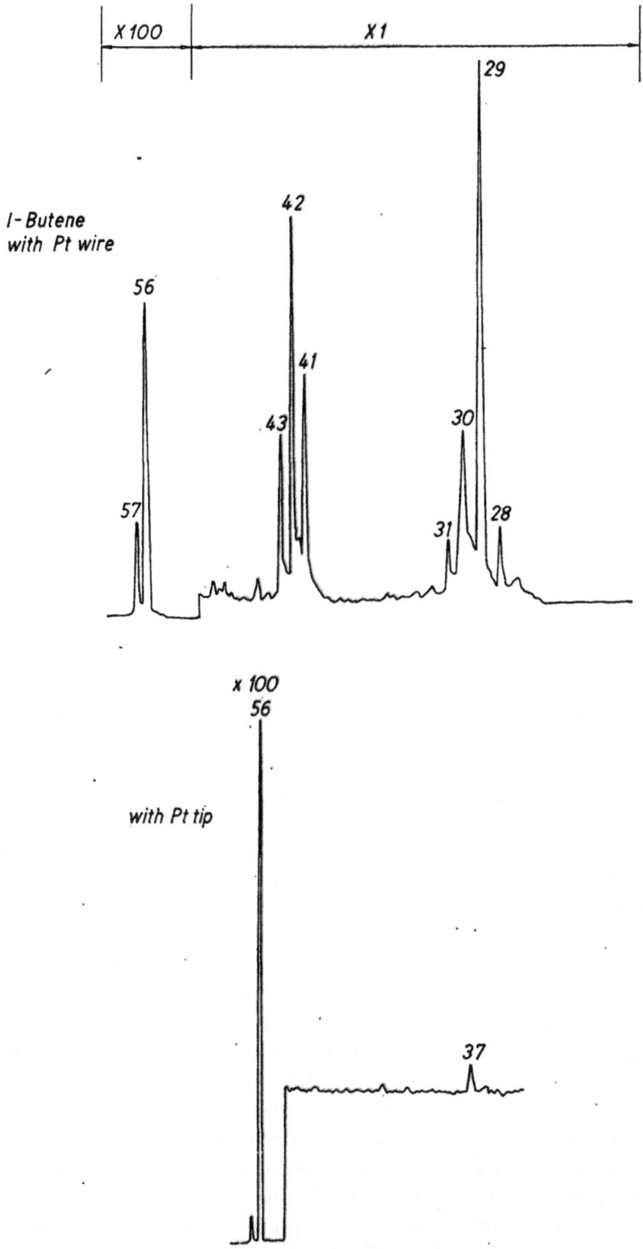

FIG. 4.10. FI mass spectra of 1-butene with Pt wire and tip.

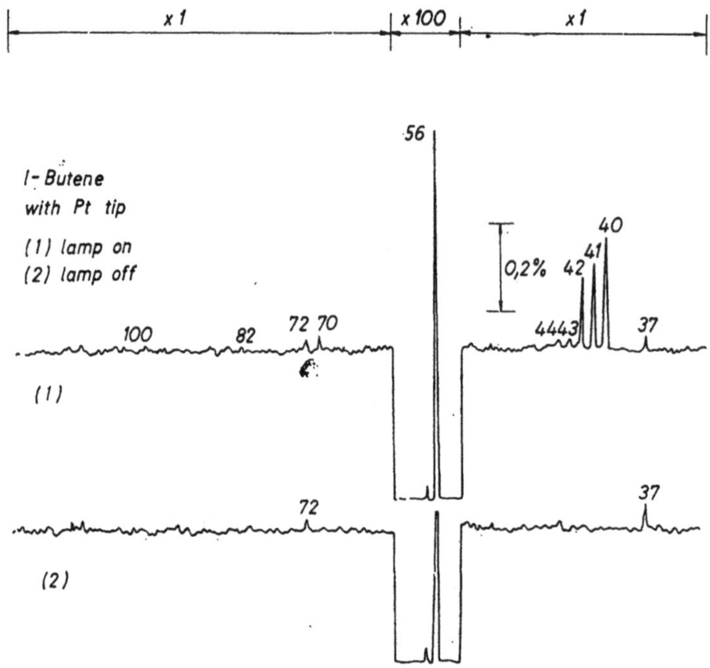

FIG. 4.11. FI mass spectra of 1-butene with Pt tip.
1) during illumination at 1 849 Å and *2*) without illumination.

4.2.2. Thermally produced free radicals

Free radicals produced by thermal decomposition of organic molecules have been detected by means of an FI mass spectrometer by Butzert and the author [199, 200]. The pyrolysis of tetra methyl lead (TML) and tetra ethyl lead (TEL) has been studied in detail. Fig. 4.12. shows the special FI source containing a thermal reactor as used for these experiments. Methyl or ethyl radicals are produced by pyrolysis of TML or TEL, respectively in a quartz tube of 5 mm diameter. The gas flow of molecules and radicals is directed on the FI emitter tip in the ion source.

Fig. 4.13. shows the FI mass spectrum of a TML — TEL mixture prior to and after pyrolysis. The intensity of the TEL parent peak was reduced by roughly a factor of nine after pyrolysis, but the intensity of the $C_2H_5^+$ radical ions ($m/e = 29$) was increased by a factor of about twelve. It was possible to derive the sensitivity coefficient of the C_2H_5 radicals as compared to that of stable molecules.

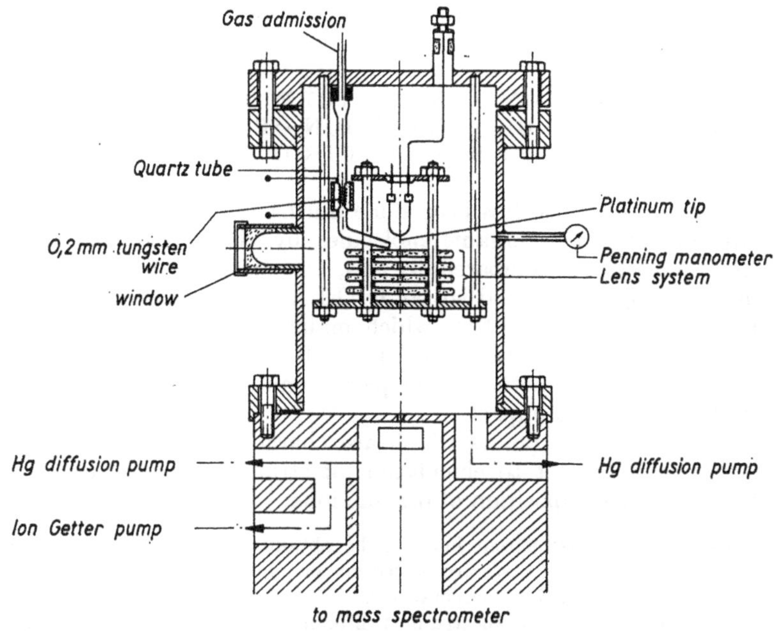

FIG. 4.12. Special FI source with a thermal reactor for pyrolysis of organic substances [199, 200].

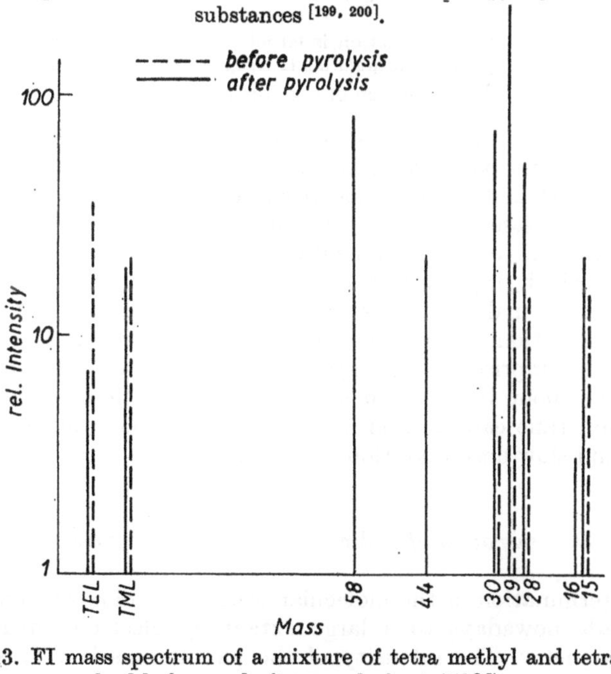

FIG. 4.13. FI mass spectrum of a mixture of tetra methyl and tetra ethyl lead before and after pyrolysis at 450 °C.

4.3. ELUCIDATION OF STRUCTURE, MECHANISM AND KINETICS OF DECOMPOSITION OF ORGANIC IONS BY FI MASS SPECTROMETRY

Tremendous work has been done during recent years on structure elucidation of organic molecules by electron impact mass spectrometry. As compared to this, the amount of information on this subject obtained so far by FI mass spectrometry is moderate. However, a feature which makes field ionization attractive for future investigations is the fact that the information obtained by this method is in part complementary to that obtained by electron impact.

There are three points which are characteristic for field ionization with respect to elucidation of structure, mechanism and kinetics of decomposition of organic ions:

1. Relatively intense parent ions can be obtained by the FI method, in many cases even if there are no parent ions detectable with low energy electron impact at low temperatures.
2. The FI fragment ion spectrum is different from the EI fragment ion spectrum because of strong polarization of the molecules in extremely high electric fields.
3. The kinetics of decomposition is largely different in FI and EI sources. Combination of both methods leads to new information on kinetics and mechanism of decomposition of organic ions.

In general, early publications on FI mass spectrometry stated that the first point described the most characteristic difference between FI and EI mass spectrometry. However, the author has recently stressed the point [201] that the third aspect is also very essential. There are fundamental differences in the time intervals within which fragments appearing at "normal" mass numbers are formed. Normal fragments are formed within max. 10^{-12} sec after field ionization, whereas normal fragments formed after electron impact comprise a large spectrum of lifetimes with an upper limit of about 10^{-6} sec. These characteristic differences in decomposition time can be used in practice for differentiating between fast and slow decomposition processes.

4.3.1. *Determination of molecular weights by FI mass spectrometry*

Determination of the molecular weight of unknown compounds is made nowadays to a large extent by electron impact mass spectrometry. The intensity of the parent ions is very small or virtually zero under normal EI conditions if the molecules are very unstable with respect to 70 eV electrons, or if they are thermally unstable (or both). In these cases, reduction of both electron energy and temperature is promising [104]. Thus in most cases molecular weight determination can be carried out successfully by electron impact mass spectrometry. However, there are a number of molecules which are unstable in such a way that molecular weight determination by EI mass spectrometry cannot be made unambiguously. In those cases FI mass spectrometry may be applied successfully. The field desorption technique seems to be even more promising.

A few examples were given at the early stage of this field [5]. Neopentane is a molecule which is electronically so unstable that the parent ion intensity is negligibly small after electron impact, even at low temperature and low electron energy. In the case of field ionization, however, the relative intensity of the parent peak is about 3% of the most intense peak if tips are used, or about 8% if wires are used as FI emitters, with the ion source at room temperature.

Another examples is that of carbohydrates. It is well-known that it is difficult to evaporate sugars like *d*-ribose without thermal decomposition. The fraction of molecules which can be evaporated undecomposed at relatively low temperatures seems to be very unstable with respect to electron impact so that the parent ion intensity is extremely small. In the FI mass spectrum of *d*-ribose, on the other hand, the parent (and parent-plus-one) peaks are the most intense ones [6] (see fig. 4.14.).

It is very probable that the parent ion intensities in the electron impact mass spectra of many unstable compounds are not exactly zero, but only very small. In these cases it would be possible to record weak parent ion signals by using low temperatures, low electron energies and a secondary electron multiplier of high sensitivity for ion detection. However, consider such a compound (a carbohydrate, for example) which contains an impurity of about 1% (an aromatic hydrocarbon, for example) having a slightly higher molecular weight and a somewhat lower ionization potential than the compound in question. It would then be possible that the parent ion of the impurity would have a much higher intensity than the compound which had to be determined. In many cases, a detailed investigation of whether the fragment ion spectrum would fit with the assumed parent ion or not could lead to a decision on the true parent ion peak [109–113]. However, if the FI method was used and if the parent ions were the most intense ones, as with the example of *d*-ribose, the parent peaks could be differentiated from impurity peaks "on the first glance", and molecular weight determination on a routine base would be facilitated.

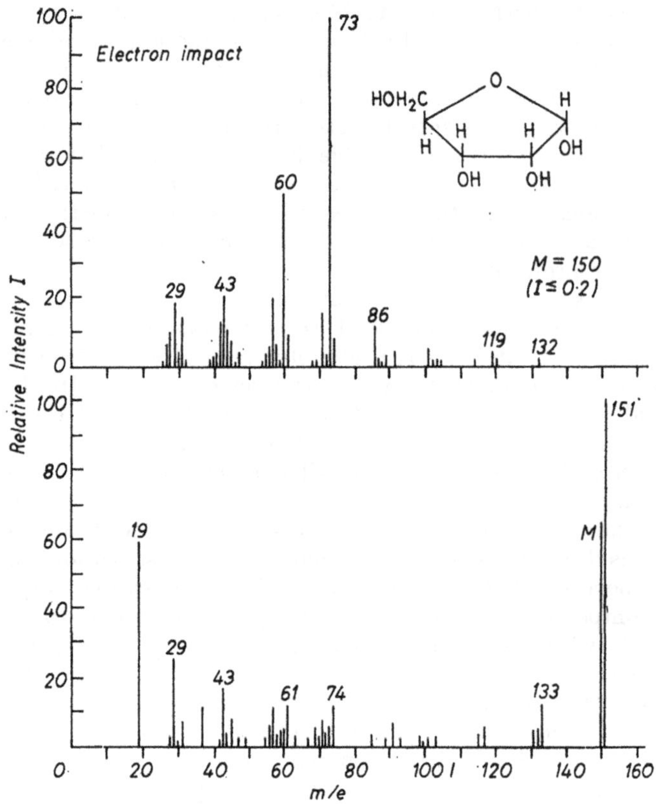

FIG. 4.14. EI and FI mass spectra of *d*-ribose.

A number of other authors have recently reported on the use of the FI method for molecular weight determination with substances where it was difficult to obtain intense parent ions with the EI method.

Brunnée [81] has compared the FI and EI mass spectra of tetradecanol (fig. 4.15.). With EI virtually no parent peak is observed because of the instability of the molecular ion. The parent peak is the most intense one in the FI spectrum. The evaporation temperature for the substance was the same for both FI and EI. Thus, the absence of a molecular EI peak cannot be explained by the thermal instability of the substance. Comparison of the FI and EI mass spectra at the same temperatures offers quite generally a means for deciding whether thermal decomposition of a substance occurs during evaporation into the sources or not.

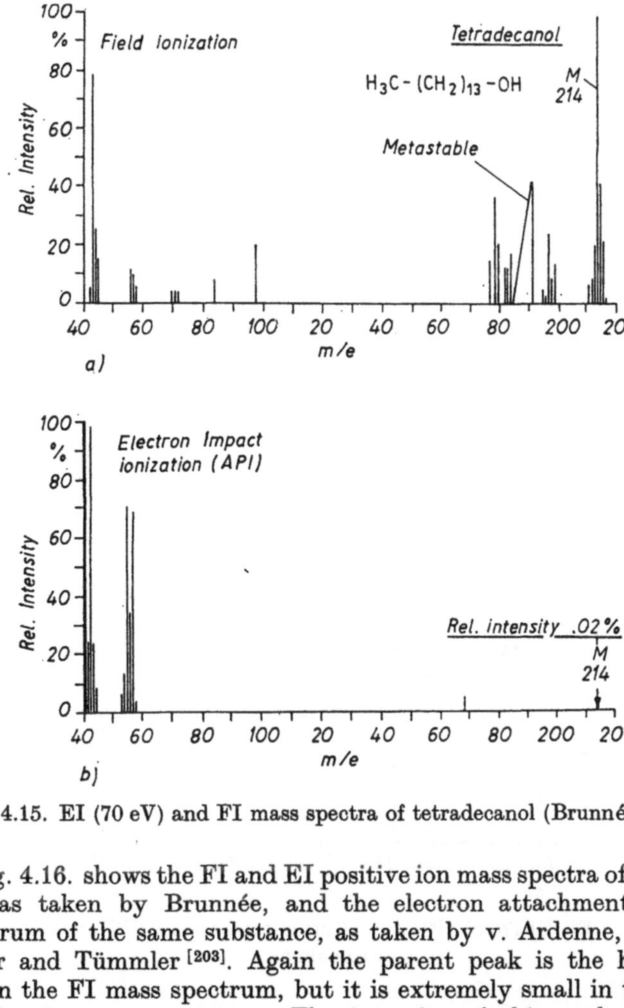

FIG. 4.15. EI (70 eV) and FI mass spectra of tetradecanol (Brunnée [81]).

Fig. 4.16. shows the FI and EI positive ion mass spectra of somaline as taken by Brunnée, and the electron attachment mass spectrum of the same substance, as taken by v. Ardenne, Steinfelder and Tümmler [203]. Again the parent peak is the highest one in the FI mass spectrum, but it is extremely small in the EI positive ion mass spectrum. The intensity of this peak in the electron attachment mass spectrum is intermediate between these two limits. (The isotope peaks are not plotted in the FI mass spectrum because the mass resolution was not sufficient. A double focusing machine would have separated the isotopic peaks.)

Chait, Shannon, Amy and McLafferty [204] have constructed a combined FI/EI source for the double focusing mass spectrometer CEC 21–110 B. Razor blades were used as FI emitters.

High resolving power was obtainable with field ionization (20000 at 10% valley definition). This allowed exact molecular weight measurement and thus determination of the elemental compo-

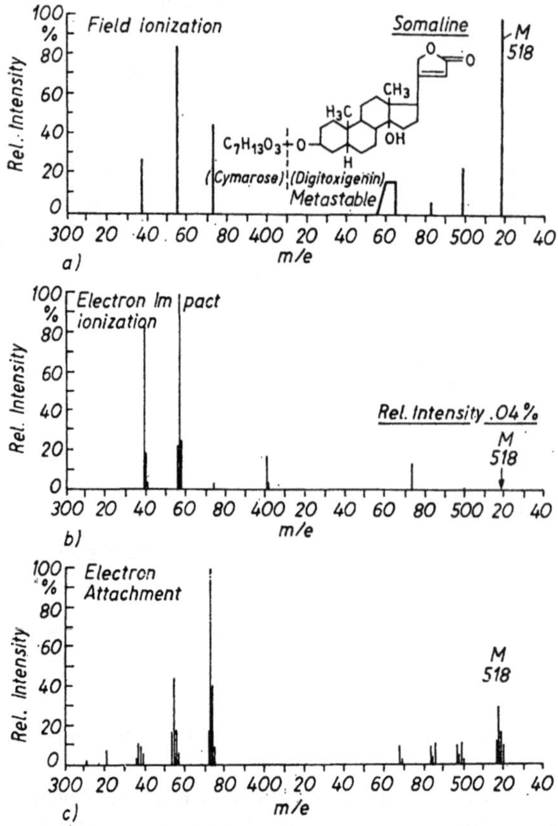

FIG. 4.16. FI, EI (70 eV) [81] and electron attachment [203] mass spectra of somaline.

sition of ions which are too unstable to give sufficiently intense parent ions by electron impact.

For example, the measured mass of serine ethyl ester, $C_5H_{11}O_3N$, was 133.0765, the calculated mass is 133.0739. ($C_8H_{16}O$ was used as a reference ion for precise mass determination.)

The FI mass spectra of more than fifty different compounds of many types have been obtained in the course of that investigation by those authors [204]. Most of the compounds show abundant molecular ions even if none are observed in the EI mass spectra. Only some halogen containing compounds show negligibly small parent ion intensities in the FI mass spectra.

Some larger molecules such as glyceryl trilaurin (MW 638), $C_{30}H_{26}SnBr_2$ (MW 664), and the cyclic peptide evalidine (MW 770) yield sizeable molecular ions. The precision of mass measurement in that range is somewhat lower because at present only a few reference lines of suitable mass are available for that range.

4.3.2. Discussion of combined FI and EI fragment mass spectra

It was pointed out earlier that FI and EI mass spectra are complementary to some extent. This was the reason for the introduction of combined FI/EI sources by the author [6] and for their initial technical development by Brunnée [81]. The first, systematic, comparative investigation of FI and EI mass spectra of a certain class of compounds was made by Hey and the author [205]. A number of mono terpenes was chosen to demonstrate the usefulness of combination of FI and EI data for structure determination.

Some common features were found in the terpene spectra which are important also for other classes of substances. These are — — among others — the following.

1. "The molecular weight of all substances could be determined unambiguously with FI, even if there was no parent peak found in the EI mass spectra".

This point has been discussed in greater detail in the preceding section and therefore needs no further treatment here. (For halogen containing compounds see section 4.3.1.)

2. "Only primary metastable peaks were found in the FI mass spectra".

A primary metastable peak is one which arises from decomposition of a parent ion. Numerous secondary metastable peaks, arising from decomposition of fragment ions, are observed in the EI mass spectra. Apparently the energy contained in fragments formed by field induced dissociation is in general not high enough for secondary metastable decomposition. In the few cases where weak secondary metastable FI fragments have been reported (for example by Patterson [188], see section 4.1.3.) the possibility cannot be excluded that the fragments from which the metastable decompositions occur were formed in the neutral state by dissociative chemisorption on the FI emitter. These fragments may be field ionized, and energy transfer by the ionization process may lead to metastable decomposition. In that case we may again speak of primary metastable processes instead of secondary ones. However, further investigations on the mechanism of formation of these weak metastable processes are required.

Elucidation of decomposition paths of organic ions is greatly facilitated by the fact that the more intense metastable ions in the FI mass spectra originate from the parent ions only.

3. "Some of the metastable ions observed in the FI mass spectra are orders of magnitude more intense than in the EI mass spectra".

This phenomenon is observed, for example, in the FI mass spectrum of α-terpineol, where metastable elimination of H_2O from the parent ion leads to a metastable fragment peak which is as intense (after integration over the peak area) as the parent peak. The reason for this phenomenon lies in the small amount of energy required for the elimination of a water molecule. This leads to very fast formation of the fragment peak after bombardement of the molecule with 70 eV electrons. The small amount of energy transferred to the molecule by field ionization shifts that process to large decomposition times, lying in the range of metastable processes. (According to the quasi-equilibrium-theory of mass spectra, the velocity of the decomposition reactions increases rapidly with increasing amount of excitation energy transferred to the molecular ion.)

4. "Ions of about double the molecular weight are frequently found in FI mass spectra if the molecule contains carbonyl or hydroxyl groups or conjugated multiple bonds in aliphatic compounds".

The intensity of these peaks is much lower than that of the molecular ion peaks (about 1% or less). Confusion with the molecular ion should therefore not be possible. These peaks on about double the molecular mass number give additional information on the nature of the molecules.

5. "Comparison of the intensities of certain fragment peaks in the EI and FI mass spectrum gives indications on whether the fragments are formed by simple bond cleavage or by more complicated mechanisms of decomposition".

This last point will be discussed in more detail in the next section. However, before doing this, a few examples of comparison of FI and EI mass spectra will be given.

Fig. 4.17.a, b show the EI [202] and FI mass spectra of menthone.

The most intense peak in the EI mass spectrum is found at $m/e = 112$ which is produced by a McLafferty rearrangement [208], as indicated in the figure. In contrast to this, the $m/e = 112$ peak is very small (0.6%) in the FI mass spectrum and it is shifted from the normal mass position as noticed by Levsen [226]. The reason for this spectacular difference of the EI and FI mass spectrum of menthone lies in the fact that this McLafferty rearrangement process requires too much time to be observed within a time interval of 10^{-12} sec which is characteristic for FI fragment peaks at normal mass position. However, the $m/e = 112$ peak is observable as a metastable FI peak. This metastable peak is the most intense fragment peak of the FI mass spectrum. Another rather intense (about 5%) metastable peak is found in the FI

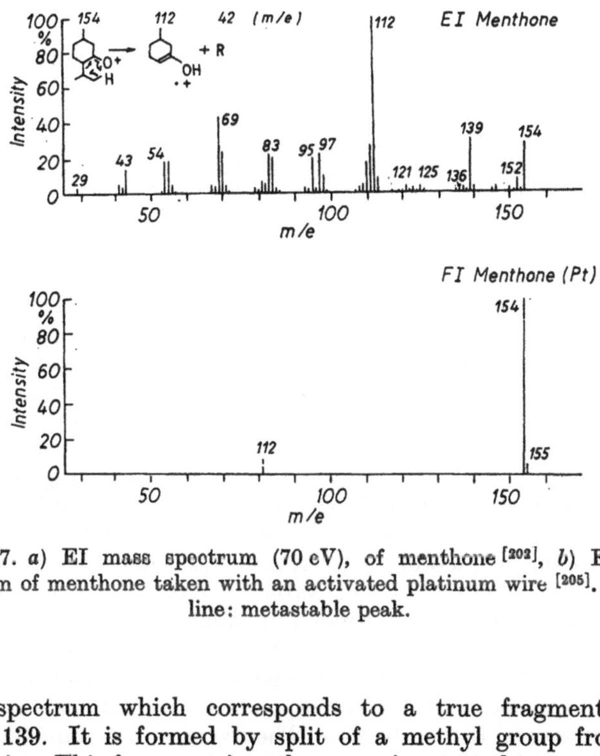

FIG. 4.17. a) EI mass spectrum (70 eV), of menthone [202], b) FI mass spectrum of menthone taken with an activated platinum wire [205]. Dashed line: metastable peak.

mass spectrum which corresponds to a true fragment mass $m/e = 139$. It is formed by split of a methyl group from the parent ion. This fragment is rather prominent at the normal mass number in the EI mass spectrum. (The metastable peaks are drawn as dashed lines at the observed mass position in the figures of this chapter. The true mass numbers are noted on top of these peaks.)

Dimeric ions are also present in the spectrum.

Table 4.7. lists some of the more important peaks of the FI and 70 eV EI mass spectra of α-terpineol.

TABLE 4.7. MASS SPECTRA OF α-TERPINEOL

Mass-number	EI	Intensities FI direct	FI metastable	Process
154	—	100	—	Molecular peak
137	4	14.0	—	Direct elimination of OH
136	31	1.7	96.4	Elimination of H_2O
121	38	—	—	Elimination of $H_2O + CH_3$
95	13	1.1	—	Direct elimination of C_3H_6OH
93	59	0.1	0.9	Indirect formation of $C_7H_9^+$
81	35	—	—	Indirect formation of $C_6H_9^+$
59	100	6.5	0.8	Direct and metastable formation of $C_3H_6OH^+$
43	33	0.9	—	$C_3H_7^+$

(Intensities related to peak areas, most intense peak = 100, FI spectrum produced by gold wire. The EI intensities are taken from a spectrum measured by v. Sydow [206].)

The molecular peak is the most intense one in the FI mass spectrum, but it is absent in the EI mass spectrum. Direct elimination of OH from the molecular ion is not pronounced in case of EI, but very intense in case of FI. (A fragment peak of a few percent rel. intensity has to be regarded as very intense in the case of FI.) The most remarkable process is elimination of a water molecule from α-terpineol. It is a very intense process in the case of EI, rather weak in the case of FI at the normal fragment mass number, but extraordinarily intense at the metastable FI fragment mass number. This shift from the normal fragment mass region to the metastable mass region indicates — as mentioned earlier — that elimination of water by molecular rearrangement requires much more time than is available for formation of an FI fragment at normal mass number (about 10^{-12} sec). The residence time of the α-terpineol parent ions in the field free region of the mass spectrometer corresponds to metastable transitions of about several 10^{-8} to 10^{-6} sec. The average rate constants of the molecules with respect to metastable decomposition in that interval of residence times are only $10^4 - 10^5$ sec^{-1}, as measured by Tenschert [130] by the retarding potential technique (see section 3.4.2.5.).

The high intensity of the fragment $m/e = 121$ in the EI, and its negligibly small intensity in the FI mass spectrum indicates a stepwise elimination of H_2O and CH_3 from the molecular ion.

The EI intensity of the fragment at $m/e = 95$ is ten times larger than that at $m/e = 93$ in the FI mass spectrum. This indicates that the first peak originates from a direct bond split whereas the second is due to a rearrangement process. This is supported by a relatively intense metastable ion corresponding to $m/e = 93$. The fragment at $m/e = 43$ is intense in the EI mass spectrum, but it can be formed only by rearrangement. Therefore, it is of rather low intensity in the case of FI.

These few examples demostrate the characteristic differences between EI and FI mass spectra.

4.3.3. *Kinetics of rearrangement reactions.*

Rearrangement reactions of organic ions are of great importance in the electron impact mass spectrometry. Although the intensities of the rearranged ions are in general smaller in the FI mass spectra, important results can be drawn from such spectra with respect to the kinetics of the rearrangement processes. These can be obtained by observation of peak shifts and broad characteristic peak shapes in the case of rearrangement reactions.

As pointed out in section 3.4, one can distinguish among three different ranges of molecular decomposition times by inspection of FI mass spectra:

I. Decomposition within about 10^{-14}—10^{-12} sec, observable as a relatively sharp fragment peak at normal mass number (termed "normal" fragment peak).

II. Decomposition within about 10^{-12}—10^{-8} sec, observable as a broad peak extending over a mass range between the normal fragment peak and the normal metastable peak. (This broad intermediate peak is termed "fast metastable" peak.)

III. Decomposition within about $10^{-8} - 3 \times 10^{-6}$ sec, observed as a somewhat broadened fragment peak at the position of "normal metastable" peaks.

Decomposition reactions within time interval (I) are mainly field induced. Molecules decomposing in the time intervals (II) and (III) dissociate statistically in the low field or zero field region of the mass spectrometer. Mainly these two regions are of importance with rearrangement reactions. The reactions can be discussed qualitatively by means of the quasi-equilibrium-theory of mass spectra (QET) [117, 118] (see section 3.4.2.3.).

Fig. 4.18.*a*) shows a plot of log k versus the internal ion energy E as obtained by the QET. (k = rate constant.) The shapes of the curves are taken arbitrarily, and no reference is made to specific formulations of the QET. The function $k_{(E)}$ has been determined experimentally for a number of substances by Hertel and Ottinger [122–124], for the case of electron impact ionization. According to one of the fundamental postulates of the QET, the curves should be independent of the method of ionization, and

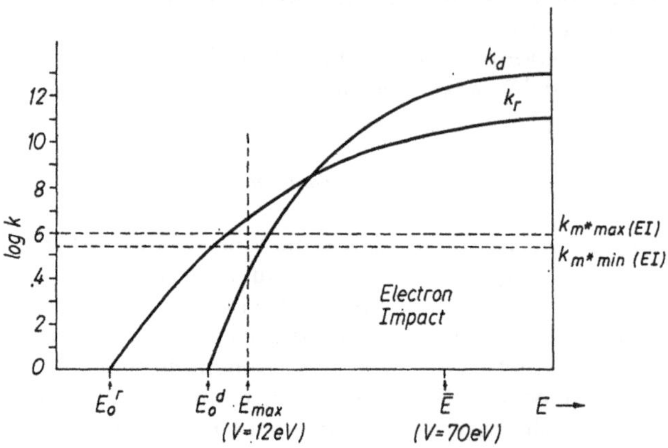

FIG. 4.18. *a*. Rate constants k_d of a direct bond fission and k_r of a rearrangement reaction as a function of the internal energy E of the molecular ion. Ionization by 12 eV or 70 eV electrons, respectively. E_0^r and E_0^d = activation energies for the two reactions. Horizontal dashed lines: Boundary condition for metastable decomposition (log k = 5.5—6.0). Log k = 6—13: normal fragmentation. E_{max} = maximum energy transferred by an electron with kinetic energy V = 12 eV. $\bar{E}$ = average energy transferred by 70 eV electrons.

hence they should be the same for EI and FI. (Except for the region of extremely high fields.) This is demonstrated by fig. 4.18.*b*.

The rate constant k_d refers to a direct C—C bond fission process, whereas k_r refers to a rearrangement reaction.

Williams and Cooks [207] and Williams [258] point out that the log k — curve should rise more slowly with the internal ion energy E with rearrangement reactions because both the activation energies and the "frequency factor" should be smaller with most of the rearrangement reactions than with direct bond fission processes. (For small frequency factors see also ref. [209].)

This can be realized by inspecting the basic equation of the QET in its earliest and simplest version [117].

$$k = \nu \left(\frac{E - E_0}{E}\right)^{s-1} \tag{4.1}$$

ν = frequency factor, E = internal energy of the parent ion, E_0 = activation energy, s = number of vibrational modes.

Eqn. (4.1) gives the rate constant for a molecule with no internal rotations. In a great number of rearrangement reactions internal rotations of the molecular ions are stopped in the activated com-

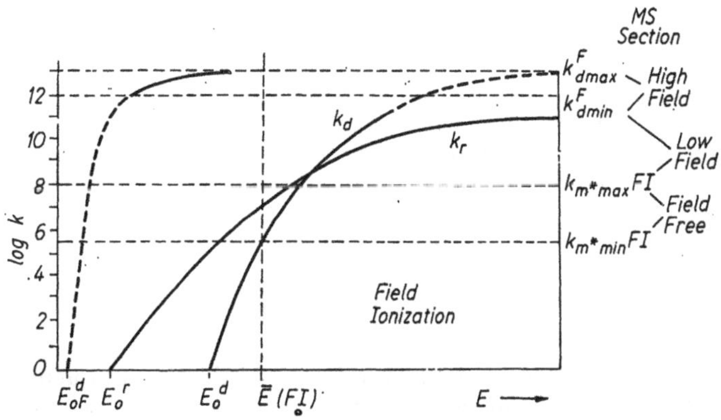

Fig. 4.18.*b*. Rate constants k_d and k_r as a function of the internal energy of the ions for the case of field ionization. The curves have the same meanings as in fig. 4.18a). The rate interval for metastable decomposition is much wider than for the case of electron impact. $k^F_{d\ min,max}$ are the minimum and maximum rate constants for detection of field dissociated fragments on the "normal" mass position. E is the internal molecular ion energy after field ionization. Log $k = 12-13$: Field induced direct bond fission with activation energy E^d_{0F}, reduced by the extremely high field. (Solid curve on the leftside.) Log $k = 5.5-8$: metastable decomposition in the field free space. $\bar{E}_{(FI)}$ = average energy transferred to the molecular ion by field ionization. The rate constants are related in the figure to the high, low and zero field sections of the mass spectrometer.

plex. Rosenstock et al. [117] have derived a rate equation for this case which, on the same first level of approximation as eqn. (4.1) is given by

$$k = \nu' \left(\frac{E - E_0}{E}\right)^{s - \frac{L}{2} - 1} (E - E_0)^{\frac{L - L^{\ddagger}}{2}} \qquad (4.2)$$

where ν' is a constant factor depending on the vibrational and rotational molecular frequencies, E, E_0 and s have the same meaning as before, and L and $L^{\ddagger}$ are the numbers of rotational modes of the parent ion and the activated complex, respectively.

Eqn. (4.2) describes the slower rise of log k with E for a rearrangement reaction in a qualitatively right manner for low energies. However, log k tends to infinity for high energies which seems to be physically unreasonable.

Therefore, eqn. (4.2) has to be replaced by another function of the argument $\frac{E - E_0}{E}$:

$$k_r = \nu_r f_{r\left(\frac{E-E_0}{E}\right)} \tag{4.3}$$

f_r depends on the number of oscillators and the number of free and stopped internal rotators. Presently we do not know the exact form of f_r, and therefore we formally continue to describe k by eqn. (4.1), but with a changed frequency factor ν_r and an effective number of oscillators s':

$$k_r = \nu_r \left(\frac{E - E_0}{E}\right)^{s'} \tag{4.4}$$

The constant ν_r of eqn. (4.4) reflects both the frequencies of oscillation and free internal rotations and also the effect of stopped internal rotations which may reduce ν_r by several orders of magnitude as compared to eqn. (4.1). With this definition of ν_r in mind we may speak of a "low frequency factor" of a rearrangement reaction. (An analog expression in the transition state theory formulation of a thermal reaction would be the enthropy of activation term, exp $\Delta S^{\neq}/R$), $\Delta S^{\neq}$ would have a large negative value for a rearrangement reaction. Let us now return to fig. 4.18.a. The horizontally dashed lines correspond to different ion residence times in different parts of an FI mass spectrometer. For example, ions decomposing in the field free region of the mass spectrometer before entering into the magnetic field ("normal" metastable ions) have residence times between about 10^{-8} and 3×10^{-6} sec. The corresponding decomposition processes are mainly those which have rate constants between 10^8 and 3×10^6 sec^{-1}.

Ions decomposing between about 10^{-12} and 10^{-8} sec do not appear on the normal fragment ion positions in the FI mass spectrum because of their reduced kinetic energies (see section 3.4.1.). These "fast" metastable ions appear as a broad intermediate peak, as described under case II at the beginning of this section. The corresponding rate constants are mainly $k = 10^8 - 10^{12}$ sec^{-1}.

Now one can draw an important conclusion from fig. 4.18.a. If the frequency factor of a direct C—C bond fission is assumed to be of the order 3×10^{13}, and if this factor is reduced for a rearrangement reaction by a factor of about 10^2, one would arrive at a maximum rate constant of $k = 3 \times 10^{11}$ sec^{-1} at high energies. Then one can derive from the figure that the maximum of the rearrangement peak will never appear at the "normal" fragment ion mass position, but will be shifted to smaller masses. These peak shifts are actually observed in most of the FI mass spectra of rearrangement reactions [259]. The situation is different for direct bond fission processes with frequency factors between 10^{12} and 3×10^{13} sec^{-1}. The bond can be ruptured either in the field free region (normal metastable peak, $k \simeq 3 \times 10^6 - 10^8$ sec^{-1}), the low field region (fast metastable peak, $k \simeq 10^8 - 10^{12}$ sec^{-1}) or in the high field region (normal fragment peak, $k \simeq 10^{12} - 3 \times 10^{13}$ sec^{-1}). In the latter case the activation energy is greatly reduced by the high electric field (see fig. 3.6.), and therefore, the log k versus E — curve is shifted to low energies, according to eqn. (4.1). This shift is indicated by the solid part of the k_d — curve, whereas the dashed parts are extrapolations which are not physically realized.

The situation is quite different for the case of electron impact (see fig. 4.18. b). The ion residence times in the field free space are only about $1 - 3 \times 10^{-6}$ sec, and hence the metastable processes occur only in the narrow interval $k \simeq 3 \times 10^5 - 1 \times 10^6$ sec^{-1}. (Dashed horizontal lines.) On the other hand, ions formed after electron impact and before full acceleration have a residence time of about 10^{-6} sec in the ion source. Therefore, all fragments formed within $3 \times 10^{-14} - 1 \times 10^{-6}$ sec appear on the normal fragment ion mass position in the EI spectrum. In other words, ions originating from rearrangement processes with small frequency factors appear at normal mass positions in the EI mass spectra, in contrast to the shifted peaks of the FI mass spectra. (The FI peaks are shifted only with gas phase rearrangement reactions, not with surface rearrangement reactions, as will be shown later.)

Summarizing all these considerations one arrives at the general result:

Reactions with a small frequency factor ($\nu <$ *several* 10^{11} *sec*$^{-1}$) *occurring in the gas phase cannot be detected in an FI mass spectrum by fragments which have peak maxima at "normal" mass positions* [201].

Those slow reactions can be detected, however, either by "fast metastable" fragment ions which are formed between the FI emitter and the cathode with shifted peak maxima at m_f^* ($m > m_f^* > m^2/M$), or by "normal metastable" ions ($m_n^* = m^2/M$).

In the electron impact mass spectra, on the other hand, very intense fragment ions originating from reactions with low frequency factors can be observed on the normal mass position. If the electron impact energy is lowered, the abundance of these fragments will be often larger than that of fragments originating from direct bond fission processes.

For these reasons it is interesting to compare for the case of

FI and EI the ratio of the intensities of fragment ions originating from low frequency factor reactions, normalized to the fragment sum intensity, $I_{\Sigma f}$. A ratio R can be studied which is defined by

$$R = \frac{(I_{nl}\, I_{\Sigma f})_{FI}}{(I_{nl}\, I_{\Sigma f})_{EI,\,12eV}}$$

where I_n are the intensities of the peak *maxima* on "normal" mass positions, and the index l refers to a low frequency factor.

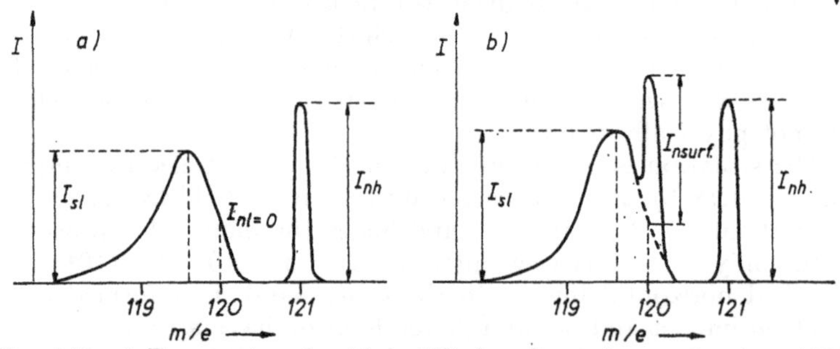

FIG. 4.19. *a*) Fragment peak with a shifted peak maximum at $m/e = 119.6$ originating from a reaction with a low frequency factor, $k = 10^{11}\ \text{sec}^{-1}$. Intensity of the shifted peak maximum: I_{sl}. There is no peak maximum at the normal mass position, i.e. $I_{nl} = 0$, $R = 0$. The peak at $m/e = 121$ is due to a direct bond fission (high frequency factor).
b) Fragment peak of the same gas phase reaction, superimposed by a surface rearrangement peak of intensity $I_{n\,surf.}$. R has a large value in this special case.

A rearrangement reaction with a small maximum rate constant (say $k = 10^{11}\ \text{sec}^{-1}$) leading to a shifted mass peak may have a nonvanishing peak tail at the normal mass position. This is because of the unimolecular decomposition kinetics: A small fraction of molecules associated with $k = 10^{11}\ \text{sec}^{-1}$ decompose already in the time interval 10^{-14}—10^{-12} sec. The intensity of this peak tail at the normal mass position is not of interest here, because the problem of interest here is the evaluation of the maximum rate constants from the shifted peak *maxima*. Only the difference between a peak maximum at normal mass position and the tail intensity of the shifted peak at the same position should be evaluated from the FI mass spectra for calculation of R. This is illustrated by fig. 4.19.a.

The definition of the R-factor was given in the early publications on that problem in a similar way, but not as clear as the defintion given here because the problem of the tail of a shifted mass line had not been discussed.

The R-factors of reactions with low frequency factors should be zero, or at least very small according to these considerations. This is actually observed with most of the rearrangement reactions occurring in the gas phase.

On the other hand, reactions with low frequency factors can also occur on the surface of a field ion emitter if the adsorption time is large enough. These reactions do not fall into the range of application of the above-mentioned rule, and thus the R-factors may become large in these cases (see fig. 4.19b). Actually some very intense rearrangement peaks which are not shifted from the normal mass position have been observed in the FI mass spectra of a number of compounds by several authors. These peaks must be due to surface rearrangement processes which do not fall into the range of application of the above-mentioned rule.

The intensities of the "normal" metastable peaks are often more intense in the FI than in the EI mass spectra by two reasons: First, the residence time interval of the ions in the field free space of the mass spectrometer is much larger in the case of FI ($10^{-8} - 3 \times 10^{-6}$ sec) than in the case of EI (about 1×10^{-6} to 3×10^{-6} sec). Second, the fraction of ions with low interval energy is larger with FI than with EI, which shifts the rate constants from high values to those low values which are characteristic for the metastable decomposition.

For comparison of reactions with low frequency factors it is interesting to introduce a further factor in addition to the R-factor:

$$Q = I_{if}/I_{im^*}$$

where I_{if} is the intensity of the integrated fast metastable peak and I_{im^*} that of the corresponding normal metastable peak. The Q-factor should be small for slow gas phase reactions.

Finally, a factor S is defined which is better suited for comparison with the Q-factor than the R-factor which should always be very small for gas phase rearrangement.reactions. S is defined in the following way:

$$S = \frac{(I_{ir}/I_{\Sigma f})\ \mathrm{FI}}{(I_{ir}/I_{\Sigma f})\ \mathrm{EI}}$$

where I_{ir} is the integrated peak intensity of the—normally shifted—rearrangement peak, corresponding to a reaction with a low frequency factor. $I_{\Sigma f}$ is the sum of the fragment ion intensities.

Damico et al. [227] found a number of intense FI peaks resulting from hydrogen transfer reactions. Very probably most of them are due to surface rearrangement reactions. Those occuring in the gas phase should have peaks which are at least slightly shifted from the normal position.

There are no principal theoretical restrictions for the integrated intensities of the shifted rearrangement peaks, and therefore S may become greater than one, in contrast to the R-factors which are much smaller than one for gas phase rearrangement reactions. These S-factors are formed in order to show a certain parallelism to the Q-factors. A reaction having a very small maximum rate constant will show an intense normal matastable peak and a weak fast metastable peak. Hence both S and Q should be small for that case.

By the same argument a fast rearrangement reaction should lead to both large S and Q-factors. These factors are listed in table 4.8 (rearrangements with H-transfer) and table 4.9) (skeletal rearrangements). These rearrangement reactions have been studied extensively by Levsen [128]). The tables show that there is a certain parallelism with respect to the order of magnitude of S and Q. The S and Q-factors are classified as "low" (l) if they are smaller than 0.1 and as "high" if they are larger than 0.1. Most of the reactions can be recognized as either ll or hh. There are also a few intermediate cases lh (salicyclic acid methylester and 2-hydroxybenzylalcohol) where the rates are between those of very fast and very slow rearrangement reactions.

4.3.3.1. Rearrangement reactions with hydrogen transfer

Hydrogen transfer reactions proceed via a cyclic activated intermediate complex in many cases. In the EI mass spectrometry the *McLafferty rearrangement* reaction is a well-known process of that type. The configuration of a six-membered ring is a very special one as compared with the variety of statistically bent open chain configurations. Therefore, the frequency factor ν_r of such a reaction as defined by eqn. 4.4 should be small. An estimation of the energetic and the sterical factors leads to an average rearrangement time $\tau_R > 10^{-11}$ sec. Therefore, the maxima of the *McLafferty rearrangement* peaks in the FI mass spectra should be shifted from the normal mass positions if the reactions occur in the gas phase. The R-factors should then be small.
Levsen [226] has studied a large number of McLafferty or related rearrangements and noticed that the peaks were actually shifted in most of the cases (and hence the R-factors were small). However, he also noticed some intense rearrangement peaks at exactly the normal position with no unusual peak broadening (large R-factors). It will be shown that those reactions are not pure gas phase reactions but surface rearrangement reactions.

Among the McLafferty rearrangement reactions which have been studied are the ketones butyrophenone, valerophenone and caprophenone [226]. The main reaction is a single rearrangement

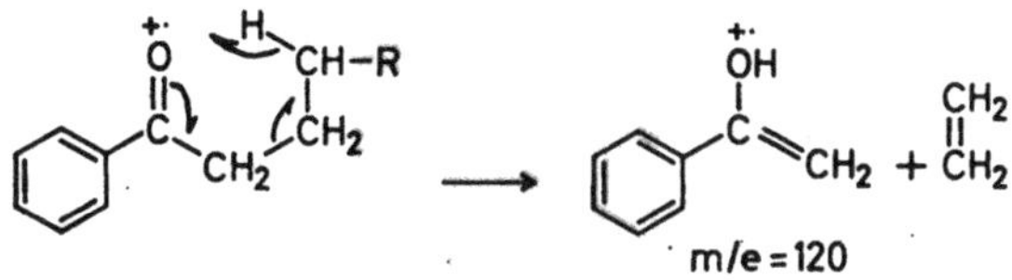

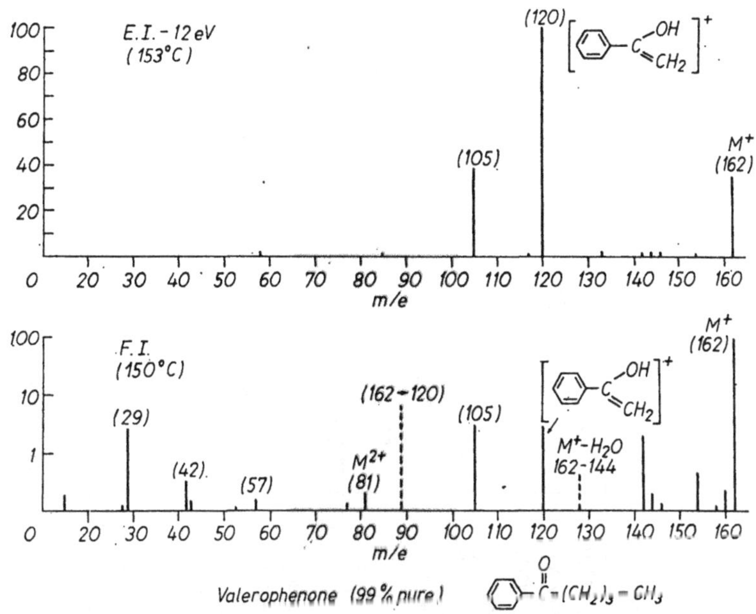

FIG. 4.20. EI (12 eV) and FI mass spectra of valerophenone (corrected for ^{13}C-contributions). Note the linear and logarithmic intensity scales for EI and FI, respectively).

reaction leading to a fragment $m/e = 120$, which had also been reported previously by Chait et al. for the case of butyrophenone.

Fig. 4.20. shows the EI and FI mass spectra of valerophenone. One of the most interesting features of the FI mass spectrum is the shift of the maximum of the rearrangement peak $m/e = 120$ by

TABLE 4.8.

Rearrangements with H-transfer

Compound (Molecular weight)	Process	I (%) F.I.	I (%) E.I. (12 eV)	$I/\Sigma I_{Fragm.}$ (%) F.I.	$I/\Sigma I_{Fragm.}$ (%) E.I. (12 eV)	I (%) F.I. Metast.	S	Q
Butyrophenone (148)	$M^+ - C_2H_4$ (120) (Mc Lafferty R.)	0,33 (150°)	49 (250°)	8	20	12	0,40	0,27
Valerophenone (162)	$M^+ - C_3H_6$ (120) (Mc Lafferty R.)	2,9 (150°)	100 (150°)	26	50	6,3	0,52	0,46
Caprophenone (176)	$M^+ - C_4H_8^+$ (120) (Mc Lafferty R.)	3,1 (150°)	100	33	76	4,6	0,44	0,67
Menthone (154)	$M^+ - C_3H_6$ (112) (Mc Lafferty R.)	0,66 (150°)	100 (150°)	48	48	0,90	1,0	0,67
Benznic acid butylester (178)	$C_6H_5-COOH_2^+$ (123) "two-hydrogen-transfer"	< 0,05 (25°)	100 (120°)	2,5	71	8,1	0,035	<0,006
	$M^+ - OH$ (161)	< 0,01 (90°)	0,9 (120°)	< 0,15	0,6	0,3	< 0,25	<0,03
Benzoic acid propylester (164)	$C_6H_5-COOH_2^+$ (123) "two-hydrogen-transfer"	< 0,05 (25°)	100 (120°)	6	70	7,2	0,08	<0,007
Butanoic acid-n-amylester (158)	$C_3H_7-CH=CH_2^+$ (70) (Mc Lafferty R.)	38,0 (130°)	100 (150°)	20	36	0,4	0,55	95
	$C_3H_7-COOH_2^+$ (89)	2,5 (130°)	74 (150°)	1,2	26	0,6	0,05	4
	$M^+ - C_2H_4$ (130) (Mc Lafferty R.)	5,0 (130°)	2,7 (150°)	2,4	1,0	0,5	2,4	4,8

Butanoic acid butylester (144)	$C_2H_9-CH=CH_2$ (56) (Mc Lafferty R.)	100 (90°)	77 (100°)	87	29	0,8	3,0	125
	$C_3H_7-COOH_2^+$ (89) "two hydrogen transfer"	0,8 (90°)	100 (100°)	0,6	38	15	0,016	0,05
	$M^+-C_2H_4$ (116) (Mc Lafferty R.)	0,3 (90°)	4,7 (100°)	0,2	1,8	6	0,11	0,05
Butanoic acid propylester (130)	$CH_3-CH=CH_2^+$ (42) (Mc Lafferty R.)	100 (100°)	8,0 (110°)	81	3,5	—	23	—
	$C_3H_7-COOH_2^+$ (89) "two hydrogen transfer"	1,3 (100°)	100 (110°)	1,0	45	7,4	0,022	0,18
	$M^+-C_2H_4$ (102) (Mc Lafferty R.)	1,0 (100°)	8,5 (110°)	0,8	3,9	4,8	0,20	0,21
Acetic acid phenylethylester (164)	$C_6H_5-CH=CH_2^+$ (104) (Mc Lafferty R.)	100 (150°)	100 (150°)	67	90	11	0,75	9
Salicylic acid methylester (152)	M^+-CH_3OH (120) (Orthoeffect)	0,03 (50°)	46 (140°)	4	90	0,13	0,04	0,2
2-Hydroxybenzylalcohol (124)	M^+-H_2O (106) (Orthoeffect)	0,06 (100°)	65 (110°)	4	68	0,25	0,06	0,2
p-Cresol (108)	M^+-H_2O (90)	< 0,01 (150°)	1,0 (160°)	< 1,2	9	0,12	<0,13	<0,08
Diphenylsulfoxide (202)	M^+-H_2O (184)	< 0,01	0,6 (130°)	< 0,2	0,8	0,11	<0,25	<0,09

several tenths of a mass unit towards smaller mass numbers. The peak shift depends on the ratio of the emitter voltage ($+U_{an}$) to the total voltage ($+U_{an} - U_{cath}$). Therefore, the peak shifts Δm are related to certain $+U_{an}$ and $-U_{cath}$ values, which are 3 and 7 kV, respectively, for the present case.

	Δm in mass units
butyrophenone	-0.47 ± 0.08
valerophenone	-0.55 ± 0.08
caprophenone	-0.71 ± 0.08

A rearrangement reaction closelly related to the McLafferty reaction is the *mass spectrometric ortho effect*. The rearrangement of 2-hydroxybenzylalcohol will be discussed here.

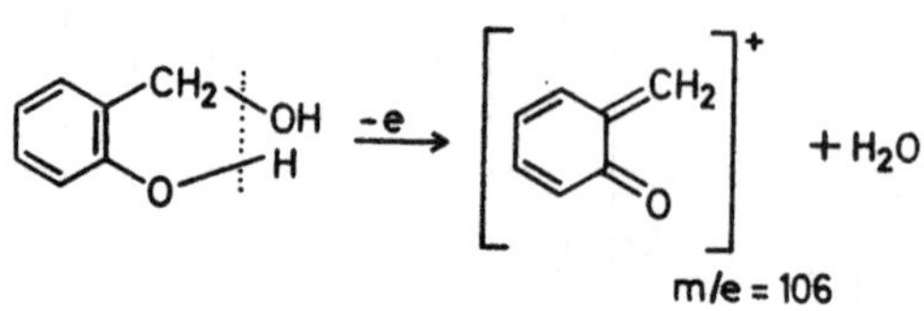

A weak $m/e = 106$ rearrangement peak is found at low temperature in the FI mass spectrum at the exact mass position. ($I_{rel} = 0.06-0.1$ at 100 °C.) The intensity of this peak increases strongly with increasing temperature. Raising the temperature to 250 °C increases the intensity by a factor of 140. This peak is entirely sharp. A shifted peak was found in addition at an about 0.5 mass units lower position if the FI emitter was covered with benzonitrile—grown micro needles. This peak is broadened (see fig. 4.21.). The intensity of this shifted peak is much less temperature dependent than the sharp $m/e = 106$ peak. The different temperature dependence and shapes of the two peaks leads to the assumption of different mechanisms of origin of the peaks. Because of the sharpness of the $m/e = 106$ peak it can be assumed that it originates from a surface rearrangement reaction. (This occurs on the hut surfaces of the ion source [226]) The minimum between the sharp peak and the shifted peak is probably due to ion optical defocusing effects.

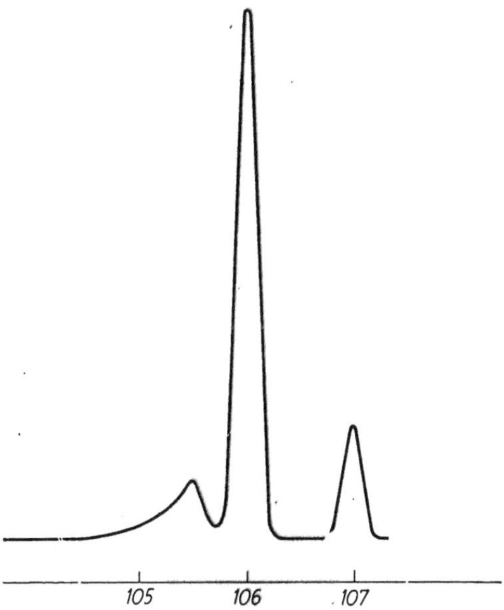

FIG. 4.21. Part of the FI mass spectrum of 2-hydroxybenzylalcohol, showing the different shapes of the gas phase and surface rearrangement peak at $m/e = 105.5$ and $m/e = 106$.

4.3.3.2. *Skeletal rearrangement reactions*

Skeletal rearrangement reactions are more complicated processes in many cases than hydrogen transfer reactions (see review article by Brown and Djerassi [228]). The large negative value of the activation entropy (configuration entropy) in the transition state formulation of the rates leads to a small frequency factor. The activation energies are in general somewhat larger than those of the hydrogen transfer reactions. Both reasons lead to the conclusion that the rate constants must be low, and therefore, one would expect small intensities of the FI rearrangement peaks at the normal or shifted mass position. The metastable rearrangement peaks would be expected to be stronger.

Skeletal rearrangements are generally defined by the equation

$$ABC \rightarrow AC + B, \qquad AC \neq \text{hydrogen}$$

The elimination of CO from diphenyl ether is a rather complicated rearrangement process of this type:

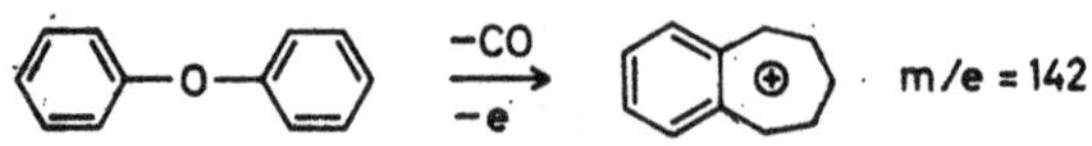

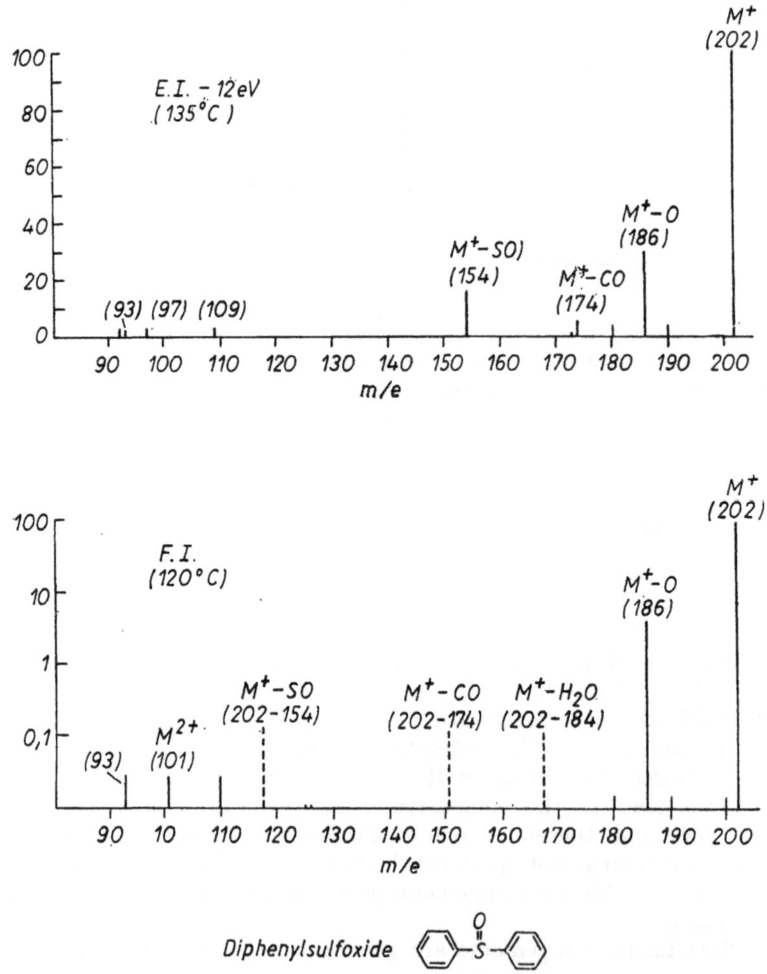

FIG. 4.22. EI (12 eV) and FI mass spectra of diphenyl sulfoxide (corrected for ^{13}C-contributions).

TABLE 4.9.

Skeletal Rearrangements

Compound (Molecular weight)	Process	I (%) F.I.	I (%) E.I. (12 eV)	$I/\Sigma I_{Fragm.}$ (%) F.I.	$I/\Sigma I_{Fragm.}$ (%) E.I. (12 eV)	I (%) F.I. Metast.	S	Q
Diphenylsulfoxide (202)	M^+-CO (174)	<0,01	4,8 (135°)	< 0,2	8	0,11	< 0,02	< 0,1
	M^+-SO (154)	<0,01	16 (135°)	< 0,2	27	0,14	< 0,007	< 0,07
p-Cresol (108)	M^+-CO (80)	<0,01 (150°)	1,6 (160°)	< 1,3	15	0,3	< 0,09	< 0,03
Diphenylether (170)	M^+-CO (142)	<0,001 (350°)	2,3 * (150°)	< 0,2	55 *	0,5	< 0,004	< 0,01
Dimethyldisulfide (94)	M^+-HS (61)	0,005 (350°)	8 (140°)	1	41	0,08	0,02	0,06
Dimethylacetylene-dicarboxylate (142)	M^+-CO_2 (98)	<0,005 (25°)	51 (150°)	< 0,5	11	1,8	<0,05	0,003
Anthraquinone (208)	M^+-CO (180)	<0,01 (290°)	8 (220°)	< 0,5	68	0,35	< 0,007	<0,03
N,N-Dimethyl-'N-phenylformamidine (148)	M^+-H_2CN (120)	<0,001 (25°)	8 * (150°)	< 0,1	8 *	0,04	< 0,01	0,03
	M^+-CH_3-HCN (106)	0,001 (25°)	9 * (150°)	0,1	10 *	—	0,01	—
p-Nitrobenzenesulfonamide (202)	M^+-SO_2 (138)	0,12 (170°)	22 (140°)	1,4	60	0,5	0,02	0,24
Benzylalcohol (108)	M^+-HCO (79)	<0,02 (250°)	14 (250°)	< 0,15	30	0,5	< 0,005	<0,04
Diphenyldisulfide (218)	M^+-HS (185)	<0,005 (190°)	4 (210°)	< 0,6	11	0,10	< 0,05	<0,05
	M^+-2S (154)	0,016 (190°)	6 (210°)	1,8	16	0,10	0,11	0,16
Dibenzylsulfoxide (230)	M^+-H_2SO (180)	<0,02 (160°)	11 (170°)	0,02	9	3,7	0,002	0,005

* 11 eV

The rearrangement peak $m/e = 142$ cannot be detected in the FI mass spectrum although its relative intensity is 2.3% in the 11 eV EI mass spectrum. The FI rearrangement peak $m/e = 142$ can be detected, however, as a metastable peak with a rel. intensity of 0.5%.

In the 12 eV EI mass spectra of diphenyl sulfoxide the rearrangement peaks $M^+ - SO$ and $M^+ - CO$ are found with a relative intensity of 27% and 8%, respectively. They are undetectable (< 0.01% of the base peak) in the FI mass spectrum. Metastable ions corresponding to these processes are found with rel. intensities of about 0.1%, indicating a very slow decomposition process. Fig. 4.22. shows the corresponding mass spectra.

These results are in accordance with observations of Damico, Barron and Sphon [227] who did not find elimination of CO in the FI mass spectra of some pesticides.

The *S*- and *Q*-factors of a large number of skeletal rearrangements are very small (with one exception), as can be seen from tab. 4.9 This is in accordance with the previous theoretical considerations. In the case of dimethylacetylenecarboxylate one finds a shifted rearrangement peak of relatively high intensity if one uses an ion source with a favourable potential distribution and geometry. The possibility cannot be excluded that some skeletal rearrangement processes with larger *S*- and *Q*-factors could be detected on the future because no serious sterical complications seem to be involved with some of these rearrangement reactions.

4.3.4. The FI emitter dipping technique

Recently we introduced a new technique for the study of thermally unstable organic compounds which may be termed "FI emitter dipping technique". This technique avoids thermal evaporation of the solid organic samples from a micro-oven to the FI emitter. For a number of substances thermal decomposition prior to ionization occurs to a large extent with the evaporation method.

The thermal decomposition can be avoided with the new method in the following way: The very large active surface area of the field anode which is produced by the above-mentioned benzonitrile activation methods establishes the possibility of direct adsorption of the sample from a liquid solution, and of subsequent field desorption of the substance from the field ion emitter.

The activated field anode is immersed in a concentrated solution to be analysed. Then the FI anode is mounted into the ion source of the FI mass spectrometer via a sliding rod in a vacuum lock. The adsorbed solvent molecules appear as a background peak in the FI mass spectrum. Solid samples of relatively high vapour pressure cause a "Field Desorption Mass Spectrum" already at room temperature. If the volatility is low, the FI wire is heated

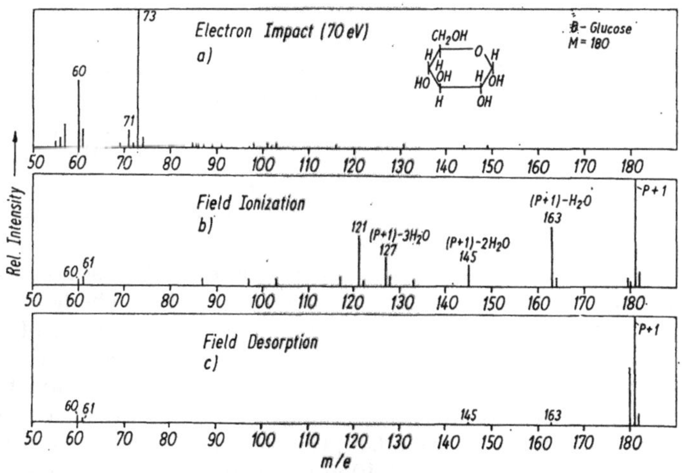

FIG. 4.23. a) EI mass spectrum (70 eV) of D-glucose. Evaporation temperature about 100 °C.

b) FI mass spectrum of D-glucose. Evaporation temperature about 115 °C. FI emitter: 2.5 μm activated platinum wire.

c) FD mass spectrum of D-glucose. Emitter wire (10 μm activated tungsten) heated by a current of 3.5 mA.

by a current of a few mA until the mass lines of the sample appear in the spectrum. The degree of thermal decomposition is much reduced by this technique as shown by fig. 4.23. The figure shows the electron impact (EI), the field ionization (FI) and the field desorption (FD) mass spectra of D-glucose. The electron impact mass spectrum shows no parent ion peak but very intense fragment peaks. The FI mass spectrum was taken in the usual manner by evaporation of the solid sample from a micro-oven to the FI emitter. The parent ion peak is now the most intense one, but strong $[(p + 1) - n\,H_2O]^+$ and $[(p + 1) - C_2H_4O_2]^+$ peaks also appear because of thermal decomposition of the sample. Finally, the field desorption mass spectrum consists virtually only of the parent ion peak, thus demonstrating the reduction of thermal degrading of the sample.

The new method avoids transfer of solid samples to the gaseous state prior to field ionization. The lattice energy which must be overcome by the evaporation process is larger than the energy necessary for elimination of smaller from larger molecules in many cases. On the other hand, if the solid sample is dissolved in a solvent, the molecular structure is normally not destroyed. The sample molecules in the adsorbed state are bound to the emitter surface only by weak van-der-Waals forces. Therefore, they can be field desorbed by little or no heat transfer. Field desorption of atoms or diatomic molecules, as described by Müller [34] and by Gomer [16], occurs at much higher field strengths. Large organic molecules can be field desorbed at lower field strengths because of their smaller ionization potentials and because of their larger size. Summarizing, one can say that a "soft" method of sample introduction has been added to the soft method of field ionization.

We have recently constructed a new field ionization source which is equipped with two sliding rod systems (fig. 2.40.). The horizontal sliding rod is used in the usual manner for introduction of solid samples into a micro-oven. The second vacuum lock serves for rapid exchange of the FI emitters. The emitters are 10 μm tungsten wires activated with benzonitrile. For removal of a sample from the wire prior to the adsorption of a new sample the wire is heated by a current of about 40 mA (about 1000 °C) in vacuo. The sliding rod can be also equipped with a new emitter within a minute. An activated blade may also be used as FI emitter.

The field desorption method can be applied as a micro-technique for analysis of organic samples. Microgram quantities of the solid sample are dissolved in a few microliters of a solvent. The solution is taken up by a 10 μl syringe, and then a small droplet of about 1 μl is pushed out of the syringe. The droplet hangs on the horizontally fixed needle of the syringe. A micro-manipulator can move the syringe in three directions. The droplet is brought in such a position that it touches the FI wire from the top. In this way the sample is adsorbed on the FI wire.

4.4. QUANTITATIVE ANALYSES OF MIXTURES

Introduction

Quantitative analyses of organic mixtures with FI mass spectrometers have been carried out so far mainly with hydrocarbon mixtures (gasolines, crude oil fractions, products of organic syntheses). One may take advantage of the FI method mainly in the case of very complicated mixtures which contain up to several hundred different components (gasolines, for example). In case of simple mixtures containing only a few components, measurement with the conventional electron impact method is accurate and simple enough.

The main advantage of the FI method as applied to analysis of very complicated mixtures consists in the fact that the mass spectra of many components consist virtually only of the parent peak. Those branched molecules tending to stronger fragmentation after FI form mainly fragments lying in the mass range below $m/e = 57$. These fragments do not overlap with the parent ion peaks of a gasoline sample, starting with butane ($m/e = 58$).

Besides these advantages of the FI method for analysis of complicated organic mixtures, there are also some disadvantages as compared with the EI method which will be discussed in detail later.

In principle, mass spectra of organic mixtures which are rather similar to FI mass spectra can be produced by low energy electrons and low ion source temperatures. As mentioned earlier, this technique has been applied successfully both to pure substances and to mixtures mainly by Spiteller [104, 113]. It is then more or less a question of whether an EI source would be available which could be operated at room temperature like the FI source. (Commercial EI sources which can be operated at room temperature are at present very rare).

Although the EI mass spectra taken at low electron energy and source temperature and the FI mass spectra of the higher boiling hydrocarbon fractions are rather similar, this is not necessarily the case for lower boiling gasoline samples which contain hydrocarbons with only a few carbon atoms. The ionization potentials of these lower hydrocarbons are different by tenths of an electron volt. One is then faced with the dilemma of whether to chose an electron energy above the highest ionization potential of a compound in the mixture or to adjust the electron energy to the lowest ionization potential of a component in the mixture. In the first case, fragmentation of substances of lower ionization potential would occur. In the second case, the substances of higher ionization potential would not be ionized at all. In the case of field ionization, fragmentation of all gasoline components is sufficiently low. The fragment masses do not interfere with parent ion masses, and the ionization efficiency is sufficiently large for all components at moderate field strengths.

4.4.1. Analyses of mixtures with tip emitters

Early measurements on hydrocarbon mixtures with FI mass spectrometers using tip emitters have been made by Wagner [219], and with wire emitters by the author [74].

Some general results of these investigations will now be discussed. If tips are used for analysis of organic mixtures, platinum tips of relatively large radius (about 5000 Å) should be used in order to obtain a sufficiently large emitter area. The relatively low field strength obtainable with these blunt tips (about 5×10^7 V/cm)

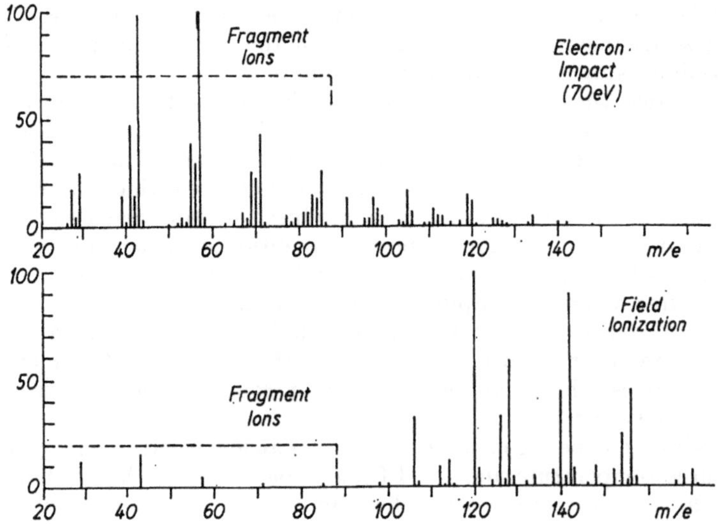

FIG. 4.24. EI and FI mass spectra of a gasoline sample. Field anode: platinum tip (Wagner [219]).

prevent strong fragmentation of the molecules which would otherwise occur at high fields. A large emitter area is desirable in order to obtain large ion intensities and low fluctuations of the ion intensities (see section 2.1.4.). Wagner has shown that the "standard deviation" of the more abundant components of a mixture is about 5% on the average. The standard deviation is defined in the following manner:

$$\delta\,[\%] = \frac{100}{\bar{x}} \sqrt{\frac{\Sigma(\bar{x} - x_i)^2}{N - 1}}$$

where $\bar{x}$ is the arithmetic mean value calculated from N measured values x_i.

The EI mass spectrum of a gasoline sample taken at 70 eV electron energy, and the corresponding FI mass spectrum taken with a platinum tip are shown in fig. 4.24. [219]. It may be seen from the figure that the EI parent ion intensities in the higher mass range are very low if the spectrum is taken under normal working conditions. (The difficulties encountered with measurement of low energy EI mass spectra of low boiling gasoline samples have been discussed before.) The parent ion intensities are large in the FI mass spectrum, and the low fragment ion peaks are concentrated entirely in the mass region below $m/e < 85$. Thus all peaks of $m/e \geqq 86$ are virtually free from overlap with fragment peaks, i.e. they represent hydrocarbons equal or higher in mass than the hexanes.

At present, thin wires or metal blades are usually taken as FI emitters instead of tips for quantitative analyses of hydrocarbon mixtures. This is because of the higher reproducibility of the FI mass spectra with wires or blades as compared with single tip emitters. The advantage of blades over thin wires is their large lifetime, during routine investigations, as mentioned in section 2.4. However, if treated with care, a 2.5 μ platinum Wollaston wire may be used for gasoline analysis over a period of more than an year, as demonstrated by Hippe of the author's laboratory. Moreover, carbon whiskers with about 20 times larger tensile strengths than platinum Wollaston wires may be used as FI emitters. The use of a thin wire has the advantage of greatest reproducibility as compared with all other FI emitters. Therefore, the following text will refer always to hydrocarbon mixtures using a thin wire.

4.4.2. Analyses of mixtures with wire emitters

A mixture of seven hydrocarbons has been analysed to get the first information about what range of accuracy is to be expected if a thin wire is used in the source of an FI mass spectrometer [74].

Fig. 4.25.a. shows the FI mass spectrum of the substances listed in table 4.10. The range of parent and fragment ions is clearly separated. Identification of the peaks at mass numbers 86, 100, 114 and 128 can be made unambiguously. They must correspond, respectively, to hexane, heptane, octane and nonane since the molecular formulae are C_nH_{2n+2}. In principle, the peak $m/e = 128$ could be attributed also to naphthalene. This possibility can be excluded by the fact that the peak $m/e = 64$ corresponding to doubly charged naphthalene is absent. (The relative intensity is about 1% for naphthalene, but zero for nonane). The mass lines 78 and 92 are to be attributed to benzene and toluene. The doubly charged ions of these molecules appear at mass numbers 39 and 46, respectively as FI emitter.

The peak at $m/e = 84$ is due to cyclohexane. Hexene would give the same peak. But the presence of hexene could be excluded by the fact that the observed relative intensity of the fragment at mass 41 and the metastable fragment at mass 37.3 is far below the theoretical limit of about 1%.

After this safe identification of the peaks, the accuracy of analysis was tested. A standard mixture containing the same seven compounds in equal mole fraction was made. The sensitivity coefficients for the different substances were obtained by measuring the FI mass spectrum of the standard mixture and by dividing the peak heights by the mole fractions of the components. Then the original mixture of the seven components with known mole fractions was taken as the unknown mixture. The measured peak heights of the components were divided by the sensitivity coefficients and the mole fractions obtained in this way were normalized to 100%.

The result is shown by table 4.10. The deviations of the results of analysis from the known mole fractions of the mixture were below ± 1 mole percent on the average.

A gas chromatographic analysis of the same mixture was taken for comparison. It turned out that the accuracy was of the same order as that obtained with the FI mass spectrometer.

TABLE 4.10. ANALYSIS OF A MIXTURE OF SEVEN HYDROCARBONS

Substance	Preset content Mole %	Analysis FI − MS Mole %	Analysis GLC Mole %	Deviation FI − MS Mole %	Deviation GLC Mole %
Hexane	26.8	27.1	27.0	+ 0.3	+ 0.2
Cyclohexane	4.3	4.1	4.3	− 0.2	0.0
Heptane	12.7	12.5	12.2	− 0.2	− 0.5
Benzene	31.6	32.8	33.3	+ 1.2	+ 1.7
Octane	7.1	7.1	7.4	0.0	+ 0.3
Toluene	11.0	11.5	11.2	+ 0.5	+ 0.2
Nonane	6.5	6.0	6.6	− 0.5	+ 0.1

(The actual content of benzene seems to be higher because the deviations of both the FI and GLC results are in the same direction and higher than the average.)

Fig. 4.25.b. shows the FI mass spectrum of a gasoline sample taken with a 2.5 μ platinum wire [6]. The olefins had been removed prior to analysis. Straightforward identification of each of the parent peaks with a group of isomeric compounds is possible. The fragment ion intensity is lower than that shown in the spectrum taken with a tip (fig. 4.24.).

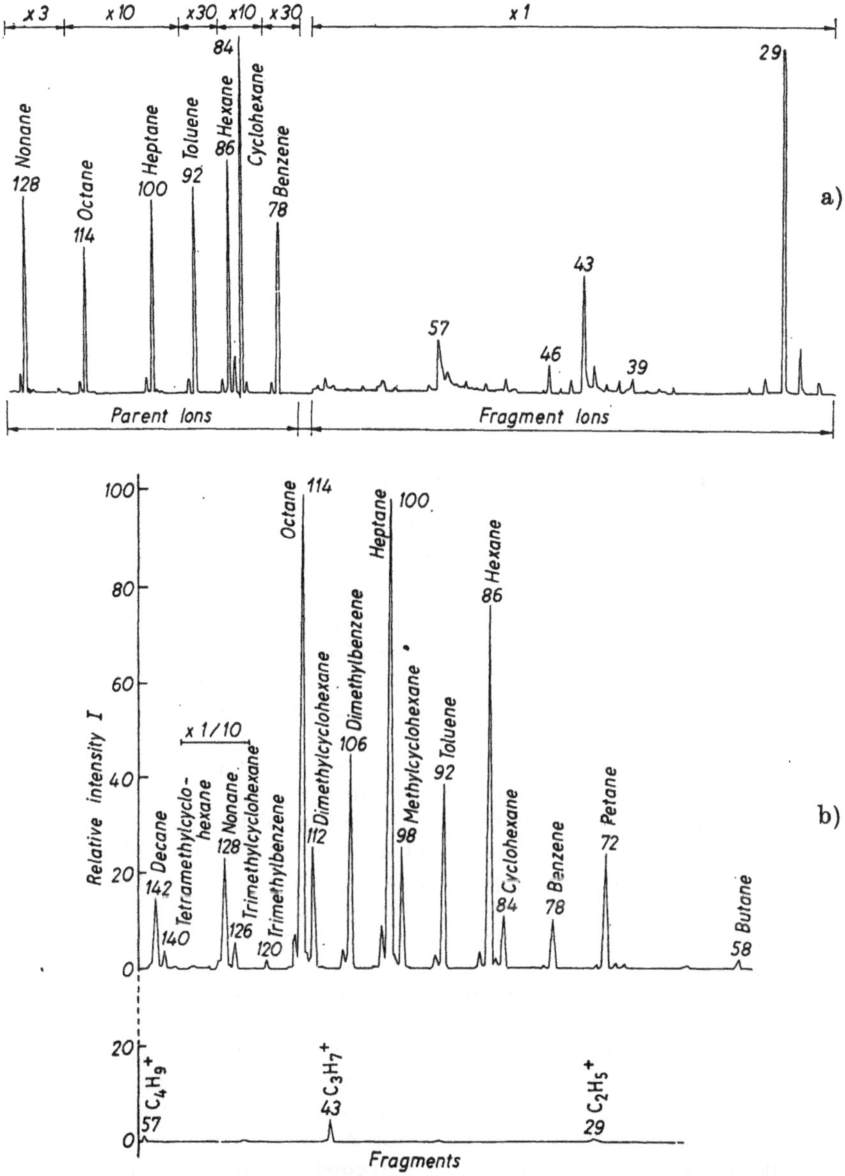

FIG. 4.25. *a*) FI mass spectrum of a mixture of seven hydrocarbons. *b*) FI mass spectrum of a gasoline sample taken with a 2.5 μm platinum wire. (One compound designation only is indicated for all isomeric compounds of a specific mass number.)

4.4.3. Foundations of gasoline sample analysis

Extended investigations on the parameters influencing gasoline sample analysis have been carried out by Hippe [221] and the author [222]. It turned out that many points have to be observed carefully in order to obtain accurate results of analyses.

Gasolines are mixtures of about 100 to 200 different hydrocarbons belonging to the classes of alkanes, alkenes, and aromatic compounds with molecular weights between about 54 and 156. Resolution and identification of nearly all components of a gasoline sample is possible by means of high resolution capillary gas chromatography. Group analysis of normal, iso, cyclo alkanes, alkenes and aromatics can be carried out by electron impact mass spectrometry. (For review see Franke [220], for example.) FI mass spectrometry evaluates the molecular weight distribution from about 54 to 156 for three groups (normal + iso alkanes, cyclo alkanes + alkenes, aromatics). This results in abundance data for about 20 different mass groups. However, FI mass spectrometry cannot differentiate between different isomers of the same mass number, insofar as only the parent peaks are used for analysis.

Thus it is seen that the information obtained by FI mass spectrometry (evaluation of about 20 mass groups) is intermediate between GLC — capillary chromatography (resolution of up to about 200 components), and electron impact group analysis (evaluation of about 5 classes of substances).

In case one would be interested in a quick and qualitative information of the composition of a gasoline sample, one would obtain this information immediately by inspection of the FI mass spectrum (see fig. 4.25.). If one would need quantitative results, calibration of the FI mass spectrum would be necessary. The basic advantage of the FI method as compared with the EI method under conventional conditions consists in the fact that only one standard mixture has to be measured in case of FI, whereas n mass spectra of the pure components of a n-component mixture have to be measured in case of EI. The difference arises from the fact that in case of EI each mass peak is composed of many contributions of fragment ions arising from different parent ions, whereas in case of FI each mass line represents only one group of parent ions.

The mathematical treatment for evaluation of the analysis with a FI mass spectrometer is extremely simple. Evaluation of a system of n equations with n variables (n being very large) necessary in case of EI mass spectra is not a principal problem if a computer is used. However, the accuracy of analysis is reduced not by mathematical but by physico-chemical reasons if a large number of substances contributes to each mass peak ("interference" effects, n is assumed to be between 100 and 200). Moreover, many of the n pure substances required for calibration in the case of a total analysis with an EI mass spectrometer are usually not available to a normal laboratory, and the time required to take the n standard spectra would be very large. In practice, such total analyses are not carried out. The amount of this work is drastically reduced if only a group analysis is carried out with an EI mass spectrometer. However, this occurs at the expense of the information obtained. (Reduction of data for 100–200 components to data for about 5 groups of substances.)

The drawback of the FI method for gasoline analyses lies in the fact that standard mixtures have to be used for determination of the sensitivity coefficients and these mixtures must be very similar to the gasoline samples to be analysed. This can be done in general only by an iteration method. At least two subsequent standard mixtures have to be made the composition of which should be increasingly matched to the sample.

This point will now be discussed in more detail. The sensitivity coefficient E_i for the i – th component of a mixture is defined as

$$E_i = H_i/g_i$$

where H_i is the height of the molecular ion peak (i) and g_i the known weight of the i – th component in the reference mixture. The weight of each component of an unknown sample is obtained by division of the peak heights of the sample by the sensitivity coefficients. The contributions of the components of the sample, g_i, in weight % are obtained by normalization. In order to test the optimum accuracy of gasoline analysis with an FI mass spectrometer, two synthetic gasoline samples, each containing 75 components, were prepared by Behling [222] and placed at disposal of the author's laboratory. The composition of the two samples was not greatly different. One of the samples was taken as a standard mixture for calibration, the other one as an unknown gasoline sample. The FI mass spectra of the samples were taken and evaluated by Hippe [221].

The result of analysis is shown by table 4.11. The analyses taken at two different days are compared, and the average deviations from the true composition of the synthetic gasoline sample are listed.

The mean deviation of the abundant components are $\pm$ 0.5 weight %, that of the components which are less abundant than 5 weight %, about $\pm$ 0.1 weight %.

The maximum deviations are observed at mass 128. The ratio

TABLE 4.11. ANALYSIS OF A SYNTHETIC GASOLINE SAMPLE

Mass	True sample composition weight %	Result of Dec. 16 weight %	Analysis Jan. 4 weight %	Mean deviation weight %
68	0.49	0.45	0.48	− 0.02
70	3.23	3.04	3.15	− 0.13
72	9.38	8.57	9.19	− 0.50
78	19.01	18.28	17.88	− 0.92
82	0.11	0.08	0.14	0
84	3.01	3.03	3.08	+ 0.05
86	6.46	6.39	6.70	+ 0.18
92	13.41	13.90	13.17	+ 0.12
96	0.11	0.11	0.09	− 0.01
98	1.63	1.70	1.65	+ 0.05
100	5.59	5.95	5.98	+ 0.38
106	18.24	18.95	18.02	+ 0.24
112	2.32	2.35	2.31	+ 0.01
114	5.91	5.93	59.7	+ 0.04
120	7.68	7.35	7.83	− 0.09
126	0.20	0.21	0.20	0
128	1.38	2.02	2.02	+ 0.64
134	1.39	1.29	1.56	+ 0.03
142	0.10	0.09	0.11	− 0.01
148	0.35	0.31	0.46	+ 0.03

Experimental data:
sample amount: 80 μl, 2.5 μ Pt-FI wire at +6 kV, cathode at −5 kV, temp. of sample reservoir: 80 °C, temp. of ion source: 20 °C.

of n-nonane to 2.2.5 — trimethyl hexane differed by a factor of three in the two mixtures. This is a good example of the fact that larger deviations of the results have to be expected if the composition of the standard samples and the unknown samples are largely different.

The variation of the sensitivity coefficients of isomeric compounds will now be discussed in greater detail. The measured peak height of a substance depends on the following parameters (all other conditions being the same):

a-priori field ionization probability (tunneling effect), supply velocity (diffusion over the emitter surface), stability of the molecule with respect to fragmentation.

The sensitivity coefficient is given by

$$E_i = \alpha \cdot \beta \cdot \gamma$$

where γ = ionization probability
β = supply velocity
α = ratio of the parent to the total ion intensity.

A relative sensitivity coefficient $E_{rel\ i}$ is introduced for a group of isomeric compounds which is defined by

$$E_{rel\,i} = E_i/E_n$$

where E_n is the sensitivity coefficient of the unbranched molecule of the isomeric group. Thus, $E_{rel\ n} = 1$.

The spectra of heptane-benzene mixtures of known composition will be discussed now in order to show the variation of the sensitivity coefficients of heptanes with the isomeric structure. The benzene serves as an internal standard because it causes a simple FI mass spectrum and it constitutes one of the main components of gasolines.

The results are indicated in table 4.12.

TABLE 4.12. IONIZATION DATA FOR HEPTANE ISOMERS IN 1 : 1 HEPTANE-BENZENE MIXTURES

Designation of compound	Boiling Point °C	$E = \frac{H_M}{g}$ $\times 10^6$	$\frac{E}{E_{n7}}$	$\frac{\delta}{\delta_{n7}}$	$\frac{\alpha}{\alpha_{n7}}$	E_{benz} $\times 10^6$	$\frac{E}{E_{benz}}$
n 7	98.43	5.48	1.00	1.00	1.00	7.15	0.77
2 *M* 6	90.05	2.88	0.53	0.58	0.90	5.04	0.57
3 *M* 6	91.85	3.79	0.69	0.73	0.95	6.98	0.54
3 *A* 5	93.47	3.33	0.61	0.64	0.95	6.26	0.53
2.3 *DM* 5	89.78	2.00	0.36	0.43	0.84	4.21	0.48
2.4 *DM* 5	80.5	1.58	0.29	0.42	0.69	3.99	0.40
2.2 *DM* 5	79.2	0.56	0.10			3.71	0.15
3.3 *DM* 5	86.06	0.53	0.10	0.51	0.19	4.45	0.12
2.2.3 *TM* 4	80.88	0.53	0.10	0.50	0.19	4.75	0.11

(High voltage: + 6 — 5 kV; emitter: acetone conditioned Pt-wire 2.5 μ; temperature of gas container: 100 °C, of ion source: 20 °C).

The sensitivity coefficient E_{benz} is not constant, but it depends on the second compound in the mixture. E_{benz} drops with increasing branching of the heptane isomers. This questions the suitability of benzene as an internal standard. However, by forming the quotient E/E_{benz}, the changing emission conditions of the FI wire are eliminated at least partially from the measurements. Measurements can be compared even if they have been obtained at very different dates.

The relative sensitivity coefficients $E_{rel\ i}$ vary from 1.00 for n-heptane to 0.1 for 2.2.3 — trimethyl-butane. This is due mainly to the strong decrease of the stability of the parent ions with respect to fragmentation with increasing branching of the isomers, as can be seen from the α/α_{n_7} — values.

Strongly branched molecules with quaternary carbon atoms are unstable not only with respect to FI and EI, but also with respect to the temperature. Therefore, their concentration in natural gasoline samples is normally small. Their contribution to the average sensitivity coefficient of the mixture is very small, and therefore, only the first six isomers of table 4.12. will be considered for the following discussions.

In principle, there exist two possibilities for setting up a reference mixture. Either one chooses one heptane isomer having an average sensitivity coefficient as a characteristic representative for the mass group 100, or one mixes all of the isomeric heptanes — excluding the species with quaternary carbon atoms — with the same concentrations. If one chooses the first possibility, one must have previously measured the sensitivity coefficients of the different pure isomers, and then one chooses the isomer having the average sensitivity coefficient.

From table 4.12. one can derive an average value for $E_{rel\ i}$ from the first six isomers which is equal to 0.58. The sensitivity coefficient of 3-ethyl pentane is closest to this value ($E_{rel\ i} = 0.61$). The maximum deviation of the average sensitivity coefficient of the mass group 100 of a gasoline sample could be $+0.39$ or -0.32 if this sample would contain only n-heptane or only 2.4 — dimethyl pentane, respectively, and no other isomers. In natural gasoline samples, the first four isomers of table 4.12. will usually dominate, and then an average sensitivity coefficient of 0.7 is adequate. The deviations of the actual average sensitivity coefficient of the sample from the value 0.7, derived under the assumption that all of the first four components are equally abundant, are expected to be of the order of $\pm$ 10%.

The ionization factor γ and the substance supply factor β may be combined to a factor $\delta = \gamma\beta$. Assuming that the ionization probabilities γ of different isomers do not differ too much, one can say that the factor δ is determined virtually only by the supply factor β.

The role which the supply factor β plays can be made clearer if the FI mass spectra of the pure isomers are compared with the spectra of the isomers with a large excess of a solvent consisting of other gasoline components. In the latter case it has to be expected that the differences of the supply factors of the individual isomers will be smeared out by the average flow of the liquid. Table 4.13. summarizes the results.

It can be seen from the table that the absolute sensitivity of the heptane isomers is greatly increased if an excess of solvent is added to the isomers (up to a factor of 14.2 in the case of 2.2 dimethyl pentane). The relative sensitivity of the isomers with respect to n-heptane is also increased considerably by the solvent. The relative fragmentation stability α is somewhat decreased by the solvent, but the transport factor δ is greatly enhanced, as was supposed before.

TABLE 4.13. COMPARISON OF THE IONIZATION DATA OF PURE HEPTANE ISOMERS AND THE ISOMERS WITH AN EXCESS OF SOLVENT CONSISTING OF FIVE COMPONENTS

Designation of isomer	Sensitivity coeff. $E_{hept} = H_M/\eta$		rel. sensitivity coeff. related to n-heptane $E/E_{n\text{-}hept.}$		rel. fragm. stability related to n-heptane α		δ-factor related to n-heptane	
	pure heptane	heptane + 5 comp.	pure heptane	heptane + 5 comp.	pure heptane	heptane + 5 comp.	pure heptane	heptane + 5 comp.
n 7	1.73	5.11	1.00	1.00	1.00	1.00	1.00	1.00
2 M 6	—	4.33	—	0.85	—	0.93	—	0.91
3 M 6	0.60	3.44	0.35	0.67	1.00	0.97	0.35	0.70
3 E 5	0.38	3.91	0.22	0.77	1.00	0.95	0.22	0.81
2.3 DM 5	0.40	3.36	0.23	0.66	0.99	9.91	0.24	0.72
2.4 DM 5	0.56	3 05	0.32	0.60	0.99	0.91	0.33	0.66
2.2 DM 5	0.08	1 10	0.05	0.22	0.64	0.35	·0.07	0.61

The composition of the heptane + 5 components was about:
10% heptane, 25% benzene, 25% xylene, 20% toluene, 10% n-hexane, 10% n-octane.

These properties are in favour of practical gasoline analyses because the differences of the sensitivity factors for different isomers are less pronounced in the presence of the very large number of other components in the gasoline samples.

It is a very interesting feature that the relative fragment ion intensities strongly increase if one proceeds from a measurement of the pure isomers to one with an excess of aromatic compounds. This can be seen from table 4.14. and 4.15.

The fragmentation probability can increase by up to a factor of 100, as in the case of 3-ethyl-pentane. A detailed theoretical interpretation of this phenomenon cannot be given at present. However, it is evident that this effect is due to adsorption properties of the organic substances on the FI emitter. Interactions between the adsorbed particles themselves and between the particles and the metal surface play an important role. This exemplifies the difficulties arising if one tries to set up standard tables for the FI mass spectra of organic substances similar to those set up for electron impact mass spectrometry [191].

Another effect showing the importance of adsorption effects is

TABLE 4.14. PERCENTAGE OF THE ADDED FRAGMENT ION CURRENTS OF HEPTANE ISOMERS IN THE PURE STATE AND WITH AN EXCESS OF FIVE ORGANIC COMPONENTS, AS LISTED IN TABLE 4.13.

	n 7	3 *M* 6	3 *E* 5	2.3 *DM* 5	2.4 *DM* 5	2.2 *DM* 5
Pure isomer	0.1	0.2	0.1	0.7	1.1	36.5
Isomer + 5 components	5.1	8.2	10.0	13.8	13.8	66.3

TABLE 4.15. FI MASS SPECTRA OF FOUR HEPTANE ISOMERS IN PURE STATE AND IN THE PRESENCE OF FIVE ORGANIC COMPOUNDS, AS LISTED IN TABLE 4.13.

Mass		*n*7		2 *M* 6		3 *M* 6		3 *E* 5	
		Pure	+5 comp.	Pure	+5 comp.	Pure	+5 comp.	Pure	+5 comp.
15	rel. peak heigth %	<0.1	<0.1	<0.1	<0.1	<0.1	<0.1	<0.1	<0.1
29		<0.1	4.3	<0.1	3.8	<0.1	3.4	<0.1	4.6
43		<0.1	0.4	0.5	6.1	<0.1	<0.1	<0.1	<0.1
57		<0.1	0.2	0.1	0.7	0.1	2.6	<0.1	<0.1
71		<0.1	0.2	0.1	0.9	0.1	2.2	0.1	5.4
85		<0.1	<0.1	<0.1	<0.1	<0.1	<0.1	<0.1	<0.1
100		100.0	94.9	99.3	88.5	99.8	91.8	99.9	90.0

the following one. If one introduces a mixture of heptane and benzene in the ratio 1 : 1 into the FI mass spetrometer, one observes sensitivity coefficients which are different from the coefficients obtained if first the heptane sample and then the benzene sample are introduced in the ratio 1 : 1. In other words, the surface state of the FI emitter greatly influences the sensitivity coefficients.

Fig. 4.26. shows that the sensitivity coefficients of n-heptane —

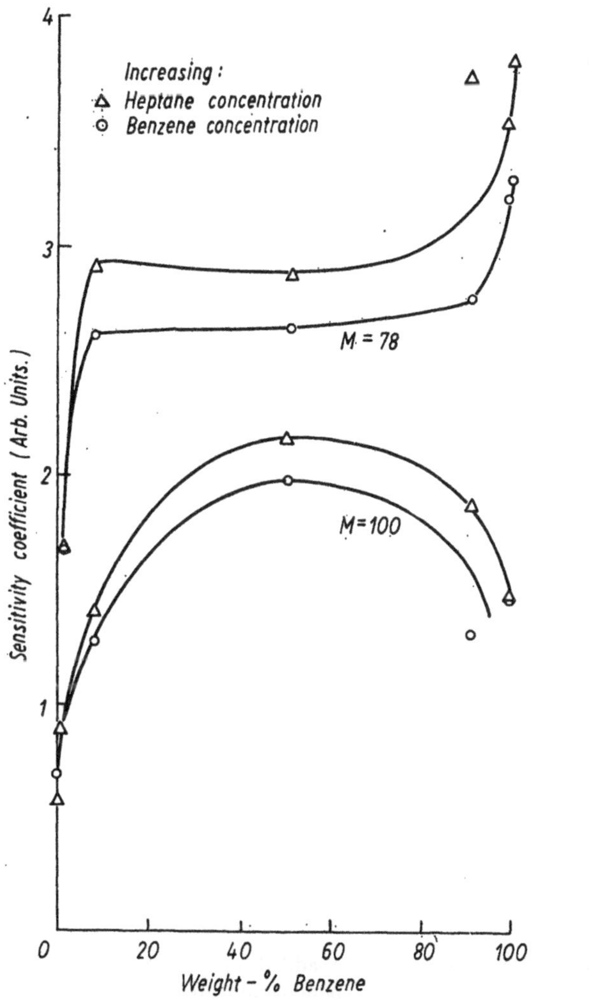

FIG. 4.26. Sensitivity coefficients of n-heptane-benzene mixtures on the mixing ratio.

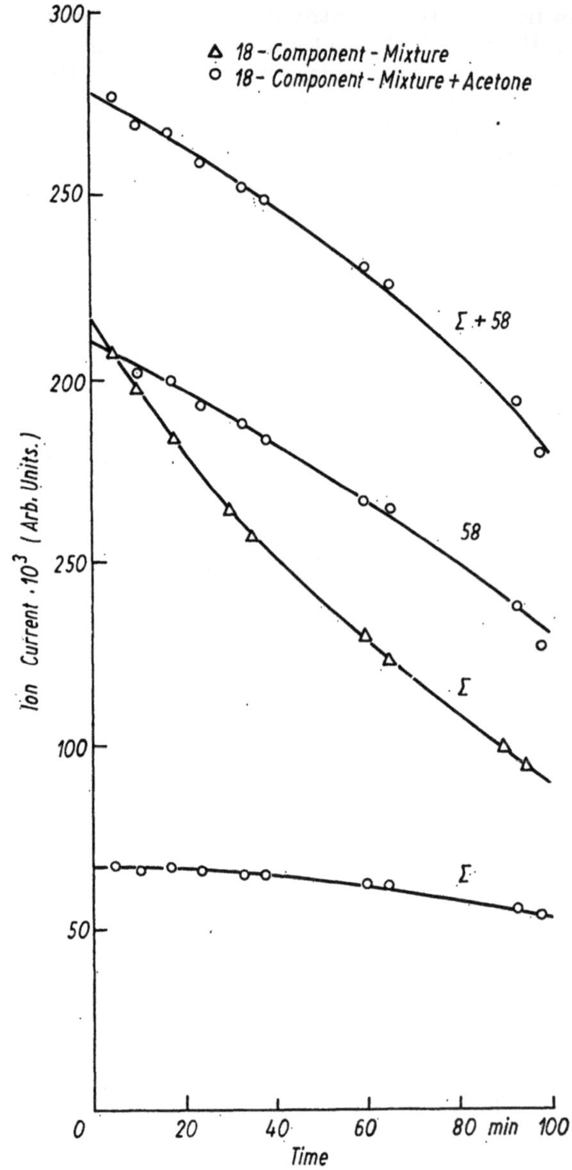

FIG. 4.27. Dependence of the added up parent ion currents of a 18-component mixture.

benzene mixtures depend both on the mixing ratio and on the order of increasing concentrations of the components. The upper curves correspond to increasing heptane concentrations, the lower curves to increasing benzene concentrations.

In addition to these effects, the time dependence of the sensitivity coefficients has to be observed. It takes some time to establish an equilibrium state of adsorption after sample introduction into the mass spectrometer.

Fig. 4.27. shows the time dependence of a mixture containing18 components (alkanes, alkenes, cycloalkanes, aromatics from C 5 to C 12). It can be seen from the figure that the intensity of the mixture without an acetone additive drops by about a factor of two within 90 minutes.

If about 15% acetone is added to the mixture, the intensity of the ion sum drops considerably however, the change of the intensity of the ion sum for the 18 component mixture with time is greatly reduced.

Fig. 4.28.a shows that the time dependence of the ion intensities is different for the different aromatics in the mixture. The heavier increase with time, the lighter decrease. After addition of acetone (fig. 4.28.b), the conditions are reversed for the C 6, C 8 and C 9 components. Summarizing all of these results obtained for gasoline analyses one can make the following statement.

The sensitivity coefficients for any component of a mixture depend on a variety of parameters in a complicated manner, which is governed mainly by adsorption phenomena. The sensitivity coefficient of one isomeric mass group depends on the following parameters:

a) on the composition of the isomeric mass group,
b) on the chemical nature of the other components,
c) on the relative concentrations of the other components,
d) on the time interval between sample introduction and measurement of the spectrum,
e) on the surface condition of the FI emitter prior to the analysis.

In spite of these difficulties and drawbacks of the FI method for quantitative analyses of mixtures, rather accurate analyses of gasoline samples can be made (as can be seen from table 4.11.) if the following precautions are taken.

Point *e*) can be eliminated if the FI emitter is conditioned by a fixed program using acetone prior to the measurement, and if the sensitivity is checked with a test substance.

Point *d*) can be eliminated if both the reference mixture and the unknown sample are measured according to the same preset time program. Points a—c) will be reduced more and more the better the reference mixture is adapted in its composition to the unknown mixture.

It is advisable to use an iteration method consisting of two steps. First, the FI mass spectrum of the unknown sample is measured. A reference mixture is then made, containing one representative component of each mass group found in the sample. The

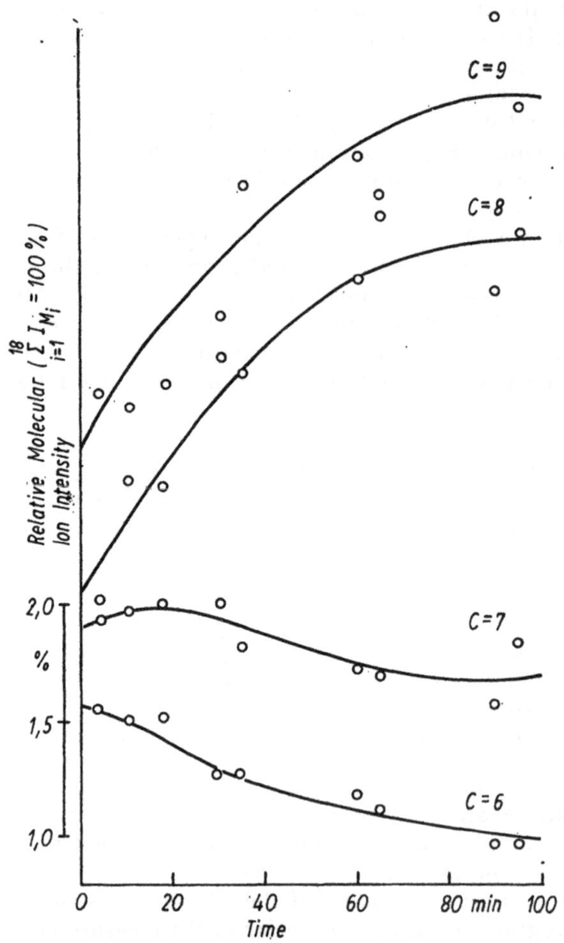

FIG. 4.28 a Time dependence of the relative parent ion currents of the aromatic components in a 18-component mixture.
a) without acetone additive.

weight percent g_i of each component of the reference mixture is adjusted closely to the unknown mixture by making $g_{i1} = H_{is}/E_{i1}$, where g_{i1} is the weight percent for the first approximation of the iteration method, H_{is} is the measured height of the ith mass line in the unknown spectrum, and E_{i1} is the sensitivity coefficient according to the first approximation. E_{i1} is obtained from previous measurements of the sensitivity coefficients of different compo-

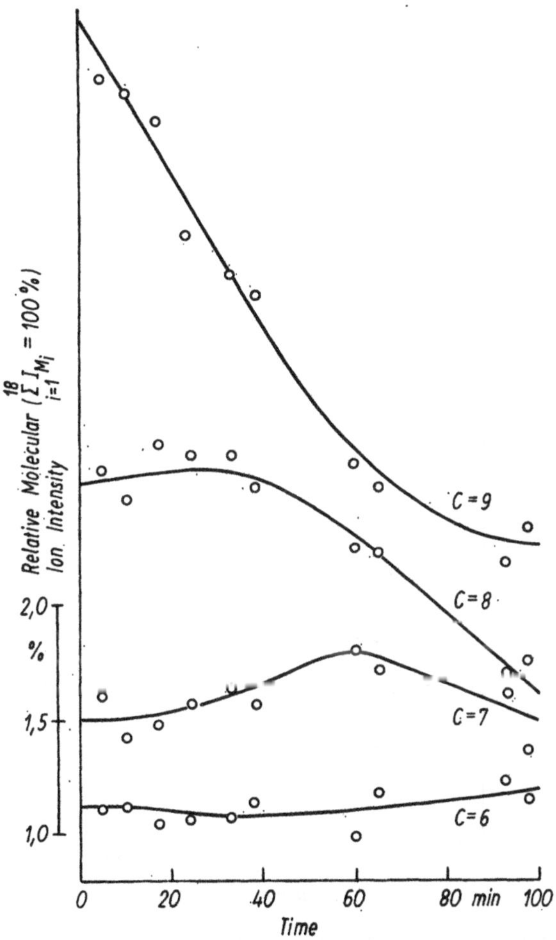

FIG. 4.28b
FI currents with acetone additive.

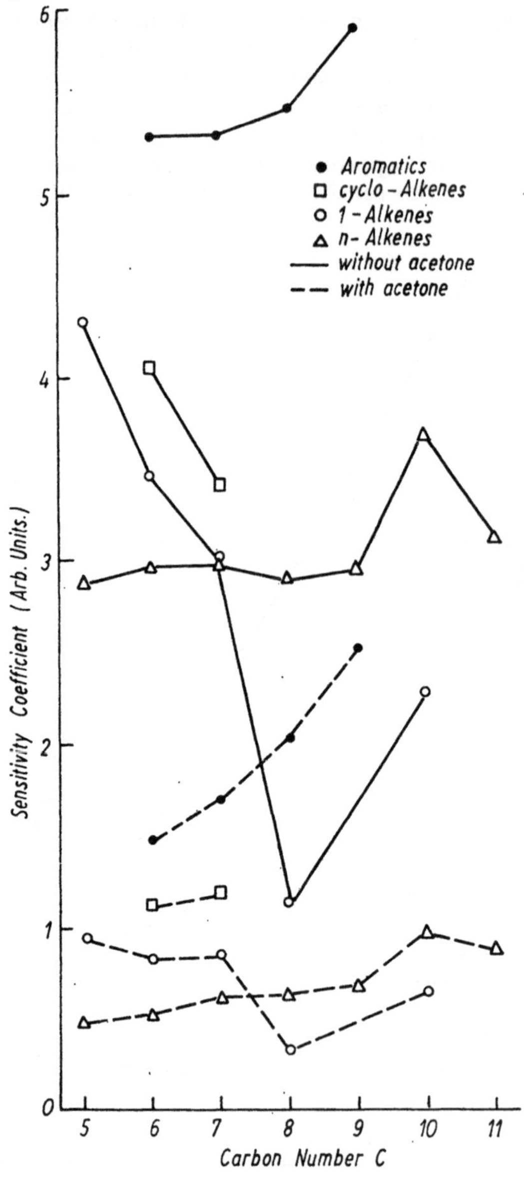

FIG. 4.29. Sensitivity coefficients of aromatics, alkenes and alkanes in a 18-component mixture.

nents in mixtures. Fig. 4.29. shows the sensitivity coefficients obtained for the components of an 18 component mixture.

For the second approximation, the sensitivity coefficients E_{i2} are derived from the measured peak height of the reference sample, H_{ir}:

$$E_{i2} = \frac{H_{ir}}{g_{i1}}$$

Then the composition of the unknown sample in weight percent g_{i2}, according to the second approximation is obtained by the relation

$$g_{i2} = \frac{H_{is}}{E_{i2}} = \frac{H_{is}}{H_{ir}} \cdot g_{i1}\,.$$

Fairly accurate results can be obtained by this method if the reference sample is adapted to the unknown sample to the same extent as in the case of the mixtures underlying table 4.11. The average deviations of the analysis from the true sample composition indicated in table 4.11. are below 1 weight %.

4.4.4. Analyses of heavy petroleum fractions (paraffin waxes)

The conditions for FI mass spectrometry of heavy paraffin waxes are much simpler than those for gasoline analyses. This is due to the fact that the mixture is composed only of one chemical group of compounds (paraffins) and that in the range of large carbon numbers of the paraffins (C_{20}—C_{40}) the differences of the sensitivity coefficients are much less pronounced than those for the molecules with small carbon numbers (C_5—C_{12}). This result has been obtained by Mead [241]. Using a razor blade as emitter in conjunction with a double focusing mass spectrometer, Mead obtained adequate sensitivities with a spectral reproducibility of the order of $\pm$ 5%.

Paraffin waxes in the 300° to 550 °C boiling range have been studied. Fig. 4.30. shows two portions of the FI mass spectrum of paraffin waxes from 350 °C to 400 °C fractions. Fig. 4.30. a is a portion of the spectrum of the sample after molecular sieve treatment. This means that the sample consists virtually of 100% normal paraffins. Fig. 4.30. b shows a portion of the spectrum obtained by applying the urea adduction technique to the residue from the sample treated with the molecular sieve (this fraction contains slightly branched paraffins and cycloparaffins with unbranched chains.) C_nH_{2n+1} and C_nH_{2n-1} fragment peaks, which are very intense in the EI mass spectra, are absent in the FI mass spectra. Thus, the $2n + 1$ peaks in fig. 4.30. b consist virtually only of isotopic contributions from the C_nH_{2n} peaks (cyclo-paraffins). The evaluation of the spectrum is then extremely simple as compared to EI mass spectra. The sensitivity coefficients were obtained in the following manner. Mead was able to show by additional experiments that the relative sensitivities of the parent peaks of individual normal paraffins are equal in the carbon number range 20 to 40, within the limit of experimental error which is usually of the order of $\pm$ 5%. Similarily, the sensitivity

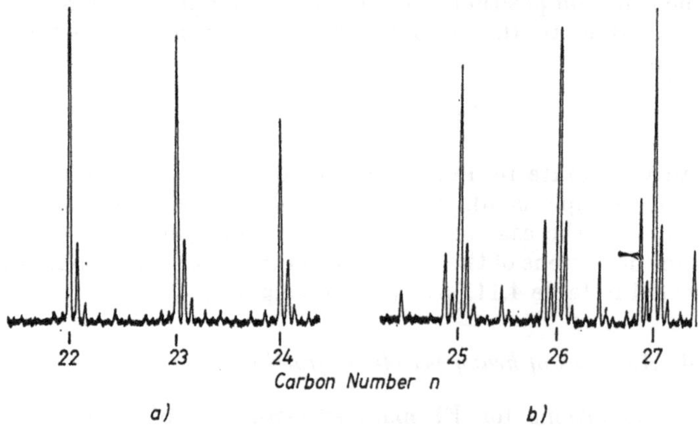

FIG. 4.30. Part of the FI mass spectrum of paraffin waxes.
a) after molecular sieve treatment.
b) after molecular sieve and urea adduction technique. (Mead [241]). Peaks designated by carbon number n originate from C_nH_{2n+2} ions.

coefficients of the cyclo-paraffins and alkylbenzenes were assumed to be equal within the limits of measuring accuracy because these coefficients varied only very slightly with several fractions covering the boiling range 300° to 550 °C. The average relative sensitivity coefficients for the different hydrocarbon types were obtained by taking the summation of the monoisotopic parent peaks for each type of hydrocarbon and by dividing them by the known concentrations which had been derived for these summations by EI mass spectrometry. Table 4.16. shows the results.

TABLE 4.16. CALIBRATION DATA ON RELATIVE SENSITIVITY OF PARENT IONS IN CARBON NUMBER RANGE C_{20} TO C_{40}

Normal paraffins	2.3
Isoparaffins	1.0
Cyclo-paraffins	2.3
Alkylbenzenes	5.6

A typical result of analysis obtained by Mead is represented by table 4.17. A wax designated ASTM wax I, the composition of which had been measured previously by eight mass spectrometry laboratories by EI techniques, was analysed by the FI technique.

TABLE 4.17. ANALYSIS OF ASTM WAX I (PERCENT)

	Cooperative analysis av. (EI)	Field ionization
Paraffins C_nH_{2n+2}	82.8	85.0
Monocyclo-paraffins C_nH_{2n}	14.2	12.1
Dicyclo-paraffins C_nH_{2n-2}	2.1	1.7
Tricyclo-paraffins C_nH_{2n-4}	...	0.3
Alkylbenzenes C_nH_{2n-6}	0.8	0.7
Indans-Tetralins C_nH_{2n-8}	0.09	0.2

There is good agreement between the cooperative and the FI analysis. Moreover, the FI analysis made clear a feature which was not evident after EI analysis. Some laboratories reported the presence of dicyclo-paraffins and indans-tetralins while other laboratories did not. The FI method clearly showed that these compounds were present, and a low concentration of tricyclo-paraffins could also be detected.

The ratio of the normal to iso-paraffins can be evaluated by quantitative removal of the normal paraffins by the molecular sieve technique.

The advantage of the FI technique for this type of analysis is that no elaborate calculations have to be used. The solution of matrices, and the making of allowances for the fragmentation of high molecular weight compounds into lower molecular weight materials — as made in EI mass spectrometry — are unnecessary. FI mass spectrometry may in the future play a role in the analysis of saturated hydrocarbon heavy fractions similar to that played by low voltage EI mass spectrometry in the analysis of aromatic fractions.

4.4.5. Analyses of the products of pyrolysis of polymers

Schüddemage and Hummel [242–245] have been the first to show that FI mass spectrometry can be applied very usefully to the field of polymer research. A special problem — among many others — is the structure of the sequence of co-polymers. Schüddemage and Hummel studied for examples co-polymers of styrene (*A*) and sulfur dioxide (*B*). They used samples with the mole fractions 2:1 and 1:1 for *A*:*B*. SO_2 cannot form co-polymers. Therefore, the 1:1 samples must consist of alternating styrene — SO_2 sequences, *ABABAB* .... The 2:1 samples could consist theoretically either only of 2:1 sequences, *AABAAB* ..., or alternatively of additional 3:1 and 1:1 sequences in the same ratio, *AAAB*, *AB*.

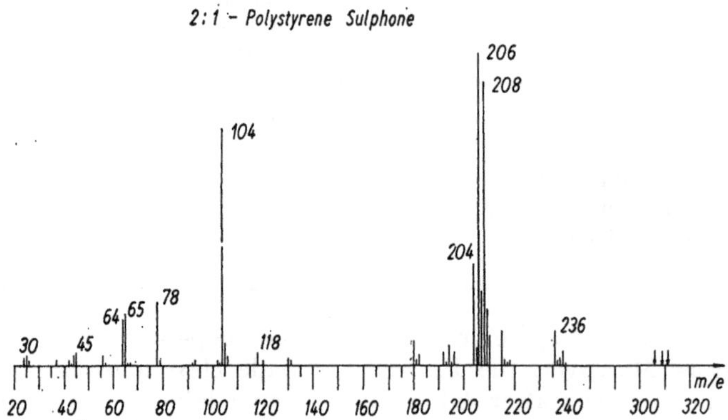

FIG. 4.31. Mass pyrogram of a 2:1 polystyrene-sulfone.

The problem can also be solved, in principle, with an electron impact mass spectrum. However, this shows a very great variety of fragmentation processes, and the evaluation is complicated. The FI method introduces very little fragmentation, and the evaluation of the spectra is straightforward.

The pyrolysis is carried out directly in the container of the heatable gas introduction system of the mass spectrometer at a temperature between 300° and 350 °C. The fragments of the pyrolysed polymers enter then directly into the FI source through a gold leak. The spectrum obtained in this manner is called a "mass pyrogram".

The spectrum is recorded quickly in order to avoid secondary reactions of the primary fragments in the gas container. Fig. 4.31. shows the mass pyrogram of a 2:1 polystyrene-sulfone. The FI mass spectrum shows preferentially dimeric species at $m/e = 204$, 206, 208 (the intensities have to be added), and monomeric species of styrene at $m/e = 104$. Trimeric species of styrene at $m/e = 312$ cannot be detected. (The peak at $m/e = 236$ is due to a secondary reaction, as proved by the time dependence of its intensity).

This experiment shows clearly that the sequence of 2:1 polystyrene-sulfones is *AAB* and that the alternative possibility

TABLE 4.18. FIELD IONIZATION METASTABL

PROBABLE PROCESS	C_3	C_4	C_5	C_6	C_7	C_8	C_9
P-15		31.88—4.2	45.13—0.51				
-16		30.41—2.3	43.56—0.22				
P-29			25.68—0.29	37.78—1.7	50.41—2.9	63.38—11.2	76.57—
-30			24.50—1.5	36.47—2.8	49.00—2.4	61.89— 3.4	75.03—
P-43							56.45—
-44							55.13—
P-57							
-58							
P-71							
-72							
P-85							
-86							
P-99							
-100							
P-113							
-114							

*) Location of metastables apply to all isomers of a given molecular weight. The abundance are for normal p chart divisions of parent ion. The field ionization conditions are defined in the experimental section.

ARAFFINS*) PRIMARY METASTABLE IONS ONLY

	C_{11}	C_{12}	C_{13}	C_{14}	C_{15}	C_{16}	C_{18}
—5.8	103.39—3.8	116.95—3.0	130.57—2.8	144.25—2.0		171.72—2.5	199.31—2.9
—1.3	101.77—0.83	115.29—0.85	128.89—0.6	142.55—0.51		169.98—0.6	197.54—
—4.5	81.85—3.8	94.88—3.4	108.05—3.0	121.34—2.7		148.18—3.5	175.28—4.2
—1.7	80.41—1.3	93.39—1.1	106.52—1.0	119.77—0.72		146.57—0.97	173.62—
—1.1	62.83—1.9	75.11—2.3	87.66—2.8	100.41—2.7		126.38—4.1	152.79—4.6
—0.73	61.56—0.94	73.79—1.1	86.28—1.0	98.99—0.90		124.88—1.4	151.24—
		57.66—0.85	69.39—1.3	81.46—1.6		106.31—3.1	131.85—3.3
		56.49—1.1	68.17—1.1	80.16—0.72		104.94—0.97	130.41—
				64.49—0.44		87.97—1.8	122.44—3.3
				35.35—0.18		86.73—0.87	111.12—
						71.37—0.63	94.59—2.9
						70.25—0.29	93.37—
							78.27—1.3
							77.17—

tive to 100

indicated above is not realized. A mass pyrogram of 1:1 polystyrene-sulfones shows that the products of pyrolysis are virtually quantitatively the monomeric species. This shows that the primary monomeric fragments do not recombine to dimeric products in the gas containers during the relatively short time of analysis. This statement is important because it proves that the dimeric species observed in the 2:1 polystyrene-sulfone mass pyrograms (fig. 4.31.) are actually formed as primary products of the pyrolysis.

This brief example has outlined the usefulness of FI mass spectrometry in polymer research. Much more detailed results have been obtained already by Schüddemage and Hummel.

4.4.6. FI metastable fragment maps

Wanless [246] has carried out a careful research on the use of metastable peaks of hydrocarbons for quantitative analyses and for correlation between the isomeric structures of the molecules and the FI mass spectra.

The fragmentation peaks of EI mass spectra possess, in general, unlimited redundancy. For example, the peaks in the C_nH_{2n+1} series can carry contributions from all isomers in a homologous series of compounds. On the other hand, FI mass spectra yield in general only primary metastable fragment ions. These possess limited redundancy. The redundancy is limited to the maximum number of isomers of a given carbon number. The EI mass spectra are overburdened with additional fragmentation which is not significant. The weak metastable EI peaks often interfere with the very strong normal fragment peaks.

In FI mass spectra, the primary metastable ions are essentially not interfered with by the weak normal fragment ions. The patterns of metastable field ions reflect the molecular structures, as can be seen from the subsequent tables. This method enables one to distinguish among isomeric structures.

Table 4.18. shows the FI metastable map for normal paraffins, as measured by Wanless. The ion source and other experimental details for these measurements have been described by Wanless and Glock [183].

The metastable ions of different isomers appear at the same mass numbers, but the relative abundances vary greatly with the isomeric structure. This had been pointed out already in the earlier papers of the author and his associates for the case of amines [189] and olefins [96]. The variation of the metastable fragment patterns with the isomeric structure is given by Wanless [246]. Finally, table 4.19. shows the metastable FI fragment map for nine iso-

					"C1" Region													"C2" Region						"C3" Region				
						(P+1 −15)	P-15			P-16								P-27		P-28	P-29	P-30			P-43	P-44		
		100	98	85	74.87	73.68	72.25	72	71	70.56	70	69.5	69	59	58	57	56	54.16	52.75	51.84	50.41	49.00	43	34.25	32.49	31.36	29.5	29.
n-Heptane (109-1)	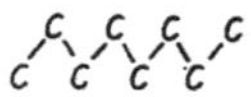	100					(0.14)			(0.30)											(1.2)	(1.2)	(0.39)					4.0
2 me Hexane (110-1)	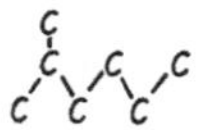	100	3.2	0.95	(1.3)	(1.8)	(7.2)			(2.9)						20					0.42	0.8	50	(0.021)	(0.029)	(0.025)		2.3
3 me Hexane (111-1)	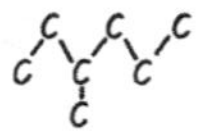	100							0.29							1.6		(0.42)		(0.81)		(1.7)	0.93					0.36
3 Ethyl Pentane (69-2)	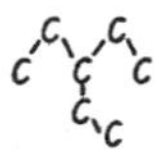	100							3.1									(0.70)	(0.40)	(1.0)		(4.5)	○					2.7
2,3 di me Pentane (70-2)	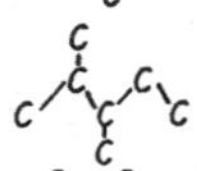	100		0.50		(2.8)			1.3		0.4		2.4	1.1	3.3	35				(0.94)	(2.3)	(4.4)	60			(0.5)		2.9
2,4 di me Pentane (62-1)	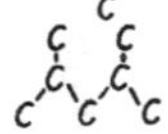	28		1.2				0.2						0.7	1.6	28							100		(0.13)	(0.50)	(0.85)	○
2,2 di me Pentane (68-1)	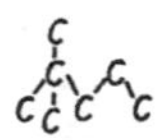	--		27					60							100							0.5					(80)
3,3 di me Pentane (73-1)	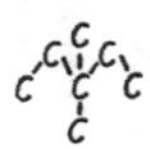	24		7.3					100			11		(0.39)	(0.44)	(0.22)				(0.44)	(0.37)	(1.0)	○					(8.0)
2,2,3 tri me Butane (72-1)	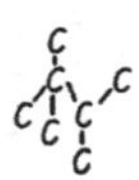	6,7		16.3										0.7	8.4	100							90					○

meric heptanes which clearly shows the structural information which one can obtain from such a map.

The unique information is encircled in the table. In addition to these metastable peaks, one can also obtain characteristic structural information from some normal fragment peaks which are also encircled in the table.

Summarizing, one can say that in the future it will probably be possible to determine the structure of isomeric compounds contained in an unknown mixture by means of the metastable FI mass spectrum.

REFERENCES

1. M. G. INGHRAM and R. GOMER, *J. Chem. Phys.* **22**, 1279 (1954).
2. M. G. INGHRAM and R. GOMER, *Z. Naturforschg.* **10a**, 863 (1955).
3. E. W. MÜLLER and K. BAHADUR, *Phys. Rev.* **102**, 624 (1956).
4. H. D. BECKEY, *Z. Naturforschg.* **14a**, 712 (1959).
5. H. D. BECKEY in "*Advances in Mass Spectrometry*" Vol. II, p. 1 Pergamon Press, Oxford 1963.
6. H. D. BECKEY, H. KNÖPPEL, G. METZINGER and P. SCHULZE in "*Advances in Mass Spectrometry*" Vol. III, p. 35 Inst. of Petroleum, London 1966.
7. M. DRECHSLER, German Patent No. 954 105 (1954).
8. W. A. SCHMIDT, *Phys. Verh.* **12**, 71 (1961), Dissertation, Freie Universität Berlin, 1962.
9. J. BLOCK, *Z. Phys. Chem.* NF, **39**, 169 (1963).
10. A. J. B. ROBERTSON and B. W. VINEY, in "*Advances in Mass Spectrometry*" Vol. III, p. 23, Inst. of Petroleum, London 1966.
11. H. RAUSCH VON TRAUBENBERG and R. GEBAUER, *Z. Phys.* **56**, 254 (1929).
12. I. R. OPPENHEIMER, *Phys. Rev.* **31**, 66 (1928).
13. F. KIRCHNER, *Naturwiss.* **41**, 136 (1954).
14. G. WENTZEL, *Z. Phys.* **38**, 518 (1926).
    H. A. KRAMERS, *Z. Phys.* **39**, 828 (1926).
    L. BRILLOUIN, *Compt. Rend.* **183**, 24 (1926).
15. R. GOMER, *J. Chem. Phys.* **21**, 1869 (1953).
16. R. GOMER, "*Field Emission and Field Ionization*" Harvard University Press, Cambridge 1961.
17. T. T. TSONG and E. W. MÜLLER, *J. Chem. Phys.* **41**, 3279 (1964).
18. E. W. MÜLLER, *Erg. exakt. Naturwiss.* **27**, 290 (1953).
19. E. W. MÜLLER, *J. Appl. Phys.* **28**, 1 (1957).
20. R. GOMER, *Rev. Sci. Instr.* **24**, 993 (1953).
21. R. A. SCHRACK, *Rev. Sci. Instr.* **27**, 172 (1956).
22. E. W. MÜLLER, J. A. PANITZ and S. B. MCLANE, *Rev. Sci. Instr.* **39**, 83 (1968).
23. T. C. CLEMENTS and E. W. MÜLLER, *J. Chem. Phys.* **37**, 2684 (1962).
24. H. D. BECKEY, *Naturwiss.* **45**, 259 (1958).
25. H. D. BECKEY, J. DAHMEN and H. KNÖPPEL, *Z. Naturforschg.* **21a**, 141 (1966).
26. M. G. INGHRAM and R. J. HAYDEN, "*A Handbook of Mass Spectroscopy*" *Nuclear Sci. Ser. Rep.* No. 14, Nat. Acad. Sci. (1954).
27. A. J. B. ROBERTSON and B. W. VINEY, *J. Chem. Soc.* A, 1843 (1966).

28. W. A. SCHMIDT, *Z. Naturforschg.* **19a**, 318 (1964).
29. H. D. BECKEY and D. SCHÜTTE, *Z. Instrkde.* **68**, 302 (1960).
30. R. SEELIGER, *Optik* **4**, 261 (1948).
31. R. SEELIGER, *Optik* **5**, 490 (1949).
32. C. F. GIESE, *Rev. Sci. Instr.* **30**, 260 (1959).
33. A. J. JASON, R. P. BURNS and M. G. INGHRAM, *J. Chem. Phys.* **43**, 3762 (1965).
34. E. W. MÜLLER, *Phys. Rev.* **102**, 618 (1956).
35. J. E. MULSON and E. W. MÜLLER, *J. Chem. Phys.* **38**, 2615 (1963).
36. D. S. BOUDREAUX and P. H. CUTLER, *Phys. Rev.* **149**, 170 (1966).
37. E. W. MÜLLER, in "*Advances in Electronics and Electron Physics*" Vol. XIII, p. 83 Academic Press, New York 1960.
38. R. H. GOOD and E. W. MÜLLER in "*Handbuch der Physik*" ed. by S. Flügge Vol. XXI, p. 176 Springer, Berlin 1956.
39. A. G. J. VAN OOSTROM, *Philips Res. Rep. Suppl. No.* 1 (1966).
40. R. HAEFER, *Z. Phys.* **116**, 604 (1940).
41. J. A. BECKER, *Bell Syst. Techn. J.* **30**, 907 (1951).
42. M. DRECHSLER and E. HENKEL, *Z. angew. Phys.* **6**, 341 (1953).
43. W. P. DYKE, J. K. TROLAN, W. W. DOLAN and G. BARNES, *J. Appl. Phys.* **24**, 570 (1953).
44. E. W. MÜLLER, *Z. Phys.* **120**, 270 (1943).
45. P. O. SCHISSEL, *J. Appl. Phys.* **33**, 2659 (1962).
46. P. SCHULZE, M. Sc. Thesis, University of Bonn 1961.
47. M. DRECHSLER and G. PANKOW in "*Proc. Internat. Conf. on Electron Microscop.*" p. 405, London 1954.
48. M. J. SOUTHON and D. G. BRANDON, *Phil. Mag.* **8**, 579 (1963).
49. M. J. SOUTHON, Ph. D. Thesis, University of Cambridge 1963.
50. W. P. DYKE and J. K. TROLAN, *Phys. Rev.* **89**, 799 (1953).
51. H. G. METZINGER and H. D. BECKEY, *Z. Phys. Chem. NF* **52**, 27 (1967).
52. R. GOMER, R. WORTMAN and R. LUNDY, *J. Chem. Phys.* **26**, 1147 (1957).
53. E. W. MÜLLER, *Z. Phys.* **108**, 668 (1938).
54. E. W. MÜLLER, ibid. p. 642.
55. G. F. TAYLOR, *Phys. Rev.* **23**, 655 (1924).
56. S. BRENNER, *J. Appl. Phys.* **27**, 1484 (1956).
57. K. LEVSEN, A. HEINDRICHS and H. D. BECKEY, *Meßtechn.* **78**, 9 (1968).
58. E. SCHMID and W. BOAS in "*Plasticity of Crystals*" F. Hughes and Comp., London.
59. LANDOLT-BÖRNSTEIN, „*Zahlenwerte und Funktionen*" 6. Ed. Vol. IV, 2b, p. 585 Springer, Berlin 1964.
60. J. NIXDORF, *Draht-Welt* **53**, 696 (1967).
61. "*Handbook of Chemistry and Physics*" 41st. Ed. 1959/60 and 45th. Ed. 1964/65 Chemical Rubber Publ., Cleveland.
62. H. BITTEL, *VDI-Zeitschr.* **97**, 486 (1955).
63. H. WIND, *Report WET* 63/009, pp. 1—14 (1963), FOM-Inst. voor Plasma-Fysica, Rijnhuizen, Netherlands.
64. D'ANS-LAX, „*Taschenbuch für Chemiker und Physiker*" 3. Ed., Berlin, Springer 1967.
65. A. J. B. ROBERTSON, B. W. VINEY and M. WARRINGTON, *Brit. J. of Appl. Phys.* **14**, 278 (1963).

66. E. STERN, B. S. GOSSLING and R. H. FOWLER, *Proc. Roy. Soc.* **A 124**, 699 (1929).
67. H. HEISING, M. Sc. Thesis, Universität Bonn 1969.
68. M. J. DUELL, B. J. DAVIS and R. L. MOSS, *Disc. Farad. Soc.* **41**, 43 (1966)
69. A. J. B. ROBERTSON and P. WILLIAMS, "*Proc. Phys. Electronics*", Annual Conf. 1967.
70. G. F. VIBRANS, *Techn. Rep.* No. 353 (U), Lincoln Laboratory, M.I.T. (1964).
71. I. BRODIE, *J. Appl. Phys.* **35**, 2324 (1964).
72. R. C. BRADLEY and L. A. D'ASARO, *J. Appl. Phys.* **30**, 226 (1959).
73. H. D. BECKEY, *Z. Instrkde.* **71**, 51 (1963).
74. H. D. BECKEY, *Z. analyt. Chem.* **197**, 80 (1963).
75. C. M. CROSS and A. J. B. ROBERTSON, *J. Sci. Instr.* **43**, 475 (1966).
76. H. D. BECKEY, H. KRONE and F. W. ROELLGEN, *J. Sci. Instr.* **2**, 118 (1968).
77. J. M. GILLILAND and B. W. VINEY, Royal Aircraft Establishment Techn. Report, 1966.
78. P. M. MORSE and H. FESHBACH, "*Methods of Theoretical Physics*" p. 100 McGraw Hill, New York 1953.
79. C. BRUNNÉE, *Z. analyt. Chem.* **217**, 33 (1966).
80. B. E. JOB, Manchester private communication.
81. C. BRUNNÉE, *Z. Naturforschg.* **22b**, 121 (1967).
82. H. D. BECKEY and F. W. ROELLGEN, *Z. Instrkde.* **74**, 47 (1966).
83. E. W. MÜLLER, *Z. Phys.* **131**, 136 (1951).
84. E. W. MÜLLER, *J. Appl. Phys.* **27**, 474 (1956).
85. B. J. WACLAWSKI and E. W. MÜLLER, *J. Appl. Phys.* **32**, 1492 (1962).
86. C. BRUNNÉE, G. KAPPUS and K. H. MAURER, *Z. analyt. Chem.* **232**, 17 (1967).
87. K. LEVSEN, Bonn, private communication.
88. J. H. BEYNON, A. E. FONTAINE and B. E. JOB, *Z. Naturforschg.* **21a**, 776 (1966).
89. H. D. BECKEY, E. HILT and H. KRONE, *EURATOM-Report* 4194d (1968).
90. H. KNÖPPEL, Dissertation, Universität Bonn 1965.
91. J. LENNARD-JONES and G. G. HALL, *Trans. Far. Soc.* **48**, 581 (1952), *Disc. Far. Soc.* **10**, 18 (1951).
92. J. C. LORQUET and G. G. HALL, *Mol. Phys.* **9**, 29 (1965).
93. J. C. LORQUET, *Mol. Phys.* **9**, 101 (1965).
94. H. D. BECKEY, *Z. Naturforschg.* **17a**, 1103 (1962).
95. H. D. BECKEY, *Z. Naturforschg.* **19a**, 71 (1964).
96. H. D. BECKEY and P. SCHULZE, *Z. Naturforschg.* **20a**, 1335 (1965).
97. A. H. SEHON and M. SWARC, *Proc. Roy. Soc.* **A 202**, 263 (1950).
98. J. R. HISKES, *Phys. Rev.* **122**, 1207 (1961).
99. J. R. HISKES, *Nucl. Fus.* **2**, 38 (1962).
100. E. TELLER, *Z. Phys.* **61**, 458 (1930); E. A. HYLLERASS, *Z. Phys.* **71**, 739 (1931).
101. H. WIND, *J. Chem. Phys.* **43**, 2956 (1965).
102. H. WIND, *Proc. Phys. Soc.* **84**, **4**, 540, 617 (1964).
103. H. WIND, Thesis, State Univers. of Utrecht, Netherlands (1965); *Culham Report* CLM-46 (1965).

104. G. REMBERG, E. REMBERG, M. SPITELLER-FRIEDMANN and G. SPITELLER, *Org. Mass Spectrometr.* **1**, 87 (1968).
105. H. D. BECKEY, *Z. analyt. Chem.* **170**, 359 (1959).
106. H. D. BECKEY, *Z. Naturforschg.* **16a**, 505 (1961).
107. H. D. BECKEY and H. KNÖPPEL, *Z. Naturforschg.* **21a**, 1920 (1966).
108. See for example: JAHNKE-EMDE-LÖSCH, „*Tafeln höherer Funktionen*" Teubner, Leipzig 1948.
109. J. H. BEYNON, "*Mass Spectrometry and its Applications to Organic Chemistry*" Elsevier, Amsterdam 1960.
110. F. W. MCLAFFERTY, "*Interpretation of Mass Spectra*" Benjamin, New York 1966.
111. K. BIEMANN, "*Mass Spectrometry*" McGraw Hill, New York 1962.
112. H. BUDZIKIEWICZ, C. DJERASSI and D. H. WILLIAMS, "*Mass Spectrometry of Organic Compounds*" Holden-Day, San Francisco 1967.
113. G. SPITELLER, „*Massenspektrometrische Strukturanalyse organischer Verbindungen*" Verlag Chemie, Weinheim 1966.
114. A. G. J. VAN OOSTROM, Dissertation, Universität Amsterdam 1965.
115. B. G. HERZBERG, "*Molecular Spectra and Molecular Structure*" Vol. I, p. 101, Van Nostrand, Toronto 1950.
116. viz. JAHNKE-EMDE-LÖSCH, „*Tafeln höherer Funktionen*" p. 108, Teubner, Leipzig 1948.
117. H. M. ROSENSTOCK, M. B. WALLENSTEIN, A. L. WAHRHAFTIG and H. EYRING, *Proc. Nat. Acad. Sci. USA* **38**, 667 (1952).
118. H. M. ROSENSTOCK and M. KRAUSS in: "*Mass Spectrometry of Organic Ions*" ed. F. W. MCLAFFERTY, Academic Press, New York 1963; H. M. ROSENSTOCK, "*Advances in Mass Spectrometry*" Vol. IV, p. 523 Institute of Petroleum, London 1968.
119. J. C. LORQUET, *Mol. Phys.* **10**, 493 (1966).
120. J. H. BEYNON, R. A. SAUNDERS and A. E. WILLIAMS, *Z. Naturforschg.* **20a**, 180 (1965).
121. J. H. BEYNON and A. E. FONTAINE, *Z. Naturforschg.* **22a**, 334 (1967).
122. CH. OTTINGER, *Phys. Letters* **17**, 269 (1965).
123. CH. OTTINGER, *Z. Naturforschg.* **22a**, 20 (1967).
124. I. HERTEL and CH. OTTINGER, *Z. Naturforschg.* **22a**, 1141 (1967).
125. M. BARBER, K. R. JENNINGS and R. RHODES, *Z. Naturforschg.* **22a**, 15 (1967).
126. N. D. COGGESHALL, *J. Chem. Phys.* **37**, 2167 (1962).
127. H. D. BECKEY, *Bull. Soc. Chim. Belg.* **73**, 352 (1964).
128. H. D. BECKEY, H. HEY, K. LEVSEN and G. TENSCHERT, *Int. J. Mass Spectrometry and Ion Phys.* **2**, 101 (1969).
129. H. D. BECKEY, A. HEINDRICHS and G. TENSCHERT, *Z. Instrkde.* **75**, 195 (1967).
130. G. TENSCHERT, Dissertation, Universität Bonn 1969.
131. H. KNÖPPEL and H. D. BECKEY, *Z. Naturforschg.* **21a**, 1930 (1966).
132. R. A. MARCUS, *J. Chem. Phys.* **20**, 352, 355, 359, 364 (1952).
133. J. BLOCK and P. L. MOENTACK, *Techn. Rep.* 36/62 Union Carbide, Europ. Res. Assoc.
134. H. D. BECKEY, *Z. Naturforschg.* **15a**, 822 (1960).
135. H. G. METZINGER, Diplomarbeit, Universität Bonn 1963.
136. P. DEBYE, "*Polar Molecules*" Dover Publ., 1929.

137. L. ONSAGER, *J. Am. Chem. Soc.* **58**, 1486 (1936).
138. J. G. KIRKWOOD, *J. Chem. Phys.* **7**, 911 (1939); *Disc. Far. Soc.* **42 A**, 7 (1946).
139. F. W. RÖLLGEN, Dissertation, Universität Bonn 1970.
140. ST. BRUNAUER, P. H. EMMETT and E. TELLER, *J. Am. Chem. Soc.* **60**, 309 (1938).
141. E. WICKE, M. EIGEN and TH. ACKERMANN, *Z. Phys. Chem. NF* **1**, 340 (1954)
142. L. ONSAGER, *J. Chem. Phys.* **2**, 599 (1934)
143. P. SCHISSEL, private communication.
144. J. BLOCK, *Z. Naturforschg.* **18a**, 952 (1963).
145. M. WIEN, *Phys. Zeitschr.* **32**, 545 (1931).
146. B. E. JOB and W. R. PATTERSON, "*Proc. Conf. on Newer Phys. Methods in Structural Chem.*" p. 129, United Trade Press, London 1967.
147. H. C. BROWN and M. GRAYSON, *J. Am. Chem. Soc.* **75**, 6285 (1953).
148. J. BLOCK and P. L. MOENTACK, *Z. Naturforschg.* **22a**, 711 (1967).
149. R. GOMER and L. W. SWANSON, *J. Chem. Phys.* **38**, 1613 (1963).
150. E. W. MÜLLER, *Z. Elektrochem.* **59**, 372 (1955).
151. R. GOMER, *J. Chem. Phys.* **31**, 341 (1959).
152. E. W. MÜLLER, *Naturwiss.* **29**, 533 (1941).
153. D. G. BRANDON, *Surface Sci.* **3**, 1 (1965).
154. J. F. MULSON and E. W. MÜLLER, *J. Chem. Phys.* **38**, 2615 (1963).
155. D. F. BAROFSKY and E. W. MÜLLER, *Surface Sci.* **10**, 177 (1968).
156. R. VANSELOW and W. A. SCHMIDT, *Z. Naturforschg.* **21a**, 1690 (1966).
157. D. G. BRANDON, *Phil. Mag.* **14**, 803 (1966).
158. E. W. MÜLLER, S. NAKAMURA, O. NISHIKAWA and S. B. MCLANE, J. *Appl. Phys.* **36**, 2496 (1965).
159. H. D. BECKEY and P. SCHULZE, *Z. Naturforschg.* **21a**, 214 (1966).
160. C. KEMBALL, *Proc. Roy. Soc.* London **190**, 117 (1947).
161. H. G. METZINGER and H. D. BECKEY, *Z. Naturforschg.* **22a**, 1020 (1967).
162. C. KEMBALL and E. K. RIDEAL, *Proc. Roy. Soc. London* **187**, 53 (1946).
163. M. H. ARMBRUSTER and J. B. AUSTIN, *J. Am. Chem. Soc.* **66**, 159 (1944).
164. G. JURA, E. H. LOESER, S. R. BASFORD and W. D. HARKINS, *J. Chem. Phys.* **14**, 117, 344 (1946).
165. R. A. W. HAUL and E. R. SWART, *Z. Elektrochem.* **61**, 380 (1957).
166. W. D. MACHIN and S. ROSS, *Proc. Roy. Soc.* London **265**, 455 (1962).
167. R. SUHRMANN, G. KRÜGER and G. WEDLER, *Z. Phys. Chem.* NF. **30**, 1 (1961).
168. J. C. P. MIGNOLET in "*Chemisorption*" ed. W. E. GARNER, Butterworth, London 1957.
169. R. V. CULVER and F. C. TOMPKINS, *Adv. Catalysis* **11**, 101 (1959).
170. R. SUHRMANN, B. HAHN and G. WEDLER, *Naturwiss.* **44**, 60 (1957).
171. H. G. METZINGER and H. D. BECKEY, *Z. Naturforschg.* **22a**, 1029 (1967).
172. J. H. DE BOER, *Adv. Catalysis* **8**, 85 (1956).
173. J. H. DE BOER, "*The Dynamical Character of Adsorption*" Clarendon Press, Oxford 1953.
174. G. DAMKÖHLER, *Z. Phys. Chem.* **B 23**, 58 (1933).
175. A. J. B. ROBERTSON and P. WILLIAMS, "*Advances in Mass Spectrometry*" Vol. IV, p. 847 Inst. of Petroleum, London 1968.
176. P. WILLIAMS, Dissertation, University of London 1966.

177. A. J. JASON, R. P. BURNS and M. G. INGHRAM, *J. Chem. Phys.* **43**, 3762 (1965).
178. A. J. JASON, R. P. BURNS, A. C. PAN and M. G. INGRAHM, *J. Chem. Phys.* **44**, 4351 (1966).
179. A. J. JASON, *Phys. Rev.* **156**, 266 (1967).
180. H. D. BECKEY and P. SCHULZE, *Z. Naturforschg.* **20a**, 1329 (1965).
181. B. STEINER, C. F. GIESE and M. G. INGHRAM, *J. Chem. Phys.* **34**, 189 (1961).
182. G. WAGNER, Dissertation, Aachen 1963.
183. G. G. WANLESS and G. A. GLOCK JR., *Analyt. Chem.* **39**, 2 (1967).
184. M. S. B. MUNSON and F. H. FIELD, *J. Am. Chem. Soc.* **88**, 2621 (1966).
185. M. S. B. MUNSON and F. H. FIELD, *J. Am. Chem. Soc.* **88**, 4337 (1966)
186. F. H. FIELD, M S. MUNSON and D. A. BECKER, *Advan. in Chem. Ser.* No. **58**, 167 (1966).
187. H. D. BECKEY, *J. Am. Chem. Soc.* **88**, 5333 (1966).
188. B. C. PATTERSON and M. SEAKINS, *Trans. Far. Soc.* **63**, 1863 (1967).
189. H. D. BECKEY and G. WAGNER, *Z. Naturforschg.* **20a**, 169 (1965).
190. J. COLLIN, *Bull. Soc. Roy. Liege* **11**, 446 (1952).
191. "*Mass Spectral Data*" Am. Petroleum Inst., Res. Proj. 44, Texas University.
192. H. D. BECKEY in "Mass Spectrometry" ed. by R. I. REED, p. 39 Academic Press, New York 1965.
193. H. D. BECKEY and W. GROTH, *Z. Phys. Chem.* NF **20**, 307 (1959).
194. H. OKABE, H. D. BECKEY and W. GROTH, *Z. Naturforschg.* **21a**, 135 (1966).
195. F. P. LOSSING, D. G. H. MARSDEN and J. B. FARMER, *Canad. J. Chem.* **34**, 701 (1956).
196. H. D. BECKEY, W. GROTH, H. OKABE and H. J. ROMMEL, *Z. Naturforschg.* **19a**, 1511 (1964).
197. G. B. KISTIAKOWSKY and P. H. KYDD, *J. Am. Chem. Soc.* **79**, 4825 (1957).
198. F. P. LOSSING, "Mass Spectrometry" p. 442, McGraw Hill, New York 1963.
199. H. BUTZERT and H. D. BECKEY, *Z. Phys. Chem.* NF **62**, 83 (1968).
200. H. BUTZERT, Dissertation, Universität Bonn 1967.
201. H. D. BECKEY, Int. J. Mass Spectry. and Ion Phys. **1**, 93 (1968).
202. E. v. SYDOW, *Acta Chem. Scand.* **18**, 1099 (1964).
203. M. v. ARDENNE, K. STEINFELDER and R. TÜMMLER, *Z. Chem* .**5**, 287 (1965).
204. E. M. CHAIT, F. W. SHANNON, J. W. AMY and F. W. MCLAFFERTY, *Analyt. Chem.* **40**, 835 (1968).
205. H. D. BECKEY and H. HEY, *Org. Mass Spectrometr.* **1**, 47 (1968).
206. E. v. SYDOW, *Acta Chem. Scand.* **17**, 2025 (1963).
207. D. H. WILLIAMS and R. G. COOKS, *Chem. Commun.* 663 (1968); see also S. MEYERSON, *Appl. Spectrometr.* **22**, 30 (1968).
208. F. W. MCLAFFERTY, *Analyt. Chem.* **31**, 82, 2072 (1959).
209. L. FRIEDMAN, F. A. LONG and M. WOLFSBERG, *J. Chem. Phys.* **27**, 613 (1957).
210. H. HEY, Diplomarbeit, Universität Bonn 1967.

211. J. DAHMEN, Dissertation, Universität Bonn 1968.
212. H. KRONE, unpublished work, Bonn 1968.
213. M. BARBER and R. M. ELLIOTT, "12th Ann. Conf. on Mass Spectrometry" Committe ASTM E–14, Montreal 1964.
214. J. H. BEYNON, R. A. SAUNDERS and A. E. WILLIAMS, *Nature* **204**, 67 (1964).
215. J. H. FUTRELL, K. R. RYAN and L. W. SIECK, *J. Chem. Phys.* **43**, 1832 (1965).
216. K. R. JENNINGS, *J. Chem. Phys.* **43**, 4176 (1965).
217. H. D. BECKEY, Int. J. Mass Spectry. Ion Phys. **2**, 500 (1969).
218. D. O. HUMMEL, T. SCHWEREN and K. KLEINHENZ, "15th Field Emission Symposium" (Abstracts) Bonn 1968.
219. H. D. BECKEY and G. WAGNER, *Z. analyt. Chem.* **197**, *58* (1963).
220. G. FRANKE in „Massenspektrometrie" ed. by H. KIENITZ, p. 323, Verlag Chemie, Weinheim 1968.
221. K. G. HIPPE, Bonn to be published.
222. H. D. BECKEY, unpublished report to R. D. Behling, B P Comp., Hamburg (1966).
223. E. W. MÜLLER, J. A. PANITZ and S. B. MCLANE in "Proc. 4th Regional Europ. Conf. on Electron Microsc." Rome 1968.
224. H. D. BECKEY, E. HILT, A. MAAS, M. D. MIGAHED and E. OCHTERBECK, Int. J. Mass Spectry. Ion Phys. **3**, 161 (1969).
225. R. BACON, *J. Appl. Phys.* **31**, 283 (1960).
226. K. LEVSEN, Dissertation, Universität Bonn 1970.
227. J. N. DAMICO, R. P. BARRON and J. A. SPHON, *Int. J. Mass Spectry. Ion Phys.* **2**, 161 (1969).
228. P. BROWN and C. DJERASSI, *Angew. Chem.* **79**, 481 (1967).
229. J. DAHMEN, Dissertation, Universität Bonn 1958.
230. J. DAHMEN and H. D. BECKEY, *Z. Naturforschg.* Series a, to be published 1971.
231. J. BIEGELEISEN and M. WOLFSBERG, *Adv. Chem. Phys.* **1**, 15 (1958).
232. H. EYRING, *Proc. Nat. Acad. Sci. USA* **19**, 78 (1933), *J. Chem. Phys.* **3**, 107 (1935).
233. C. KEMBALL, *Proc. Roy. Soc.* **190**, 117 (1947).
234. Y. A. EIDUS, *Russ. Chem. Rev.* **5**, 338 (1967).
235. E. K. RIDEAL, "*Concepts in Catalysis*" Academic Press, London 1968.
236. W. M. MCFADDEN, M. LOUNSBURG and A. L. WAHRHAFTIG, *Can. J. Chem.* **36**, 990 (1958).
237. W. BENZ and K. BIEMANN, *J. Am. Chem. Soc.* **86**, 2375 (1964).
238. C. G. MACDONALD, J. S. SHANNON and G. SUGOWDZ, *Tetrahedron Letters* **13**, 807 (1953).
239. S. MEYERSON and L. C. LEITCH, *J. Am. Chem. Soc.* **86**, 2555 (1964).
240. L. D. LANDAU and E. M. LIFSHITZ, "*Quantum Mechanics*" p. 70 Addison-Wesley, Reading 1958.
241. W. L. MEAD, *Analyt. Chem.* **40**, 743 (1968).
242. D. O. HUMMEL and H. D. R. SCHÜDDEMAGE, *Kolloid Z.* and *Z. f. Polymere* **210**, 97 (1966).
243. H. D. R. SCHÜDDEMAGE and D. O. HUMMEL ibid. **210**, 103 (1966).

244. D. O. HUMMEL, H. D. R. SCHÜDDEMAGE and U. POHL, ibid. **210**, 106 (1966).
245. H. D. R. SCHÜDDEMAGE and D. O. HUMMEL, ibid. **220**, 133 (1967).
246. G. G. WANLESS in "*Advances in Spectroscopy of Fuels and Related Materials*" Plenum Press, New York 1970, p. 15.
247. M. J. DUELL, Proc. Roy. Soc. **A 302**, 473 (1968).
248. H. D. BECKEY, H. HEISING, H. HEY and H. G. METZINGER, in "*Advances in Mass Spectrometry*" Vol. IV Inst. of Petroleum, London 1968, p. 817.
249. E. M. CHAIT, T. W. SHANNON, W. O. PERRY, G. E. VAN LEAR and F. W. MCLAFFERTY, Int. J. Mass Spectry. Ion Phys. **2**, 141 (1969).
250. I. V. GOLDENFELD, V. A. NAZARENKO and V. A. POKROVSKII, *Doklady Phys. Chem. USSR* **161**, 276 (1965).
251. I. Z. KOROSTYSHEVSKII and I. V. GOLDENFELD, *Doklady Phys. Chem. USSR* **172**, 153 (1967).
252 E. L. RICHTER and W. A. SCHMIDT, *Z. Naturforschg.* **20a**, 317 (1965).
253. W. K. ROHWEDDER, "15th and 16th Annual Conf. on Mass Spectrometry" ASTM Committee E–14, 1967 and 1968.
254. M. DRECHSLER, *Z. Elektrochem.* **61**, 48 (1957).
255. W. L. MEAD, A. R. WEST, J. S. HALLIDAY, A. RIDDOCH and H. HOWARTH, "15th Annual Conf. on Mass Spectrometry" ASTM Committee E–14, Denver 1967.
256. G. C. ELTENTON, *J. Chem. Phys.* **15**, 455 (1947).
257. E. W. MÜLLER and R. D. YOUNG, *J. Appl. Phys.* **32**, 1472 (1961).
258. D. H. WILLIAMS and R. G. COOKS, Chem. Commun. (12) 663, (1968).
259. H. D. BECKEY and K. LEVSEN, in "Proc. of the Summer School on Mass Spectrometry" Gordon + Breach, 1970.
260. F. W. RÖLLGEN and H. D. BECKEY, Surf. Science, **23**, 69 (1970)
261. I. V. Goldenfeld and V. A. Nazarenko, Int. J. Mass Spectry. Ion Phys. **5**, 197 (1970).
262. W. A. Schmidt, O. Frank and J. Block, 16 Field Emission Symposium, Pittsburgh, Pa (1969).
263. F. W. Röllgen and H. D. Beckey, paper submitted to Surface Science, 1970.

6. INDEX